The Changing Arctic Environment

This accessible and engagingly written book describes how national and international scientific monitoring programmes brought to light our present understanding of Arctic environmental change and how these research results were successfully used to achieve international legal actions to lessen some of the resulting environmental impacts. David P. Stone was intimately involved in many of these scientific and political activities. He tells a powerful story, using the metaphor of the "Arctic Messenger" – an imaginary being warning us of the folly of ignoring Arctic environmental change. This book will be of great interest to anyone concerned about the Arctic's fate, including lifelong learners interested in the Arctic and the natural environment generally; students studying environmental science and policy; researchers of circumpolar studies, indigenous peoples, national and international environmental management and environmental law; and policy makers and industry professionals looking to protect (or exploit) Arctic resources.

DAVID P. STONE received a degree in zoology from the University of Aberdeen in 1973 and a PhD in oceanography from the University of British Columbia. From 1977 to 1980, David worked as an oceanographic scientist in the Canadian Arctic. From 1980 to 2004, he managed environmental research for the Northern Affairs Programme of the Canadian government, becoming director of Northern Science and Contaminant Research. In 1989, he became heavily involved in the development of circumpolar cooperation under the Arctic Environmental Protection Strategy that subsequently became the Arctic Council. One of its key activities was the establishment in 1991 of the Arctic Monitoring and Assessment Programme (AMAP), where he served as Canada's delegate on the governing working group until 2004 and as its chair between 1993 and 1997. During this time, AMAP set up a circumpolar monitoring capacity and produced its first circumpolar assessment of the state of the Arctic environment. From 1990 to 1994, David co-chaired the task force on persistent organic pollutants (POPs) under the United Nations Economic Commission for Europe's (UNECE) Convention on Long-Range Transboundary Air Pollution (CLRTAP) and led the working group that prepared the negotiating text used by the convention's governing executive to negotiate a legally binding protocol under the convention. He participated in the negotiation of the Stockholm Convention on POPs, and after retirement, he was retained by the convention secretariat to assist in the development of a process to measure through global monitoring whether environmental levels of POPs are falling as a result of the new international controls. In 1997, he was instrumental in developing a virtual University of the Arctic based on existing circumpolar institutions. The Arctic Council formally announced the creation of the university in September 1998. In 2001, David received an award from Jean Chrétien, the then–prime minister of Canada, in recognition of his work on developing and managing Arctic contaminants research and on using the results to achieve negotiation of global agreements.

The Changing Arctic Environment

The Arctic Messenger

DAVID P. STONE

CAMBRIDGE
UNIVERSITY PRESS

CAMBRIDGE
UNIVERSITY PRESS

32 Avenue of the Americas, New York, NY 10013-2473, USA

Cambridge University Press is part of the University of Cambridge.

It furthers the University's mission by disseminating knowledge in the pursuit of education, learning, and research at the highest international levels of excellence.

www.cambridge.org
Information on this title: www.cambridge.org/9781107094413

© David P. Stone 2015

First published 2015

Printed in the United States of America

A catalog record for this publication is available from the British Library.

Library of Congress Cataloging in Publication Data
Stone, David P. (Oceanographic scientist)
The changing Arctic environment : the Arctic Messenger / David P. Stone (former chair of the Arctic Monitoring and Assessment Programme (AMAP)).
    pages cm
ISBN 978-1-107-09441-3 (Hardback)
1. Arctic regions–Environmental conditions.   2. Environmental monitoring–Arctic regions.   3. Climatic changes–Arctic regions.   4. Pollution–Environmental aspects–Arctic regions.   5. Environmental impact analysis–Arctic regions.   I. Title.
GE160.A68S76 2015
363.700911'3–dc23   2014042076

ISBN 978-1-107-09441-3 Hardback

For Thérèse, who insisted that the Arctic tale be told, and for Dáithí and Scellig, who continue to enlighten our lives

Contents

Acknowledgements

It is a humbling experience to reflect on the multitude of people who developed and assessed the knowledge on which the Arctic Messenger's tale is based. Even if I could list all the people, too many would still be unacknowledged.

I relied very heavily on my own memory and files for the historical and scientific context of how our understanding of the Arctic environment has changed over the last thirty years and of how governments have reacted to this knowledge. However, I often needed help to check facts, find ways to fill in gaps or quite simply to kick my memory into life. The kind cooperation of all the staff at the AMAP secretariat in Oslo proved invaluable. I was given access to AMAP archives dating back to twenty-five years ago, which not only revived my memories but also added fresh details from this period. Margaret Davis spent hours anticipating my needs by retrieving and organising historical files. Her help saved me many hours of time and likely of frustration. I also much appreciated the help of Grethe Øksnes and Inger Utne, who effortlessly provided any practical assistance I needed at the secretariat.

On the science side, the manuscript was much improved through technical reviews provided by Robert Corell, Martin Forsius, Derek Muir, Lars-Otto Reiersen, Dáithí Stone, Per Strand and Simon Wilson. All are experts in one or more of the main topics covered in the book. I am much indebted to all these people for pointing out errors or suggesting improvements and also for their support for the book being published.

An early draft of the manuscript contained three climate chapters that were reviewed by Dáithí Stone. He promptly gave his father a much-needed lesson in humility by telling him to throw the chapters away and to start again, which indeed I did! Thanks are also due to AMAP for providing permission to adapt figures used in the book from

AMAP sources and for carrying out the work to modify existing AMAP graphics for their new life. I also much appreciate being able to use two previously unpublished figures provided by Dáithí Stone and one by Robert Corell.

I am deeply grateful to Caitríona Uí Ógáin and Sophia Kakkavas (Philomel Productions) for graciously providing their permission to use three poetic fragments from *Footsteps From Another World* by the late Dáithí Ó hÓgáin. Dáithí was a much-respected folklorist, author and poet. *Footsteps From Another World* was his only collection of poetry in English. It gives those of us who do not speak Irish Gaelic a tantalizing glimpse of what we have missed. Anyone who is tempted to seek out this collection in English will not be disappointed.

The last step in bringing a book to the public is of course publication and I am much indebted to all of the staff at Cambridge University Press and their contractors who were involved in this process. In particular, I would like to acknowledge the help of Matt Lloyd, for his early encouragement and decision to promote publication by Cambridge; Holly Turner, for never being impatient with a novice author; Rituleen Dhingra for valuable advice; and Paul Smolenski, who managed the entire production process. I am also grateful for the kind attention of Britto Fleming Joe and his colleagues at SPi Global.

I very much appreciated the help of Christopher Stolle for his efficient and sensitive copyediting. He made this process an educational and enjoyable experience. I will never again look at a text in quite the same way and will always be on the lookout for a "dangling modifier".

Finally, my overriding debt is to Thérèse, my wife of more than forty years. She insisted that this book be written and never relaxed in her enthusiasm. On several occasions, my confidence in the project came close to vaporizing. By one way or another, Thérèse allowed me no escape. She also prepared the list of acronyms and, more importantly, the index. When the book returned from copyediting and for final examination of the proofs, Thérèse was seated at her desk, carefully scrutinizing every word.

Acronyms

AAR:	AMAP Assessment Report
AARI:	Arctic and Antarctic Research Institute
ACAP:	Arctic Contaminants Action Programme
ACIA:	Arctic Climate Impact Assessment
ADHD:	Attention Deficit Hyperactivity Disorder
AEAM:	Adaptive Environmental Assessment and Management
AEPS:	Arctic Environmental Protection Strategy
AMAP:	Arctic Monitoring and Assessment Programme
AMDEs:	Atmospheric Mercury Depletion Events
AMOC:	Atlantic Meridional Overturning Circulation
AO:	Arctic Oscillation
AOGCM:	Atmosphere-Ocean Global Climate Model
APECS:	Association of Polar Early Career Scientists
AR4:	Fourth Assessment of the IPCC
AR5:	Fifth Assessment of the IPCC
AVHRR:	Advanced Very High Resolution Radiometer
BAT:	Best Available Technique
BC:	Black Carbon
BEMP:	Beaufort Environmental Monitoring Programme
BMDL:	Benchmark Dose (Lower Confidence Limit)
CAFF:	Conservation of Arctic Flora and Fauna
CCMA:	Canadian Chemical Manufacturers Association
CDUT:	Canada/Déline Uranium Table
CFC:	Chlorofluorocarbon
CFI:	Canada Foundation for Innovation
CLRTAP:	Convention on Long-Range Transboundary Air Pollution
COP:	Conference of the Parties
CTD:	Conductivity Temperature and Depth

DDE:	Dichlorodiphenyldichloroethylene (a derivative of DDT)
DDT:	Dichlorodiphenyltrichloroethane
DEW Line:	Distant Early Warning Line
DHA:	Docosahexaenoic Acid
EI:	Energy Intensity
EL:	Emission Level
ELA:	Equilibrium Line Altitude
EMEP:	European Monitoring and Evaluation Programme (also known as the Cooperative Programme for Monitoring and Evaluation of Long-Range Transmission of Air Pollutants in Europe)
ENSO:	El Niño Southern Oscillation
EPA:	Environmental Protection Agency
EPPR:	Emergency Prevention Preparedness and Response in the Arctic
EURT:	East Ural Radioactive Trace
GAPS:	Global Atmospheric Passive Sampling
GEA:	Global Energy Assessment
GEF:	Global Environment Facility
GHG:	Greenhouse Gas
GP:	Global Prosperity
GPL:	Global Population Level
GRACE:	Gravity Recovery and Climate Experiment
GWP:	Global Warming Potential
HBCD:	Hexabromocyclododecane
HBFC:	Hydrobromofluorocarbon
HCB:	Hexachlorobenzene
HCFC:	Hydrochlorofluorocarbon
HCH:	Hexachlorocyclohexane
HFC:	Hydrofluorocarbon
IAEA:	International Atomic Energy Agency
IAM:	Integrated Assessment Model
IASAP:	International Arctic Sea Assessment Programme
IASC:	International Arctic Science Committee
ICARP:	International Conferences on Arctic Research and Planning
ICC:	Inuit Circumpolar Council
ICES:	International Council for Exploration of the Seas
IFCS:	Intergovernmental Forum on Chemical Safety
IGY:	International Geophysical Year

IIASA:	International Institute of Applied Systems Analysis
IMO:	International Meteorological Organization
INSROP:	International Northern Sea Route Programme
IOMC:	Inter-Organization Programme for the Sound Management of Chemicals
IPCC:	Intergovernmental Panel on Climate Change
IPCS:	International Programme on Chemical Safety
IPY:	International Polar Year
IRIS:	Integrated Regional Impact Study
KMI:	Korea Maritime Institute
LOAEL:	Lowest Observed Adverse Effect Level
MEA:	Multilateral Environmental Agreement
MODIS:	Moderate Resolution Imaging Spectroradiometer
MVP:	Moving Vessel Profiler
NAO:	North Atlantic Oscillation
NCE:	Networks of Centres of Excellence
NCP:	Northern Contaminants Programme
NEFCO:	Nordic Environment Finance Corporation
NILU:	Norwegian Institute for Air Research
NIP:	National Implementation Plan
NMDA:	N-Methyl-D-Aspartic Acid
NOAEL:	No Observed Adverse Effect Level
NORM:	Naturally Occurring Radionuclide Material
NRPA:	Norwegian Radiation Protection Authority
NSIDC:	National Snow and Ice Data Centre
OPEC:	Organization of Petroleum Exporting Countries
PACE:	Polynyas in a Changing Arctic Environment
PAH:	Polycyclic Aromatic Hydrocarbons
PAME:	Protection of the Arctic Marine Environment
PBDEs:	Polybrominated Diphenyl Ethers
PCBs:	Polychlorinated Biphenyls
PCSP:	Polar Continental Shelf Project
PFAS:	Polyfluoroalkyl Substances (also known as PFCs: Perfluorinated Compounds)
PFCA:	Perfluorinated Carboxylic Acid
PFOA:	Perfluoroctanoic Acid
PFOS:	Perfluorooctane Sulfonate
POPs:	Persistent Organic Pollutants
PTS:	Persistent Toxic Substances
PUF:	Polyurethane Foam
PUNES:	Peaceful Underground Nuclear Explosions

QA/QC:	Quality Assurance/Quality Control
RAIPON:	Russian Association of Indigenous Peoples of the North
RCP:	Representative Concentration Pathways
REACH:	Registration, Evaluation, Authorisation and Restriction of Chemicals
RTG:	Radioisotope Thermoelectric Generator
SAICM:	Strategic Approach to International Chemicals Management
SAO:	Senior Arctic Official (in the context of the ACIA and the Arctic Council)
SAON:	Sustaining Arctic Observing Networks
SLCF:	Short-Lived Climate Forcers
SMB:	Surface Mass Balance
SOAER:	State of the Arctic Environment Report
SRES:	Special Report on Emissions Scenarios
SST:	Supersonic Transport
SWIPA:	Snow, Water, Ice and Permafrost in the Arctic
TEAP:	Technology and Economic Assessment Panel
TENORM:	Technically Enhanced Naturally Occurring Radionuclide Material
UNCED:	United Nations Conference on Environment and Development
UNECE:	United Nations Economic Commission for Europe
UNEP:	United Nations Environment Programme
UNFCCC:	United Nations Framework Convention on Climate Change
UNSCEAR:	United Nations Scientific Committee on the Effects of Atomic Radiation
UV:	Ultraviolet
VCPOL:	Vienna Convention for the Protection of the Ozone Layer
VECs:	Valued Ecosystem Components
VOCs:	Volatile Organic Compounds
WCED:	World Commission on Economic Development
WCRP:	World Climate Research Programme
WGE:	Working Group on Effects
WHO:	World Health Organization
WMO:	World Meteorological Organization

1

Personal Beginnings

The White Rabbit put on his spectacles. "Where shall I begin, please your Majesty?" he asked.
"Begin at the beginning," the King said gravely, "and go on till you come to the end: then stop."

Lewis Carroll, Alice's Adventures in Wonderland

The beginning for me was summer 1970. I sat in a tent with two other students on Pabbay, a small uninhabited island in the Outer Hebrides off the north-west coast of Scotland. Andrew Ramsey, Norman Macdonald and I were working for Operation Seafarer, the first complete census of seabirds of Britain and Ireland. As part of this effort, I had trudged after Andrew to other islands in the Sound of Harris, to Auskerry in the Orkneys and for several visits to Canna, a wonderful Hebridean island to the west of Rùm and to the south of Skye. For close to 1,000 years, Pabbay was home to a small settlement. Now all that remains are the foundations of the village, some lazy beds (those patches in which vegetables were cultivated by sowing them on terraces created from rotting kelp and peat) and the roofless walls of a small church. The former islanders rest under its protection, lying beneath little mossy hummocks and peacefully metamorphosing into luxuriant pink splashes of sea thrift. There were fulmars sitting atop these natural gardens, brooding on their nests with an air of great and inner sanctity unless you approached them too closely. Then, you were welcomed with a warm vomit of fishy oil.

We were marooned. There were no birds uncounted and it was hard to find one that was not shyly sporting a shiny new ring on a leg. Our work was over, a north Atlantic gale was blowing, and our boatman in North Uist was wisely prudent. Pabbay, situated in the Sound of Harris, is completely exposed to the Atlantic. Our boatman had decided to stay at home. As we watched our daily food supply dwindle with the wind and rain doing its best to flatten our tent, we amused ourselves with my ragged volume of Robbie Burns and imagined where in an ideal world we would like to be the following summer. It was difficult to dry off and we rarely ventured outside, but we did have a short moment of excitement. Andrew headed out one morning with spade and toilet paper. A little later, he exploded into the tent, shouting "Quick – get the nets." A brightly coloured bunting had fluttered around him just as he settled into his moment in the lazy beds. Soon, the bunting was in our mist nets and identified as a yellow-breasted bunting a long way from home.

When the seascape reached new heights of foam-streaked anger, shouts came from the beach – and there was the boatman at last. Our accommodation in the boat – a lifeboat recycled from an old freighter – was beneath the roofed-over forward deck. All we could see was our man, silhouetted starkly in front of alternating sky and walls of frothy water, racing horizontally eastward. I had often been in small boats around Scottish seabird colonies and in seas that seemed to me to be pretty rough. Any sense of adventure was always deflated by a glance at the helmsman, nonchalantly steering with elbows on the tiller or foot on the wheel. This is how it had been on our way out. Now the boatman gripped the tiller with both hands, with his gaze riveted on whatever was happening to the west. I still remember the aroma and taste of the fresh bread he kindly brought for us and that we munched as we wedged our feet against the little hull. I am not a great sailor, but my stomach never had a fleeting thought of giving up that bread. It was on the foam-flecked journey on the Sound of Harris that next year's destination was decided. It would be the Svalbard archipelago.

I never really expected to see Svalbard, but Andrew had the energy of A. A. Milne's Tigger, and before long, he had persuaded Chalmers Clapperton, a geologist faculty member at Aberdeen University, to organise a small expedition. Andrew then departed for graduate work at the University of Manitoba and Norman became the president of the student union at Aberdeen University. Two years later, however, Sandy Anderson and Bill Murray (researchers at Aberdeen), Robert Swann (a fellow student at Aberdeen University) and I were working amidst the seabirds of Kongsfjorden.

We were a year late, but I never complained because in the interim summer, I met Thérèse Ní Ógáin while we worked on either side of a conveyor belt in a smoked herring factory on the Shetland

Islands. Thérèse had a B.A. from University College Dublin and was en route to an archaeological dig in Brittany.

Our main camp in Svalbard was just outside the small settlement of Ny-Ålesund, but Robert and I had another small camp at the head of the fjord. In front of our tent lay raised beaches of shin-high moss, where hummocks of Arctic-alpine flora celebrated their brief moment in the sun. Amidst the flowers, eider duck females sat with maternal pride over their young, and in the small tundra lakes, red-throated diver females and phalarope males imitated their eider neighbours. Above and behind us was a small mixed-species seabird colony, while immediately to our right was the huge ice front of the merged Kongsvegen and Konebreen glaciers. Kongsvegen is an outlet glacier, descending at the breakneck velocity (for a glacier) of about 2 metres a day from an upper ice field. With this speed, it is heavily crevassed and is almost continually calving small icebergs into the fjord. The ice front was a pandemonium of seabird activity. Our task was to chart the diurnal behavioural pattern of the kittiwakes (the most common species) feeding in the twenty-four-hour cycle of sunlight intensity. It turned out that the birds were much more interested in the intensity of ice calving than they were in the passage of the sun. The glacial front floats on the fjord, and if there was a correlation with anything, it was with the tidal cycle.

Robert and I were concerned about polar bears and our lack of a gun was considered by the staff at the Ny-Ålesund research station to be the height of folly. We explained confidently to them that we did not need one, as the experts at Aberdeen had told us that bears in summer always follow the retreating sea ice with its cargo of seals. After a few moments of quiet and reflective thought, our sage in Ny-Ålesund responded in that deliberate Norwegian way: "Ja, that is true, but the bear that stays behind – he is a hungry bear!" Of course, he was right. On the way north between Nordkapp and Longyearbyen, our ship made a supply stop at Bear Island, where a radio operator had been killed the previous autumn by a bear that had come ashore several months before the arrival of the ice and seals. Seven years later, when doing an oceanographic section by ship, we encountered a bear swimming exactly halfway between Disco (Greenland) and Baffin Island. It was more than 150 kilometres from the nearest land and (it being late summer) was at least 300 kilometres from the closest pack ice. The bear did not seem to be lost or disorientated and kept swimming in a straight line towards the west, sublimely undisturbed by the ship circling around.

It is constantly surprising how quickly time glides by. I have not seen Robert for 40 years. He became the geography teacher at the school in Drumnadrochit, not far from Loch Ness, and later at Tain. His enthusiasm for birds obviously never waned. From time to time, I read about him accompanied by bands of lucky schoolchildren working on birds all over the highlands and islands but especially on Canna. Here, he has maintained a summer ringing (banding in North America) and census programme of the bird population that must be one of the most valuable ornithological records in Scotland. I often think about the learning experiences provided by Robert, along with Andrew Ramsey, Peter Macdougal and Alastair Duncan (three other teaching friends from my Aberdeen days), and by my wife, Thérèse, who also became a teacher. Their exciting extracurricular environmental activities with young minds must have encouraged a public awareness of environmental issues in adult life.

Glaucous gulls preyed heavily on the young seabirds behind our Kongsfjorden tent. The adult glaucous gulls are magnificent, but from a purely human viewpoint, they probably have few friends. They will murder anything they can. However, with them having not evolved into raptors, it is a clumsy and cruel business. A few days before our return to Aberdeen, I noticed one close to our tent. It was behaving in a distressed way, and soon after, it was dead. I was puzzled. It looked so healthy, with a lot of fat and no obvious parasite problems. I took some organ and fatty tissue samples and added them to the puffin material collected earlier by Sandy Anderson. All the samples were passed to Bill Bourne, who was then working at Aberdeen University. Later, we learned that the puffins contained similar levels of PCB derivatives as were found in other auks from the Scottish coast despite their apparent isolation in Svalbard. However, my glaucous gull had the highest levels of PCBs that the analyzing laboratory had encountered up to that time.

I do not believe the glaucous gull levels were ever published. It was a single bird and of no statistical significance. However, it was a personal milestone for the rest of my life and for the theme of this book. How could an Arctic bird, whose winter migrations would rarely (if ever) take it into industrialized waters, carry such a large burden of a toxic chemical and what is the underlying message for the Arctic and global environments?

I made a museum skin of the gull. It remained a barely tolerated guest at the home of my aunt in southern England. One night, after an aggressive spring-cleaning operation, my bird was tossed over the estate wall of some local gentry whom she particularly disliked.

What is it about the Arctic that is so seductive? People whose ancestors have lived there through hundreds if not thousands of generations will have a perspective that outsiders such as myself can never experience or fully understand. What has it been for me? I cannot really say, but I do know it is a magic that has never weakened since those days in 1972 amongst the flowering dryas with the utterly unstoppable Kongsvegen glacier thundering into Kongsfjorden and the feeding orgy of the clamorously shrieking kittiwakes. It is a landscape that engenders the true measure of how insignificant one really is. To paraphrase from a poem by Chief Dan George, your spirit soars.

Whatever it was and still is, the spell was cast. That summer beside the Kongsvegen became a turning point for the rest of my life. Returning to Aberdeen, I wondered what research topic would get me back into the Arctic. The seabirds at Kongsfjorden were obviously feeding at the ice front. What were the conditions that gave them such a feast? I wrote a doctorate proposal and sent it to every university I thought would be interested. The first reply came from Alan Lewis, a professor at the Institute of Oceanography at the University of British Columbia in Vancouver, Canada. I arrived there in late August 1973 and spent the next four years working on plankton ecology in Knight Inlet, a very long glacial runoff fjord about a 24-hour sail north of Vancouver.

These were wonderful years. Thérèse came to Vancouver just before my first Christmas there and we were married two days later before five guests and the officiating priest. Our two sons, Dáithí and Scellig, were born in Vancouver. After many years living and working in Nova Scotia, the Arctic and Ottawa, we now have a home in retirement on a small island in the Salish Sea. On a clear day, from the windows of our house, we can see Vancouver and the forest that surrounds the university. Hidden in those trees is the little church in which we were married.

Knight Inlet is not the Arctic, but my research inevitably led me north once the doctorate was completed in 1977. I spent the next three years as a biological oceanographer working mainly between Greenland, Baffin and Ellesmere islands and Lancaster Sound. That was the end of my experience as "a bench and field scientist". For the rest of my career, I worked as a manager of Arctic environmental science programmes. Initially, the scope was restricted to the Canadian Arctic. However, Arctic science is incredibly expensive and the topics of interest have no respect for international boundaries. Therefore, as time passed, the work took on ever-growing levels of circumpolar cooperation and I was soon also involved in intergovernmental actions to protect the Arctic and global environments.

Those of us given these roles are very fortunate. We could gaze at the landscape of interdisciplinary Arctic environmental knowledge as it evolved over the past 35 years. We could see emerging issues on the Arctic's overall health and do our best to set up international cooperation and funding to move the science frontiers forward. Perhaps most importantly, we were placed in the privileged positions of bringing knowledge on the deteriorating Arctic environmental situation to circumpolar and global political levels and to argue for governmental mitigation.

Despite the huge expansion of Arctic environmental knowledge, very little has penetrated beyond the small circle of Arctic specialists. At science planning meetings over the last few years, the challenge has been thrown out many times to the dwindling band that can trace a history back to the 1970s: "You people should write a book." Well, this is one response.

This book is not an intellectual review of the last 40 years of Arctic environmental science. Neither is it a summary of Arctic international environmental cooperation. What it contains is a very personal selection (almost a memoir) of some key developments and events over the last 40 years that have contributed to our present knowledge about the Arctic's general health and the Arctic's role as part of the global ecosystem. I have concentrated on the physical, chemical and toxicological parts of the story that mark the beginnings of the environmental and their associated human health problems now coming to pass in the Arctic. Until the recent advent of climate warming, there was a general perception that threats to the Arctic environment are largely the result of human activity within the Arctic (such as hydrocarbon exploration and production). This is not the whole story and this book will concentrate instead on the insidious impacts experienced in the Arctic resulting from human activities located at much lower latitudes.

Perhaps an analogy of the Arctic tale as being a theatre tragedy in three acts would help clarify this. The physical, chemical and toxicological parts of the story are the actors who star in Act One. In Act Two, they are still present on the stage of life and are progressively eroding the well-being of the marine, freshwater and terrestrial ecology of the Arctic as well as the cultural survival of its indigenous peoples. In Act Three, we – the audience – make our entrance. We have been shifting guiltily in our seats throughout Act Two as we recognize our destructive roles as paymasters to the Act One cast. We also come to realize that we

are also being drawn into an unhappy destiny because the Arctic interacts with the physical, chemical and biological elements of our own
lower-latitude ecosystems. Not only do we have a moral responsibility to
change our ways, but it is also in our own interests to do so and we
will examine the adequacy of our circumpolar and global political
responses. To continue the analogy, this book deals with Acts One and
Three. The wildlife and indigenous cultural aspects of Act Two are
equally as important, but they are completely outside my area of expertise. You will find here the barest Act Two threads necessary to connect
the first and last acts. It remains the work of other writers to do justice
to the subjects of Act Two.

If you have been irretrievably lost in my meandering excuses,
suffice it to say that the general framework I have chosen to follow is
to describe our growing understanding of Arctic change, the Arctic
environmental interconnections with ecosystems elsewhere and the
reasons why we and our politicians should be paying close attention to
what is happening in the far north.

Many remarkable people from around the world have played vital
roles in the Arctic story. They are Arctic indigenous peoples and their
leaders, scientists from almost every discipline imaginable, medical
and nursing practitioners, school teachers, international diplomats,
"bush" pilots who routinely land planes on snow, ice and water (and
sometimes, it seems on all three at the same time) and ships' crews who
are masters of improvisation. Most will remain anonymous, as I expect
they would wish to be. A few people are identified for a number of
reasons – in most cases because of a blinding scientific, managerial
or political contribution that cannot be told without attribution. In this
vein, you will learn that I have several personal heroes and heroines for
my story. Finally, in some cases, names appear simply to add some life
and depth to the story or to help locate key references.

In writing this book, I have aimed at the nonspecialist but have
taken every opportunity to fully exploit the reader's curiosity. My hope
is that university undergraduates will take a close look at this story.
The issues facing the Arctic are extraordinarily diverse and offer wonderful career opportunities. These range, for example, from international politics and the challenges of managing Arctic science and
monitoring to atmospheric chemistry and physics, oceanography, wildlife, human health and toxicology. To young students, I could shout:
"The Arctic needs you!" For young Arctic indigenous peoples, I would
shout even louder: "We need *you* most of all!" Some parts of the book
will not be easy going. If a section is just too information laden or

interminably boring, my advice is to skip it until you land on something tastier. At the end of each science chapter is a short summary of the main points to remember. I will be very happy if readers are encouraged to dive into the more detailed texts suggested in the bibliography to better understand what I have been trying to say. Failure on my part will be if readers never reach the last chapter and if the book fails to inspire further interest from young people in Arctic studies.

A few brief notes on the book's organisation: In Part 1, we will take a quick look at Arctic environmental change, with the result (I hope) that the reader will be tempted to read on. In Part 2, we will look at how the abrupt end of the Cold War enabled countries to work together and to set up ways to study the health of the Arctic environment that would previously have been impossible. Part 3 consists of a thematic summary of the present state of knowledge dealing mainly with persistent organic pollutants, mercury and climate change. Finally, in Part 4, we take stock of where we are.

In addition to the main story are two vignettes that provide the brain with a little rest. They are generic in nature and not exclusive to any particular theme. These commentaries appear as Chapters 5 and 9.

This is my first (and probably only) book. Getting started is not an easy task, but this is enough procrastination and we are ready to meet the "Changing Arctic".

> *Biting my truant pen, beating myself for spite;*
> *"Fool," said my Muse to me, "look in thy heart and write."*
> Sir Philip Sidney, *Astrophel and Stella*

PART I

The Changing Arctic

When my grandfather was born in 1877, the Arctic environment appeared to be in much the same condition as it was when our younger son was born exactly 100 years later. Today, it is known that even in 1877, change was under way and now it is unequivocal that these changes are beginning to happen much more quickly. We could be utterly amoral and say: "Well, that's too bad, but not many people live there." However, even if we had the moral turpitude to sacrifice such a unique ecosystem with its irreplaceable human cultures, we would be unforgivably ignorant of what these changes mean to the globe as a whole. We are now beginning to understand the towering import of the role played by the Arctic in moderating the global climate. If the Arctic climate continues to follow its present rate of change (it is actually exceeding projections), the implications for the rest of the globe are ominous. The words of John Donne written 400 years ago were never as apt as they are today:

> No man is an island entire of itself; every man
> is a piece of the continent, a part of the main . . .

That is one way of capturing the stark and naked message the Arctic is giving the world.

2

The Arctic Messenger

More dangerous is my whisper than the roar of a hundred men ...
Dáithí Ó hÓgáin, "Harsh Words Spoken" from
Footsteps From Another World

The phrase "Arctic Messenger" is taken from a conference held in Copenhagen in May 2011 and organised by the Arctic Monitoring and Assessment Programme (AMAP), the University of Copenhagen and Aarhus University. In this book, I imagine the Arctic Messenger as a living entity – a harbinger that possesses omnipotent consciousness. It is of enormous age and experience, akin to the Sumerian Utnapishtim or the biblical Methuselah. Our messenger is able to tell us about its (the Arctic's) well-being and to warn us that in the past, the global ecosystem has always been astonishingly sensitive to geophysical changes in the Arctic.

This chapter takes a quick look at the breadth of environmental issues that are eroding the Arctic that was and the Arctic that is now. It is therefore an introduction to the Arctic Messenger and a rough summary of what is to come in later pages. I hope it will tempt you to read on.

About 10–13 million people live in the entire Arctic region. Between 1.5 million (Arctic Council Indigenous Peoples Secretariat) and 0.4 million (United Nations Permanent Forum on Indigenous Issues) are indigenous.[1] The Arctic environment is their home, their source of food and the foundation of their culture and spirituality. It is part of an extended soul that binds atavistically through their ancestors into the deep past and outwards into the living world. It is a psyche that

---

[1] The range reflects different definitions of the circumpolar Arctic and of the term *indigenous*.

outsiders can admire and respect but never acquire. Arctic indigenous peoples are, quite simply, an intimate part of their natural Arctic ecosystem. However, they do not enjoy an existence disconnected to the rest of the globe, and for the last 60 years, they have faced an incremental tide of environmental dangers flooding from the industrial world.

Early tangible and visible dangers faced by the Arctic environment and its peoples began to appear in the 1960s from such northern resource development as hydrocarbon exploration and production and mining. By "tangible", I mean that one could see the activities themselves and make mental connections to any visible environmental degradation that may have occurred. Arctic peoples could possibly make a stand against such activities in their homelands, as happened with the proposals for a Canadian Mackenzie Valley oil pipeline in the late 1970s and of certain historical hydroelectric power development projects in northern Norway. Alternatively, one could argue that most of the potential environmental impacts from northern development would be local and with proper management and regulation could be effectively mitigated. If Arctic peoples themselves could invest in and be genuinely involved in Arctic resource management, perhaps such activities could occur with strong local economic benefit and social, cultural and environmental impacts could be controlled even if in practice this may be difficult. For example, the settlement of comprehensive land claims in much of Canada above the tree line has greatly strengthened the role of Arctic peoples in deciding if and how their northern resources may be developed. Issues of Arctic economic resource development remain current today, but we will say no more about them here. Instead, we will concentrate on the agents of a changing Arctic environment that are not tangible, not visible and whose origins lie in human activities originating far from the Arctic. Figure 2.1 will help orientate us as we embark on the Arctic story.

Our story begins with the Cold War, when the nuclear powers tested many of their weapons by exploding them in the atmosphere. Some of the unstable radioactive nuclear products (radionuclides) resulting from the tests remained close to the test site, but a large portion was injected into the stratosphere, where they would take about a year to reach the troposphere below. From here, the radionuclides fell to Earth as dry or wet deposition (rain or snow). Monitoring results of Sami people in northern Europe were published in 1961 and revealed far-higher body burdens of radionuclides than seen in other European populations. Further circumpolar studies quickly clarified that similar body burdens existed in other circumpolar people whose diet heavily depended on local terrestrial foods. The source was attributed to the nuclear bombs that exploded over Hiroshima and Nagasaki in 1945 and, more substantially, to atmospheric nuclear weapons testing. These results played a major role in decisions being taken that ultimately resulted in the complete abandonment of atmospheric weapons testing in 1980. The peak of radioactive fallout in the Arctic from global weapons testing occurred between the late 1950s and 1963. Since that time, levels have significantly decreased – except for a short burst caused by the 1986 Chernobyl accident.

It is not surprising that after 1961, Arctic peoples who harvested caribou and reindeer found themselves facing difficult dietary decisions – decisions related not just to their diet but also to the basis of their culture. They knew that they as a people had played no role in causing the root cause of their predicament. But the potential effects of the root cause to their health and particularly to their children were frightening and invisible.

Acid rain presented a variation on the same theme. Awareness of how sulphur dioxide and nitrogen oxides emitted from combustion sources can react with water in the atmosphere, resulting in acid rain far from such sources, began to be well understood in the 1960s and early 1970s. In this case, many of the combustion source areas were visible, such as the industrial heartlands of north-eastern North America and of the United Kingdom and Eurasia. Likewise, the effects on forest growth and freshwater ecosystems in areas of intense acid rain could also be recognized and monitored. These areas were mainly south of the Arctic, but there were several very important exceptions. Much of Norway, Sweden and Finland, including their Arctic regions, lies in the midst of the atmospheric stream moving from the United Kingdom and from western continental Europe. Furthermore, their underlying geology is rich in acid rocks, which leaves their soils and lakes with little protective buffering capacity. In addition, several smelters in the eastern and central Arctic of the former Soviet Union, such as in the Kola Peninsula at Nikel and Monchegorsk and between the Yenisei River and the Taymyr Peninsula at Norilsk, had resulted in devastated landscapes of lifeless forests and lakes.

The acidification story is important in an Arctic context, even though the most significant impacts (with the exception of those resulting from smelters located within the Arctic) are found south of the Arctic. This is because it demonstrated how an activity in one geographic region can – through the transport of pollutants in the atmosphere – lead to significant environmental effects in a different and

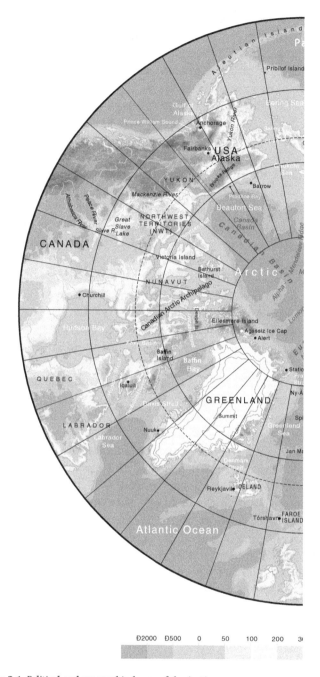

Figure 2.1 *Political and geographical map of the Arctic*

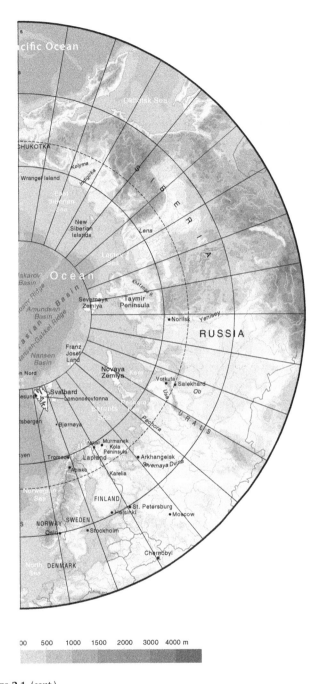

Figure 2.1 (cont.)

distant region. The international collaboration that took place to address this problem also acted as a forerunner to several actions taken later to address other pollution issues of universal importance to the Arctic.

A more insidious and invisible dimension to our story began in the 1980s, with confirmation that the so-called spring "stratospheric ozone hole" first observed in the Antarctic now had an Arctic twin. It was not as strong as its southern sibling but had the potential to expose larger populations of people to increased ultraviolet (UV) radiation. Ozone is produced naturally in the stratosphere, where it absorbs much of the incoming solar UV radiation. This is of great biological significance because all wavelengths of UV radiation are sufficiently energetic to cause disruption of molecular bonding. In 2011, the International Agency for Research on Cancer of the World Health Organization classified the entire UV spectrum as a Group 1 carcinogen. The stratospheric ozone layer is therefore vital to protecting life on Earth's surface. The root cause for stratospheric ozone loss is the release into the atmosphere of several families of chemicals known as ozone-depleting substances that were formerly used, for example, in refrigeration, as pesticides and as aerosol propellants. When ozone-depleting substances are released into the atmosphere, they are rapidly dispersed and are very resistant to decomposition in the troposphere. However, when they are exposed to solar UV radiation in the stratosphere, they break down. Their chlorine and bromine atoms are released, which then attack the surrounding ozone molecules. Ozone-depleting substances are uniformly distributed in the stratosphere, but their ozone-depletion activity is much more intensive at the two poles, where very cold stratospheric temperatures at the end of the polar winter are ideal for the chemical processes of ozone loss.

This time, the source of the danger was global. Wherever spray cans were used, ozone-depleting substances were entering the atmosphere. Wherever leaking refrigerants were used, ozone-depleting substances were entering the atmosphere. There were no point sources, such as smokestacks or mines, and no well-known military activities.

In the early 1980s, the health risks posed by the increased exposure to UV radiation in the Arctic were poorly understood, but enough was known to realize that such risks were real and serious. For indigenous peoples in the Arctic, there was little they could do but adjust their behaviour to protect against cataracts and melanoma. What of the caribou and other wildlife that do not even have the option to reduce their exposure to increased UV radiation by adjusting their behaviour?

Once again, the activities of humanity throughout the world were resulting in an impact focused on the two polar regions.

There was more bad news to come. Also in the 1980s, researchers studying Arctic marine and freshwater food chains began to find high levels of a class of organic chemical that became collectively known as *persistent organic pollutants* (POPs). In some Arctic communities, particularly those that heavily depend on the marine environment for their diet, human exposure had reached levels categorized by health authorities as indicating "concern". Most of the chemicals involved were pesticides and industrial products, such as polychlorinated biphenyls (PCBs). Very few of the pesticides had ever been used in the Arctic, but a few of the industrial chemicals were used to a small degree. By mechanisms then incompletely understood, the chemicals were being carried to the Arctic from low and mid-latitudes by the atmosphere and by ocean currents and were being biologically accumulated in some segments of the Arctic ecosystem. Several heavy metals were also identified as contaminants to the Arctic and one of these – mercury – showed similarities to the observations for POPs. In this case, the situation is even more complex because there are natural as well as anthropogenic sources of mercury. It was a devastating situation for the indigenous communities involved. Their traditional diet – the core of their social, cultural and biological well-being – had been invisibly compromised. They had played no role whatsoever in the cause, and as with the ozone situation, there was no prospect for a "quick fix".

Details of the POPs and mercury stories will form an important portion of this book, but at this stage, I want only to note that international action has been taken over the last 20 years to curb the release of ozone-depleting substances, POPs and anthropogenic sources of mercury from entering the global environment. Therefore, although the ecosystem is slow to purge itself of these substances, Arctic indigenous peoples could be forgiven if – at the turn of the century – they felt some cautious optimism for their future. Sadly, this was not to be and the most difficult challenge of all began to emerge.

At the same time that the United Nations was establishing the Framework Convention on Climate Change (UNFCCC) in the late 1980s, indigenous peoples in the Arctic began to speak about their observations on weather and climate. There were common themes across the circumpolar Arctic. They could not predict the weather in the same way they could in earlier years. In addition, times of snow buildup and melt seemed to be changing, as were the behaviours of freshwater and marine ice. Even the quality of snow suitable for igloo construction was changing. Weather was becoming more variable and extreme conditions seemed to be occurring with greater frequency. Summers appeared to have many more warm days and to be generally wetter. However, because of melt patterns, in some regions, lakes were growing smaller. The list went on.

Since then, there has been an explosion of scientific research devoted to climate change. This has built a widespread scientific consensus that the global climate system has been warming since at least 1950. The observed warming is attributed to increased atmospheric concentrations of carbon dioxide and certain other gases since the onset of the industrial era in about 1750. As we will see in later chapters, this information is periodically reviewed and assessed by the Intergovernmental Panel on Climate Change (IPCC)[2].

A large part of this book is devoted to the Arctic dimension of climate change and we will notice a remarkable convergence of the observations coming from indigenous traditional knowledge and that being generated by classical science. It turns out that the Arctic is much more sensitive to climate change than most other regions. The impact on the Arctic as we know it will be very substantial indeed. In fact, it has already begun. However, there is even more significance in Arctic warming because the Arctic is a major link in the global climate system that redistributes heat around the world. Wherever they live, all people on Earth should be concerned about how the Arctic is reacting to global warming.

These issues illustrate a historical trend in terms of the relationship between the location of the activity that caused concern (source) and the locality of potential environmental impact (receptor). In the case of Arctic resource development, the two are side by side. If there is an impact, it is not going to creep up unnoticed, and in theory, the potential for effective remedial action is always a possibility and could be in the hands of local people. With acid rain, the source area is geographically separated from the receptor by a distance of perhaps 1,000 kilometres or more. However, although the areas affected

---

[2] The Intergovernmental Panel on Climate Change (IPCC) is an ongoing scientific assessment process tasked with providing information to support negotiations on international policy for dealing with climate change conducted under the United Nations Framework Convention on Climate Change (UNFCCC). See Appendix I for a brief description of the UNFCCC and of the IPCC.

(receptor areas) can include a number of countries, they are not significantly circumpolar or global and it is possible to locate the sources.

In the case of radioactivity, there have been only a limited number
of global source localities (weapons test sites or accidents) where anthropogenic radionuclides were released into the troposphere and stratosphere. Other point sources, mainly on land, have released soluble
radionuclides directly or indirectly (via rivers) to the marine environment. In both cases, the receptor can extend very far from the source
and may even be hemispheric, if not global. A particularly sensitive
receptor element is the Arctic terrestrial environment that provides
habitat and food for caribou and reindeer. With stratospheric ozone
depletion, POPs and mercury, we move to a situation where the source is
diffuse because the substances concerned have entered the environment
all over the industrialized and agricultural world, while the Arctic
source has been minimal. However, for complex reasons that were not
recognized until quite recently, the most sensitive receptors are located
in the Arctic.

Finally, in our evolutionary progression, we reach climate change.
The sources of greenhouse gas emissions (considered by the IPCC as
being "very likely" to have caused most of the global warming observed
over the last 50 years) are also diffuse and global. The receptor is the
entire global environment. However, not all parts of the world are
equally sensitive to climate change. In fact, the Arctic and parts of
the Antarctic are probably the most sensitive regions on our planet.
The Arctic has warmed (particularly over the Greenland ice sheet) more
than twice as much as the global average over the last 50 years. Clearly,
this is of great concern to Inuit, whose way of life depends on sea ice.
However, there is more to the story. The Arctic is a major element in a
complex atmospheric and oceanographic system that moves heat from
the tropics towards the poles and keeps our climate in a happy state of
homeostasis. Later, we will look at how some of this works, what
elements of it are susceptible to change, whether there are any signals
to suggest that our global homeostatic mechanisms are in trouble and,
if so, what the consequences may be. This knowledge should be catching
the attention of our politicians. Our increased understanding of
Arctic environmental science enables us to listen to the Arctic
Messenger. In the last chapters, we will examine the extent to which
the Arctic Messenger has been heard.

As already mentioned, this book is a personal selection of evidence
to illustrate how the Arctic environment is changing and a brief
explanation of what the international political community has done or

not done in response. To be practical, I have concentrated on climate change, POPs and mercury and, to a lesser degree, radioactivity, stratospheric ozone depletion and acid rain. Others may have made different selections. I simply offer a personal reflection and I hope colleagues who have travelled along the same or parallel roads but have different memories of the details will forgive me. However, I am confident we would all share the same goal: to help make accessible to the general public the new knowledge of how the Arctic environment is changing and the global implications of such change.

To see how the Arctic Messenger gained a voice, we need to look back as far as the late 1980s. This will form the subject of the next chapter.

PART II

Working Together

When the Cold War dominated world politics, the Arctic held the longest common boundary or front line between East and West. It was an area of great military significance and sensitivity. If there had ever been a major nuclear exchange between the Soviet Union and the United States, missiles and aircraft would have passed over the Arctic, which was therefore encircled by radar stations and other military installations. Cooperation to investigate the health of the Arctic environment was not high on the agenda of Arctic country governments.

In this section, we look at how quickly this hiatus changed after Mikhail Gorbachev came to power in the Soviet Union, an event that at last provided real and meaningful opportunities for all the Arctic countries to work together. In particular, we will recognize the foresight of the Finnish (Rovaniemi) Initiative that began in 1989 and resulted in the Rovaniemi Declaration of 1991. The declaration and its accompanying Arctic Environmental Protection Strategy (AEPS) were adopted by all eight Arctic countries (Canada, Denmark/Greenland, Finland, Iceland, Norway, Sweden, Russia and the United States). From the perspective of this story, the most important element of both documents was the creation of the Arctic Monitoring and Assessment Programme (AMAP). This new institution was charged with providing Arctic governments with comprehensive and reliable information on the state of and threats to the Arctic environment. In addition, AMAP was to provide advice on actions to help Arctic governments put in place appropriate preventative and remedial environmental actions. Today, AMAP remains a key element of the Arctic Council (which subsumed the AEPS in 1996).

AMAP is quite a remarkable organisation, deserving several pages to describe how it was built and how it operates. It has essentially no guaranteed operational budget to conduct monitoring. It functions primarily due to the skill of its secretariat supported by Norway and the dedication of hundreds of scientists and indigenous peoples who volunteer to prepare AMAP assessments. When you have reached the final page of this book, I think you will share my appraisal.

3

The Arctic Messenger Gains a Voice: The Arctic Monitoring and Assessment Programme

Remember that to change your mind and follow him who sets you right is to be none the less free than you were before.
Marcus Aurelius, Meditations

Before the late 1980s, any attempt to work cooperatively in the circumpolar Arctic had to deal with the grim realities of the Cold War. The Warsaw Pact and the NATO countries obsessively distrusted one another. Any suggestions that topics of mutual interest might exist and could be tackled cooperatively were met with deep suspicion and formidable hurdles on both sides of the Iron Curtain. It was a poisonous milieu and certainly not one likely to welcome a proposal to monitor and assess the health of the circumpolar Arctic environment. Imagine suggesting to either the United States or the Soviet Union before the arrival of Mikhail Gorbachev that cooperative studies should be put in place to evaluate the distribution of radionuclides around such potential sources as sunken nuclear submarines or military and civilian marine radioactive waste sites! But this is exactly what happened in the Gorbachev years.

During the Cold War, there were some bright spots. In the years immediately following World War II, the United Nations divided the world into a number of economic commissions – one of which became the United Nations Economic Commission for Europe (UNECE). Despite its misleading title, the UNECE was composed of the countries of Europe north of the Mediterranean, the Soviet Union, Canada and the United States. This membership remains the same today – except we now have the "new" states that emerged from the disappearance of such confederations as the Soviet Union and Yugoslavia.

In the 1970s, the East and West realized that the transboundary atmospheric transport of sulphur and nitrogen products of power generation and metal smelting was leading to widespread acidification and the degradation of forest and freshwater aquatic ecosystems. This led to the successful negotiation in 1979 of the UNECE Convention on Long-Range Transboundary Air Pollution (CLRTAP) and of its successive legally binding protocols,[1] which detail the actual actions to be taken by parties to combat acidification. What is really interesting about the CLRTAP acidification protocols is that their implementation has evolved to become responsive to results coming from the European Monitoring and Evaluation Programme (EMEP). The CLRTAP/EMEP experience served as a tantalizing example of what could be done in a rational world. It was, however, a lonely example. In addition, with the very important exception of northern Scandinavia and north-western Russia, the CLRTAP was not active in the Arctic. There was no Arctic Messenger.

In the mid-1980s, the Soviet economy was moribund when (in 1985) Mikhail Gorbachev was appointed general secretary of the Communist Party, a post he held until 1991. It quickly became obvious that he was not of the same mould as his predecessors. Almost immediately, he embarked on a dramatic paradigm shift of policy that was grouped under two broad umbrellas: *perestroika*, which consisted of a total restructuring of the Soviet economy, and *glasnost*, which referred to new levels of openness and transparency within the Soviet political and governmental system. One of the first signs of the changes ahead came immediately. Discussions began between Canada, Denmark, Finland, Iceland, Norway, the Soviet Union, Sweden and the United States about the need for a nongovernmental forum to encourage and facilitate international cooperation in Arctic science. Eventually, this led to the creation of the International Arctic Science Committee (IASC) in 1990. Then, in 1987, Mr. Gorbachev visited the north-western port of Murmansk to confer the Order of Lenin on the city. The event is remembered today for a speech that is generally recognized as the first bold step towards a new cooperative Arctic. It is a curious hybrid of rhetoric – typical of either protagonist in the Cold War East-West divide – followed by six proposals for the Arctic that are akin to a "Damascus road" vision. The first two proposals called for a nuclear weapons–free northern Europe and for a reduction in military activities in the region, accompanied by the promotion of military trust. Next were proposals for the

---

[1] In order to avoid confusion, I have not followed the custom used by such international organizations as the United Nations and in technical literature that use an uppercase initial letter whenever an existing convention, protocol, or organization is referred to but its full title is not spelled out.

cooperative development of Arctic resources and for opening up the north-eastern sea route to international shipping. Finally came two proposals aimed to protect the Arctic environment, which included the development of a Pan-Arctic comprehensive plan to bring it about and the setting up of a corresponding environmental monitoring programme.

Political speeches usually have their origins in text fragments provided by several authors. The final mosaic may not be settled until moments before the politician speaks. I have often been a part of this scramble, and when one looks at translations of the Murmansk speech, one is struck by the contrast in style and content between the first and second parts. I often wondered where did the material for the second part come from? There were rumours of a story that should soon be confirmed. In the 2013 IASC bulletin, it was announced that Odd Rogne from Norway is working on a document that includes the committee's early history. Here is an enticing quote from the announcement: "Wording in the IASC planning papers found its way into President Gorbachev's Murmansk speech (1987), a speech that changed the Soviet Arctic policy fundamentally." It will be fascinating to read Odd's document.

But what happened after Murmansk? If I had been asked at any time until January 11, 1989, I would have said that very little changed. I would have been wrong. People in Helsinki had been very busy. On January 12, I was sitting comfortably in my office in Ottawa when Garth Bangay (my boss at the time) bounded in with his usual enthusiasm. He threw a file on my desk and shot off down the corridor, yelling over his shoulder "Come over in fifteen minutes and tell me how you think Canada should respond." The file was from our Department of Foreign Affairs and inside was a letter from Kalevi Sorsa and Kaj Barlund, the then-ministers of foreign affairs and environment for Finland, respectively. In a nutshell (although they did not mention Murmansk), they were floating a plan to implement the environmental elements of Mr. Gorbachev's 1987 proposals. It was all very simple. In the brief letter, they devoted a paragraph to determine that the Arctic environment was not in a healthy state and a few more paragraphs to judge that it was the collective responsibility of the eight governments of countries that surround the Arctic to do something about it. As a first step, they proposed holding a consultative meeting in short order in Finland. Attached to their letter was a four-page working paper that added a little more flesh to their ideas. For example, the working paper explicitly noted that "effective protection of the

Arctic region requires development of intergovernmental cooperation, scientific research and monitoring of the ecosystems of the Arctic region" – words that most probably reflected Finland's satisfaction with the role of the EMEP in the CLRTAP protocols.

The paper also showed that Finland foresaw an intergovernmental process that would lead to coordinated action to protect the Arctic environment, not just a statement of common objectives. The instrument to be used to implement such proposed actions could be a declaration, a convention or some other form of multilateral arrangement.

At the time, I was inheriting responsibility for a fledgling Canadian environmental monitoring programme in the Arctic from Garth. We wanted to measure the levels of certain industrial and agricultural persistent organic pollutants (POPs), such as DDT (dichlorodiphenyltrichloroethane) and polychlorinated biphenyls (PCBs), in Arctic air, water, biota and people. We were perplexed because early results showed that some chemicals were at much higher levels in the Arctic than in southern Canada, even though a specific chemical may never have been used in the Arctic or in Canada. We suspected long-range atmospheric transport from sources in mid-latitudes, but we had little data to evaluate our hypothesis. A single scan of the letter from Finland was enough to set me dreaming of a sparkling circumpolar necklace of pollutant monitoring stations. In less than the 15 minutes provided by Garth, we were both in his office, drafting a supportive reply from Canada to Finland.

It has become the norm to trace a succession of developments from Mr. Gorbachev's speech. It is true that some interchange of ideas was crossing the Iron Curtain well before 1987, such as the setting up of the IASC, but there was nothing on the scale of intergovernmental commitment proposed by Mr. Gorbachev and carried forward under the leadership of Finland three years later.

In September 1989, the first consultative meeting of what was then called the "Finnish Initiative" was held in northern Finland at Rovaniemi. It was agreed that no international arrangements existed with the capacity to coordinate effective environmental cooperation in the Arctic. A work plan was developed to decide what should be done. First, six "state of the Arctic environment reports" should be prepared on (1) acidification, (2) heavy metals, (3) marine underwater noise, (4) oil pollution, (5) organochlorine pollutants and (6) radioactivity. Second, Norway and Russia were to prepare a review of existing environmental monitoring activities in the Arctic. Finally, Sweden was to prepare a discussion paper on practical early actions and Canada was to develop a similar paper on possible common objectives and principles for a future Arctic sustainable development strategy. Jill Jensen prepared the report on organochlorine pollutants. Things were off to a flying start with a momentum that was sustained right until the arrangements for cooperation were signed in 1991. Garth was involved throughout the negotiation process, and over the following years, I learned from Nordic diplomats just how much his boundless ebullience contributed to the process.

During the 1989–1990 winter, I first began to hear the name of Lars-Otto Reiersen, the Norwegian who was leading the Norway-Russia responsibility of reviewing arrangements for Arctic environmental monitoring. I was anxious to meet Lars-Otto and I tried to arrange a detour to Oslo while in Europe attempting (at that time without success) to generate interest in establishing an international agreement to control POPs (such as DDT and PCBs) that were of growing concern in the Canadian Arctic. Then, in March 1990, I was in Moscow, Yekaterinburg (then called Sverdlosk), Murmansk and finally Leningrad (the city did not revert to its former name of Saint Petersburg until 1991), trying to develop working relationships with Russian environmental chemists for the joint monitoring of POPs in the Arctic. I knew that the Arctic and Antarctic Research Institute (AARI) in Leningrad could be fertile ground because I had been there twice before since 1988. When I got back to my hotel, there was a message from Garth via our embassy telling me that Lars-Otto was in Leningrad at the AARI and was looking for me. He was staying at the Hotel Olympic. I could not believe my luck.

I learned that Lars-Otto's meeting at the AARI was over, but we had to quickly find each other's hotels because I was leaving for Geneva the following afternoon. Like me, Lars-Otto had been told that I was booked at the Hotel Leningrad, but in those days, the Soviet travel office (Intourist) put you where they felt like it at the moment when you happened to walk in to claim your booking. There was no such thing as a single hotel desk at a Soviet hotel. It was a succession of office windows that gradually brought you closer to the Amazonian key lady on your assigned floor. I had "failed" at the first window of the Hotel Leningrad and been reassigned to the Hotel Moscow. I knew that he would probably never find me. Therefore, it all depended on my finding the Hotel Olympic. No one had a clue where it was. It was a failed rendezvous. When I finally did meet Lars-Otto, he explained that the Hotel Olympic was the accommodations part of the *Henrik Ibsen* that had been placed on a large barge and moored in the Neva River.

The *Henrik Ibsen* had been a sister to the *Alexander Kielland*, an offshore oil accommodations' platform that had capsized with the loss of 123 lives at the Ekofisk Oilfield in March 1980. Immediately after the accident, the *Henrik Ibsen* was removed from the North Sea and recycled to help with an unanticipated impact of perestroika: a huge demand for hotel space in Russian cities.

Arctic governments held two more Finnish Initiative preparatory meetings: one in Yellowknife (Canada) in April 1990 and the other in Kiruna (Sweden) in January 1991. During this period, the proposed Arctic sustainable development strategy evolved into the Arctic Environmental Protection Strategy (AEPS) and its component activities and programmes began to take shape. The key development for Arctic environmental monitoring took place in November 1990 at an expert meeting hosted by Norway at the Hotel Bristol in Oslo. It was there that I finally met Lars-Otto and it was immediately obvious to me that we had an outstanding champion for the Arctic environment. He had previously worked on the North Sea Task Force, an international science programme set up to measure the state of health of the North Sea. He could easily identify the key elements of cooperative monitoring and the challenges we could expect. It was a long and difficult meeting. There will be more on Lars-Otto's unique skills later, but they include the ability to focus on an objective while listening carefully and sympathetically to what others have to say. He patiently guided us through the agenda, and at the end of the week, we had a plan for the Arctic Monitoring and Assessment Programme (AMAP) that was ready for review at the January 1991 Finnish Initiative meeting in Kiruna. Today, the Finnish Initiative is more often called the Rovaniemi Process.

Much of our negotiation of the structure of AMAP took place in Moscow and Leningrad/Saint Petersburg during the painful economic change that accompanied the transition from Soviet Union to Russia. After driving past endless queues of freezing people hoping to buy a loaf of bread on snow-covered streets in the winter of 1990–1991, I was embarrassed by my own comfort. We easily forget how fragile our civilizations can be.

In June 1991, ministers for the environment of Canada, Denmark, Finland, Iceland, Norway, Russia, Sweden and the United States met again in Rovaniemi and signed on to the Declaration on the Protection of the Arctic Environment, which laid out their long-term plans and vision. Accompanying the declaration was the more comprehensive AEPS, which provided details of the arrangements that would be established to put their plans into action.

The workhorses of the AEPS were identified as AMAP, tasked with monitoring and assessing the state of the Arctic environment; the Conservation of Arctic Flora and Fauna (CAFF), tasked with exchanging information and the coordination of research on species and habitats; the Emergency Prevention, Preparedness and Response in the Arctic (EPPR), tasked with providing a framework for cooperation with environmental emergencies; and the Protection of the Arctic Marine Environment (PAME), tasked with reviewing measures regarding Arctic marine pollution. A working group supported each activity. The ministers agreed they would meet at regular intervals (two years) to review progress and to further develop the strategy. In 1996, a new working group on sustainable development was set up, and in the same year (through the Ottawa Declaration), the Arctic Council was inaugurated. At that time, the Arctic Council subsumed within itself all activities of the former AEPS but did not hold its first meeting until 1998 in Iqaluit, Canada.

The two major changes in this development were, firstly, that the underlying raison d'être became sustainable development rather than environmental protection, and consequently, national representation moved from environment ministers to foreign affairs ministers. This change made ministerial declarations potentially more powerful because a minister of foreign affairs speaks for all ministries within a country[2]. Secondly, a new category of representation in all activities of the Arctic Council was created – known as *permanent participants* and composed of the indigenous peoples of the Arctic. These are the Russian Association of Indigenous Peoples of the North (RAIPON), representing about 250,000 people mainly from Siberia; the Inuit Circumpolar Council (ICC), representing about 150,000 Inuit in Alaska, Canada and Greenland; the Sami Council, representing more than 100,000 Sami people within Norway, Sweden, Finland and north-western Russia; the Arctic Athabaskan Council and the Gwich'in Council International, representing about 32,000 people and 9,000 people each in north-western Canada and south-eastern Alaska, respectively; and the Aleut International Association, representing at least 2,000 people from the Aleutian Islands. Arctic indigenous peoples' organisations had already been involved in the work of AMAP (since the latter was formed),

[2] At the time of writing (2013–2014), Canada is chairing the Arctic Council. Curiously, the minister assigned to this role is Canada's minister for the environment, a change that some Arctic Council countries fear may diminish the council's status.

and for simplicity, I have used the term *permanent participants* to refer collectively to these organisations before and after the creation of the Arctic Council.

This was a very significant step because, in theory, circumpolar indigenous peoples could now participate at all levels of activity within the Arctic Council. However, although the permanent participant organisations receive funding to attend meetings of the Arctic Council and are provided with a secretariat now located in Tromso (Norway), there is no blanket funding to support their active participation in Arctic Council projects or programmes. In some cases, a national programme – the purpose of which is closely aligned to that of an Arctic Council working group (such as exists between the Canadian Northern Contaminants Programme and AMAP) – may provide some project funding or the working group secretariat may be able to organise funding. Otherwise, significant funding for participation in actual Arctic Council work tends to occur only when one or more of the permanent participants can carry out a component of that work. The latter occurred with the 1997–1998 AMAP *Assessment Reports* and the 2005 *Arctic Climate Impact Assessment Report*, where the permanent participants or their representatives prepared or significantly contributed to entire chapters. These two examples provide powerful glimpses of the contributions that could be made by the permanent participants if a more stable funding arrangement could be found.

The timing of the Rovaniemi and Ottawa declarations was serendipitous. They slipped into that short period of global optimism when East and West politics were emerging from their mutual isolation and when the work of the Brundtland Commission and the 1992 Conference on Environment and Development Summit appeared poised to move the world economies towards a more sustainable future. If you may be tempted to learn more about the aspirations, disappointments, triumphs and national rivalries of those who strove to create the AEPS and the Arctic Council, I recommend the highly readable *Ice and Water* by John English. It provides a clear historical overview, particularly from the viewpoint of the Canadian protagonists.

The architecture of the AEPS and the subsequent actions embedded in work plans of what is now the Arctic Council provided the Arctic Messenger with not just a voice but potentially a very powerful one. This is how the responsibilities of AMAP and the ministerial levels of the AEPS and the Arctic Council have been described (quotation marks indicate direct extracts from the Rovaniemi Declaration and the AEPS):

1. AMAP is established to set up arrangements "to monitor the levels of, and assess the effects of, anthropogenic pollutants in all components of the Arctic environment and since 1991" and, more recently, "to include effects of increased UV-B radiation due to stratospheric ozone depletion, and climate change on Arctic ecosystems" with "special attention on human health impacts and the effects of multiple stressors".

2. AMAP is implemented by a task force (soon renamed as a working group) and is supported by a secretariat provided by Norway.

3. The AEPS/Arctic Council ministers receive regular State of the Arctic Environment Reports summarizing the results from AMAP. The assessments report on status and trends in conditions of Arctic ecosystems; identify possible causes for changing conditions; detect emerging conditions, their possible causes, and the potential risk to Arctic ecosystems including indigenous peoples and other Arctic residents; and recommend actions required to reduce risks to Arctic ecosystems.

4. The ministers assimilate the information they have received.

5. The ministers commit "to full implementation" (of the Rovaniemi and Ottawa declarations) "and consideration of further measures to control pollutants and reduce their adverse effects on the Arctic environment".

6. The ministers resolve "to pursue together in other international fora those issues affecting the Arctic environment which require broad international cooperation".

More recently, the Arctic Council approved an AMAP Strategic Framework 2010+ document that summarizes the responsibilities of AMAP as follows (omitting references to geographical scope):

> AMAP has a mandate to monitor and assess the status of the Arctic region with respect to pollution (e.g., persistent organic pollutants, heavy metals, radionuclides, acidification and petroleum hydrocarbons) and climate change issues by documenting levels and trends, pathways and processes, and effects on ecosystems and humans, and by proposing actions to reduce associated threats for consideration by governments. This mandate is fulfilled through the implementation of a circumpolar monitoring and assessment programme as outlined in this strategic framework and in a separate monitoring plan document. . . .

AMAP's primary function is to provide sound science-based information to inform policy and decision-making processes in relation to issues covered by its mandate. AMAP aims to make effective use of up-to-date information and results from monitoring and research activities, and to promote and harmonize activities under relevant national and international programmes that can support AMAP assessments.

If all these elements function as intended, this design enables the Arctic Messenger to inform Arctic Council governments and the general public on its state of well-being. If the news is not good, the governments are expected to act individually using their domestic environmental legislation and, if necessary, to work together internationally when regional or global action is needed. How well this scheme has succeeded in directly linking science to policy and action is a question that will gradually unfold as we move to the end of this story. In some cases, it has managed quite well, but in others, it has been less successful. What can be said without reservation is that Arctic governments have regularly received unequivocal scientific assessments.

AMAP is not alone in having responsibilities relating to Arctic environmental science. Other organisations include the IASC, the Arctic Ocean Science Board and, within the Arctic Council, the work undertaken by CAFF. However, I know of no other programme that combines such a wide environmental science and protection mandate with a process that includes political assimilation, domestic action and real leverage to promote international remedial action. AMAP's monitoring is by definition reiterative, thus enabling the detection of environmental trends over time and the setting up and adjustment of remedial actions.

My close involvement with AMAP began when Heikki Sisula from Finland and I were elected as the first chair and vice chair, respectively, at the inaugural meeting of the AMAP working group in 1991. In 1993, Heikki stepped down and I became the chair, with Lars-Erik Liljelund from Sweden as the vice chair. In 1998, Lars-Erik became chair. Lars-Erik had already held senior management positions in Sweden dealing with acid rain and he later became director general of the Swedish Environmental Protection Agency. While we worked together on AMAP, he was director of the Swedish government's Environmental Advisory Council. His office was adjacent to that of his minister, Anna Lindh. As I had come from the stuffy confines of government in Ottawa, it was a revelation to meet such a vibrant minister who arrived at work on her bicycle, would join us for a coffee break and was sometimes accompanied by her preschool child. Her premature death at the hands of an assassin was a tragedy not just for Sweden but also for Europe. Lars-Otto Reiersen was appointed as executive secretary to the AMAP secretariat in Oslo in 1991. He quickly recruited two outstanding deputy secretaries: Simon Wilson and Vitaly Kimstach. Simon had worked with the International Council for Exploration of the Seas (ICES) on the North Sea Task Force and had accumulated extensive knowledge on data quality control, storage and interpretation. Vitaly was the former deputy head of the Russian Federal Service for Hydrometeorology and Environmental Monitoring and carried immense experience, particularly, of course, on Russian environmental programmes. The seven years during which we built AMAP were the most rewarding of my working life.

Our first task was to set up the monitoring programme that the AEPS ministers had made clear should be based on existing national and international programmes. The basic process that was used to assemble and evaluate the contributing programmes was essentially the same as the process that is now used to maintain it. An idealized AMAP monitoring plan was developed, which is periodically adjusted to reflect new knowledge, evaluations of environment needs or fresh direction from the AEPS/Arctic Council ministers. Countries submit their National Implementation Plans (NIPs), which detail their existing or new activities that can contribute to the overall monitoring plan. This may be supplemented by additional contributions from non-Arctic countries and regional and international organisations. The next step was to establish how they could be harmonized in terms of such matters as chemical analysis and quality assurance/quality control (QA/QC). This is absolutely essential because many of the pollutants of greatest concern are measured at (and may have toxicity at) levels of parts per billion or even parts per trillion. Such harmonization is not easy, largely because any long-running monitoring programme is likely to be reluctant to make major changes in methodology due to the fear that analytical results after the change will not be comparable to those made earlier. This could jeopardize the possibility of such programmes being able to detect environmental change over time.

One of the practices used to understand these issues is to organise periodic "round robins", where a subdivided test sample and/or a certified "standard" is analyzed blindly by participating laboratories and the results compared by an independent party. When a new contributing project enters into AMAP, the project organisers often now decide to use an existing laboratory that is already a part of the programme in order to simplify the harmonization process.

The contributing programmes gathered through the NIPs invariably reveal gaps in coverage relative to the idealized plan. In some cases, an Arctic country has been able to adjust its contributions to add new monitoring capacity. This was done on a large scale by Denmark/Greenland with the initiation of a programme of contaminant monitoring in Greenland and the Faroe Islands that is still in existence today. In other cases, the AMAP secretariat has been able to organise funding to enable knowledge gaps to be filled. An example of the latter occurred when the first AMAP report of 1997–1998 revealed a large gap in understanding the level of exposure and human health aspects of persistent toxic substances (PTS) in eastern Arctic Russia. Here, the term *PTS* was used in order to include POPs, polycyclic aromatic hydrocarbons and several heavy metals, such as mercury. With financial support from such organisations as the Global Environment Facility and the Nordic Council of Ministers and most of the Arctic Council countries, a large project was organised to address the situation. It was called Persistent Toxic Substances, Food Security and Indigenous Peoples of the Russian North and was coordinated by the AMAP secretariat with assistance from RAIPON. The results were then incorporated into subsequent AMAP reports and were also published separately in 2004. It was the first of many projects where funding has been raised by the secretariat, which has then usually provided management of the activity. The contributions of Pavel Sulyandziva (from RAIPON) and Vitaly Kimstach were crucial to the success of this project.

A different type of project coordinated by the AMAP secretariat that has become more common over the years has been the production of reports and analyses that are supportive of the aims of the Arctic Council but which are actually completed at the request of a different organisation. For example, the secretariat has produced a number of reports for the Nordic Environment Finance Corporation (NEFCO) to support the evaluation of proposals for remedial actions at pollution "hot spots" in the Russian Barents Sea region.

To bring information together for the preparation of assessment reports, the secretariat set up the AMAP Project Directory, which contains a description of national projects and programmes that contribute to the overall AMAP programme. In addition, over time, it arranged for the establishment of AMAP Thematic Data Centres, where results from contributing projects are compiled. They are then made available to scientists participating in AMAP assessments under a protocol that protects the publication rights of those who undertook the original research or monitoring. Two other important roles of the data centres are to record such matters as the quality assurance regimes that have been applied to each set of data and to provide the foundation for AMAP's plans to maintain long-term archives of information to support work on environmental change over time. At present, AMAP Thematic Data Centres include the Norwegian Institute for Air Research (NILU) in Norway that provides this service for atmospheric data, the ICES in Denmark for marine data and the Norwegian Radiological Protection Agency (NRPA) in Norway for radioactivity data.

The next big headache was to decide what the first full assessment report (due to be completed in 1997) would look like and how it would be written. We quickly settled on a basic report that would be a detailed and fully referenced document prepared by a team of scientists, together with experts nominated by the permanent participants. It would be called the *AMAP Assessment Report [AAR]: Arctic Pollution Issues* and would provide the foundation for a second document to be written in plain language and suitable for use by politicians and the general public. With an embarrassing lack of imagination, we called this report "Arctic Pollution Issues: A State of the Arctic Environment Report" (SOAER).

The AMAP working group was composed of managers of national science programmes. These people were exactly what we needed to help set up the content and operational practices of AMAP. However, to coordinate writing the scientific AAR, we needed the best active Arctic scientists we could find. To fulfill this function, we created an assessment steering group chaired by Lars-Erik. The separation of the AMAP working group from the SOAER also served to distance the drafting of the scientific AAR from any possibility of political interference.

The AAR itself was a single volume containing 12 chapters. Eight environmental issues were identified and each provided with an individual assessment chapter: (1) contaminant transport to and within the Arctic, (2) persistent organic pollutants, (3) heavy metals (mainly, lead, cadmium and mercury), (4) radioactivity, (5) acid rain, (6) petroleum hydrocarbons, (7) climate change and stratospheric ozone depletion and (8) pollution and human health. The four remaining chapters framed the geophysical, ecological and social conditions of the Arctic, therefore serving to put the assessment chapters into context. For each chapter, one or more "editors" gathered a team to prepare the assessment. In addition to writing significant parts of the assessment themselves, the editors were assisted by up to 60 "authors", who also contributed text but did not compile the chapter. Finally, "contributors" provided information or data that was unpublished at the time of the assessment. Each chapter was then subjected to an independent scientific review. The latter was not always easy to organise because it was often difficult to find specialists who were not involved in some way with the chapter.

The "social" chapter was actually called "Peoples of the Arctic: Characteristics of Human Populations Relevant to Pollution Issues". The permanent participants, who also provided most of the membership of the assessment team for this topic, nominated Henry Huntington as the editor. The chapter provided an essential resource for the assessment of human health risk from pollutant exposure as undertaken by the human health, radioactivity and climate assessment teams – a factor that led to a substantial interchange amongst experts between the drafting groups. Janine Murray from Ottawa spent some time in Oslo at this stage, helping the AMAP secretariat create a harmonious AAR document from the mosaic of individual inputs and also coordinating chapters on the Arctic's physical and ecological characteristics.

The plan was for the "plain language" SOAER to be prepared by a professional science writer, Annika Nilsson, who would have the scientific AAR in front of her as she wrote. However, we were committed to present the SOAER at the Fourth AEPS Ministerial Conference in Alta, Norway, in 1997. Some of the AAR chapters were not completed in time, and in these cases, Annika had to resort to a combination of using drafts and interviews with editors and authors. The final check on the veracity of the SOAER to science was the review by AAR drafting teams of the draft SOAER. Annika met her 1997 Alta deadline and the AAR was presented to the First Ministerial Meeting of the Arctic Council in Iqaluit, Canada, in 1998. The impact of these reports in providing the key political motivation for regional and global actions to control POPs and heavy metals under the conventions of the United Nations will be explained in the following pollutant chapters of Part 3.

An amusing event occurred in Alta. Just before the AEPS/Arctic Council ministerial conferences, senior Arctic government officials (SAOs) meet to review the key findings and recommendations coming from the working groups. They then negotiate amongst themselves the wording of the declaration and work plan of the meeting for ministerial endorsement and signature. Of course, in order to preserve unanimity, the text provided to ministers tends to represent the most progressive steps that the least progressive country can accept. In general, this is common practice for most intergovernmental meetings and it continues to be the practice in the Arctic Council today. To be fair, there may be no other practical way. It is a depressing thought when you consider the pressing imperative for bold actions to limit climate change. In Alta, the very first AMAP plain language assessment report (SOAER) was the most significant item on the agenda and the text prepared by the SAOs to map out the collective ministerial response was very cautious and disappointing for most of us with a concern for the Arctic. However, we were not alone with this judgement. The AEPS at that time was chaired by Norway. When the ministers arrived, they took about 10 minutes to compare the AMAP report with the response prepared by their senior officials before deciding that they wanted an action plan to be proud of. What is more, they set about writing it themselves. It was a sweet moment!

The next comprehensive AMAP assessment was published in 2003–2004 as four separate reports dealing individually with updates in knowledge on POPs, heavy metals, radioactivity and human health. In addition, there was a fifth report that originated from an initiative of Robie Macdonald, a Canadian senior oceanographic chemist who we will encounter later. Towards the end of the 1990s, he pointed out to the working group that climate change could influence the way in which contaminants are transported to and from the Arctic and how they move between media (such as the atmosphere and water – a process known as *partitioning*).

Following their review of the climate chapter contained in the 1998 AAR, Arctic Council ministers requested AMAP, together with CAFF and the IASC, to complete an Arctic Climate Impact Assessment (ACIA). This was a massive undertaking. The plain language overview "Impacts of a Warming Arctic" was presented at the Fourth Arctic Council Ministerial Conference (Reykjavik, Iceland) in 2004 and the full 1,030-page scientific ACIA report followed a year later. It proved to be a much-cited document. It firmly established functional linkages between AMAP and the Intergovernmental Panel on Climate Change (IPCC) and brought the Arctic dimension of climate change to the attention of the United Nations Framework Convention on Climate Change (UNFCCC). All these reports (except the ACIA overview) were written and peer-reviewed following the basic format set up for the 1998 AAR.

The AMAP assessments of 1997–1998 and 2002–2004 were very successful. However, this model had a voracious appetite for human and financial resources that could not be sustained. It was also slow and lacked sufficient flexibility to quickly refocus effort towards new and changing priorities. In 2005, I was asked to suggest a new business strategy for AMAP. The new strategy is founded on the decision to produce three types of assessment output:

1. Outputs prepared at the request of the Arctic Council or for international organisations with whom specific cooperative activities have been identified by Arctic Council ministers as being crucial for protecting the Arctic environment. Examples include:

   • Comprehensive circumpolar AMAP *Assessment Reports*, such as those of 1997–1998 and 2002–2004 but now produced at roughly 10-year intervals or as required. The circumpolar assessments are produced in the same way as the 1997–1998 plain language SOAER and the scientific AAR – except that as with the 2002–2004 assessments, each issue or report is now produced individually. Frequently, the scientific AAR would form the basis of a series of linked papers published in a special issue of a well-known and respected scientific journal. Recently, AMAP has been experimenting with using this approach in place of the AAR. It reduces redundant efforts on the part of scientists and increases the exposure of the assessment in the global scientific community.

   • AMAP reports on "Issues of Concern". These reports are not individually requested by ministers but enable AMAP to rapidly inform ministers of key information resulting from the work of AMAP. If necessary, they may be produced at a greater frequency than the two-year cycle of ministerial meetings.

   • Assessment information prepared to assist in evaluating the effectiveness and sufficiency of agreements for protecting the Arctic environment, including the POPs and heavy metals protocols to the CLRTAP, and the Stockholm Convention on POPs under the United Nations Environment Programme (UNEP).

2. Reports or contributions to reports produced and normally funded by international organisations and linked to existing AMAP activities. Examples since 2004 include: The AMAP/UNEP *Background Report to the Global Atmospheric Mercury Assessment* (2008); the follow-up AMAP/ UNEP *Technical Background Report for the Global Mercury Assessment* (2013); the UNEP/AMAP report on *Climate Change and POPs: Predicting the Impacts* (2011); and AMAP reports on the *Impact of Short-Lived Pollutants on Arctic Climate* (2008), on *Sources and Mitigation Opportunities to Reduce Emissions of Short-Term Arctic Climate Forcers* (2008), on *Combined Effects of Selected Pollutants and Climate Change* (2011) and on *The Impact of Black Carbon on Arctic Climate* (2011). Work of this

nature is only undertaken when approved by the AMAP working group and following notification of senior officials of the Arctic Council governments. They are prepared by experts who are organised and coordinated by the secretariat.

3. Fact sheets, web-based information products, videos and leaflets are developed to communicate AMAP information and results to Arctic residents and the general public.

How well this strategy will serve AMAP and the Arctic Council in the next decade remains to be seen, but at the moment, it seems to be providing a satisfactory and practical framework.

This is all I want to say at this stage about the birth and evolution of AMAP. Four people have now completed terms as AMAP chair since the formative years with Heikki Sisula (Finland; 1991–1993), me (Canada; 1993–1997) and Lars-Erik Liljelund (Sweden; 1997–1998). They are Hanne Petersen (Denmark; 1998–2001); Helgi Jensson (Iceland; 2001–2004); John Calder (United States; 2005–2009); and Russel Shearer (Canada; 2009–2013). The present chair is Morten S. Olsen from Denmark.

I hope this chapter has provided a basic insight into the breadth of Arctic environmental issues being tackled by AMAP through its monitoring and assessment activities. At the same time, it is important to understand that AMAP reports to the Arctic Council. The potential impact of this relationship is that the council has political influence. It can take steps within Arctic countries in response to adverse environmental news from AMAP and can call for such actions beyond the Arctic. It truly has provided the Arctic Messenger with a voice.

PART III

What Is the Present State of Knowledge?

The chapters in this section provide a summary of what we have learned from the Arctic Messenger about ongoing changes in the state of the Arctic environment and of the significance of such changes not only to the Arctic itself but also to the global ecosystem. The choice of themes is personal and not all are covered with the same degree of detail. The main criterion I used to select a theme is the notion of an anthropogenic (human) activity causing an adverse environmental impact at a far-distant location. Consequently, I have not included such possible themes as chronic and acute oil spills. This is not a statement about the environmental damages that can be caused by such incidents but is a recognition that such spills do not generally result in circumpolar, hemispheric or global effects.

4

Radioactivity

Ophelia: "Rich gifts wax poor when givers prove unkind."
                                    William Shakespeare, Hamlet

Before starting to draft this chapter, I gave myself a short revision of school physics and chemistry. It could be useful to pass it on now to any reader whose memory is as poor as mine.

Fundamentals

The nucleus of every atom of an element has a fixed number of protons, which is equivalent to its atomic number. For example, the nuclei of hydrogen, helium, oxygen and uranium have 1, 2, 8, and 92 protons, respectively. Each proton carries a positive electric charge. Surrounding the nucleus are orbiting electrons, each of which carries a negative electric charge. Because the number of protons is nearly always equivalent to the number of electrons, the atom is neutral (carries neither a positive nor a negative charge). It is the number and arrangement of electrons (especially those in the outer orbits) that determine the chemical characteristics of the element.

In addition to its protons, the nucleus also contains neutrons. These have the same mass as the proton but do not carry an electric charge. Therefore, the same element can come in a number of different varieties depending on its total number of neutrons. These varieties are called *isotopes* or *nuclides*. There is a subtle difference in the definition of these two words, but we need not bother about it here. Isotopes (nuclides) of a substance are chemically identical but have a different mass. For example, uranium-238 (^{238}U) has 92 protons and 146

43

neutrons, giving it an atomic mass of 238. Uranium-235 (^{235}U) has the same number of 92 protons but has 143 neutrons.

Some nuclides are stable, but others are said to be unstable because they constantly show a tendency to change (decay) into something else, which may then itself also undergo further decay. For example, the nucleus of ^{238}U has a tendency to lose two protons and two neutrons and therefore transforms into thorium-234 (^{234}Th), consisting of 90 protons and 144 neutrons. Thorium-234 is itself highly unstable, with a half-life of only 24 days. Finally, after 14 transformations, the decay series of ^{238}U arrives at the stable isotope of lead-206 (^{206}Pb). In the middle of the ^{238}U decay series lies radon-222 (^{222}Rn), while the penultimate member in the chain is polonium-210 (^{210}Po). We will revisit both later. Each transformation step involves the release of energy known as radiation. This could be through:

- The emission of two protons and two neutrons (a helium nucleus), which is called an alpha (α) particle. We saw this in the decay of uranium-238 to thorium-234. This form of radiation (α) does not have much penetrating power and is unlikely to pass through the outer dead layers of the skin. However, biological damage can occur (such as an increased risk of cancer) if alpha particle emitters get into the body and expose tissues to alpha radiation.
- The emission of a beta (β) particle. These are physically equivalent to electrons but originate from inside the nucleus of some radioactive atoms. Such emission occurs, for example, when thorium-234 decays to protactinium-234. Some familiar sources of β radiation are tritium, cobalt-60, strontium-90, technetium-99, iodine-129 and iodine-13, phosphorus-32 and caesium-137. β radiation is able to penetrate 1–2 centimetres of skin, but inhaled or ingested beta particle emitters carry the greatest risk of causing biological damage at the molecular and cellular levels.
- A burst of energy called *gamma (γ) radiation*. This form of very high-energy ionizing radiation has no mass and no electrical charge. It can pass through many materials, including the human body. The high energy of gamma radiation can lead to biological damage directly and indirectly by causing secondary ionization of atoms with which they interact. A common naturally occurring source of gamma radiation is potassium-40, which is found in soil, water and foods with a high potassium content, such as bananas.

Nuclides that release energy are called *radionuclides*. The degree of instability of a radionuclide is expressed as its half-life. Looking again at the first three nuclides in the natural series – from uranium-238 to thorium-234 to protactinium-234 (^{234}Pa), we have half-lives of 4.47 billion years, 24.1 days and 1.17 minutes, respectively.

There are three categories of natural radionuclides: primordial radionuclides, which date from when Earth was formed (such as ^{238}U); those that are decay products of primordial radionuclides (such as ^{234}Th); and cosmogenic radionuclides produced when high-energy cosmic radiation has kicked a nucleus into an unstable condition, such as carbon-14 (^{14}C). Finally, there are fission products, such as caesium-137 (^{137}Cs), strontium-90 (^{90}Sr) and iodine-131 (^{131}I), and activation products, such as cobalt-60 (^{60}Co), which are all anthropogenic. Fission products have atomic masses in the range of 70 to 170 and are formed by the thermal fission of heavy nuclei, such as ^{238}U. Activation products result from neutron capture by stable nuclei in artificial high-neutron conditions, such as reactor cores and accelerators. They generally decay by γ ray and β particle emission.

Radiation that is so energetic that it can overcome the binding energy of electrons to atoms or molecules creates ions – a situation where the electric charge of the nucleus is no longer balanced by the charge of the electrons. Ionizing radiation is capable of disrupting the internal structure of biological molecules, including DNA. Gamma (γ) ray and alpha (α) and beta (β) particle emissions are all examples of ionizing radiation.

A Note on Units of Measurement and Concepts of Radiation Dose

Now it is time for an apology. The units and definitions used by radiation scientists can be quite confusing (to me anyway). I am going to offer the standard definitions mainly to help if you do some further reading. Do not be intimidated by this terminology. I will do my best to avoid it as much as possible in this chapter, but the definitions do help explain some of the concepts of radioecology and radioprotection.

The becquerel (Bq): This is the most common unit of ionizing radiation. One becquerel (Bq) is equal to one disintegration or nuclear transformation per second (1 Bq = 1/s or 1 s^{-1}). It replaced the older curie (Ci) unit.

The gray (Gy): This is used to quantify the amount of radiation energy that is absorbed per gram of biological tissue. This is known as the *absorbed dose*.

The sievert (Sv): The gray (absorbed dose) does not tell us all we need to know about dose because the same dose of α radiation is more biologically damaging than one of β or γ radiation. Therefore, a weighted dose has been devised – known as the *dose equivalent* – that is measured in sieverts. It is also equal to 1 J/kg. In addition to being used as a measure of dose equivalence, the sievert is also used to measure another important concept of dose known as the *effective dose equivalent.* This introduces a weighting factor (W_T) to account for differences in radiation sensitivity of different tissues and organs. The stomach is more radiosensitive than the oesophagus. If the stomach (W_T = 0.12) and bladder (W_T = 0.05) are exposed separately to radiation and the equivalent doses to the tissues are 100 and 70 mSv, respectively, the effective dose equivalent is (100 mSv × 0.12) + (70 mSv × 0.05) = 15.5 mSv. In other words, the risk of harmful effects from this radiation is equal to 15.5 mSv received uniformly through the whole body.

For radiological protection purposes, 0.001 sievert (1 millisievert, or mSv) is used as a yearly dose limit of exposure for members of the public to all anthropogenic radiation. It amounts to an increased risk of fatal cancer of 0.005%.

The man sievert (man-Sv): If you add all the individual effective dose equivalents of a group of people, you have what is known as the *collective effective dose equivalent,* which is measured in man-Sv.

Finally, if you are still with me, many radionuclides have half-lives measured in thousands of years. Therefore, we have the concept of *collective effective dose equivalent commitment,* which is the collective effective dose equivalent that will be delivered to many generations of people.

In the rest of this chapter, we will frequently hear about the effective dose of radiation resulting from a particular source and the reader will probably be wondering what the health implications of such a dose may be. In fact, this is a very complicated question with no simple answer. However, I hope the following will help. To investigate the long-term effects of radiation, scientists speak about stochastic effects. These are radiation-induced health effects that occur without a threshold level of dose. The probability of the expected effect is proportional to the dose, but the severity is not. Radiation-induced cancers and hereditary effects are examples of stochastic radiation effects. They characteristically have a late appearance measured in years (for cancer) or in generations (for hereditary effects). Using this approach, the International Atomic Energy Agency (IAEA) conducted a life span study on survivors from Hiroshima and Nagasaki and calculated a fatal solid cancer mortality risk of 0.005% per mSv.

There will be no more new units of measurement to remember in this section!

Arctic Monitoring and Assessment Programme (AMAP) Assessments

Assessments of the state of Arctic environmental radioactivity began in August 1993, when the Norwegian Radiation Protection Authority (NRPA) and the Department of Radiation Physics in Lund, Sweden, organised an international conference on the topic, held in Kirkenes, Norway, in August 1993. This was one of the first meetings that made extensive Russian information available for international review. The proceedings were published in December of the same year. As a legacy from the then-fading Cold War, Russia communicating with and giving environmental radioactivity information to Norway was more acceptable at that time to both of the old Cold War protagonists than was direct liaison with a group of countries.

Subsequently, AMAP has published comprehensive assessments of radioactivity in the circumpolar Arctic in 1998, 2002 and 2009. The driving force behind each of these reports was Per Strand of the NRPA. Several reports were also co-produced by AMAP and other organisations usually associated with practical remedial projects to address specific radiological issues within the Arctic.

To simplify the summarizing of present knowledge, I have roughly followed the lead of AMAP and have categorized our survey into historical, existing and potential anthropogenic sources, along with natural sources of radiation in the circumpolar Arctic.

Fallout From Atmospheric Testing of Nuclear Weapons, Peaceful Nuclear Explosions and the Chernobyl Accident

The main overall source of anthropogenic radiation in the circumpolar Arctic has been fallout from atmospheric nuclear weapons testing. Between 1945 and 1980, 520 such tests were carried out. Arctic fallout was in particular dominated by Northern Hemisphere tests, including the 88 tests that took place on the Arctic island of Novaya Zemlya. The radionuclides most commonly used to study this fallout are the fission products caesium-137 (^{137}Cs), strontium-90 (^{90}Sr) and iodine-131 (^{131}I).

Apart from distance from test locations, the actual geographical distribution of fallout depends mainly on patterns in air movement (which roughly translates to increasing fallout with increasing latitude) and the degree of wet deposition (that is, with rain). The latter essentially means that rain and snow tend to wash out pollutants, including radionuclides, from the atmosphere. Therefore, when the 1998 AMAP assessment team estimated the circumpolar ground distribution of nuclear fallout in the Northern Hemisphere from 50 to 90 degrees latitude, they found relatively low deposition in High Arctic areas that receive low levels of precipitation, such as northern Greenland (less than 250 ^{137}Cs Bq per square metre), and high levels in very wet regions, such as western North America (including southern Alaska) and the Atlantic coastal margins of Europe, including Ireland and southern Iceland (above 2,500 ^{137}Cs Bq per square metre). Much of the remaining western Eurasian Arctic received deposition in the range of 1,000 to 2,500 ^{137}Cs Bq per square metre.

AMAP has also reported a consistent time trend of decreasing deposition of nuclear fallout that is independent of geographic location. It directly reflects the falling frequency of Northern Hemisphere nuclear testing in the atmosphere from the 1960s to the global curtailment of testing in 1980. Therefore, in 1963, there was a combined wet and dry deposition rate of more than 500 Bq per square metre for ^{137}Cs and ^{90}Sr in Finland and Greenland, but by 1985, rates had fallen to well below 10 Bq per square metre. In 1986, the accident at the Chernobyl nuclear power plant in Ukraine resulted in the injection of a new inventory of fission products into the atmosphere. The weather patterns present at that time resulted in a substantial increase in ^{137}Cs and ^{90}Sr deposition rates in Fennoscandia. However, this proved to be a temporary setback to the long-term trend of diminishing deposition. Within five years, annual deposition of ^{137}Cs and ^{90}Sr at Arctic monitoring sites was back to pre-Chernobyl levels.

What is more important from environmental and human health perspectives is the fate of deposited radionuclides. Several characteristics of Arctic terrestrial ecosystems lead them to conserve certain radionuclides rather than rapidly burying them in the organic layers of soils. For example, lichens, mosses and mushrooms are major components of the Arctic flora. They are slow growing, obtain much or all of their water directly from the atmosphere and are consequently much more efficient collectors and retainers of atmospheric fallout than are rapidly growing mid-latitude plants and agricultural crops.

In winter, they form a vital food source for caribou and reindeer, to which they transfer their burden of radionuclides. The berries of slow-growing Arctic shrubs, such as cloudberry, bilberry and cranberry, also tend to accumulate and conserve radionuclides. Reindeer in north-western Europe and caribou in North America, together with wild berries, make up a large portion of the traditional diet of Arctic peoples who live in these regions. Therefore, looking back, it is not surprising that by the early 1960s, several studies reported that Sami peoples in northern Europe were accumulating body burdens of ^{137}Cs and ^{90}Sr that were of concern to national health agencies. The Sami subsequently found themselves facing difficult decisions when dietary advisories recommended restrictions on how much reindeer meat should be consumed. Such decisions related not just to their diet but also to the basis of their culture that is so closely linked to reindeer.

In comparison, AMAP assessments found that marine pathways were far less impacted by ^{137}Cs and ^{90}Sr deposition. Freshwater food chains (particularly those involving oligotrophic lakes) were subject to a greater level of contamination, but this was well below that involving reindeer, caribou and mushrooms.

It may be of small compensation for the Sami, but global recognition of their predicament led to a much wider interest around the world in the exposure of the general population to radioactive fallout from atmospheric weapons testing. In 1963, the Soviet Union, the United Kingdom, the United States and a large number of non-nuclear states signed the Limited Test Ban Treaty. It led to the progressive decline in atmospheric fallout we have just been reviewing. We will come back to this and several other agreements related to radioactivity at the close of our story. The annual collective doses attributable to fallout reached a peak in 1963 when they amounted to about 7% of the equivalent exposure to natural sources. By the early 1980s, this had decreased to 1%. However, these global averages hide regional variation, such as we have been reviewing in certain Arctic populations in the ^{137}Cs lichen-reindeer/caribou food chain. Here, accumulated radiation doses were estimated to have been between 100 and 1,000 times higher than natural background sources. Grass is also an efficient collector of atmospheric contaminant deposition and radionuclides can therefore be rapidly moved from grass to milk. Figure 4.1 shows the time trend of decreasing ^{90}Sr and ^{137}Cs activity in northern Finland between 1960 and 2008.

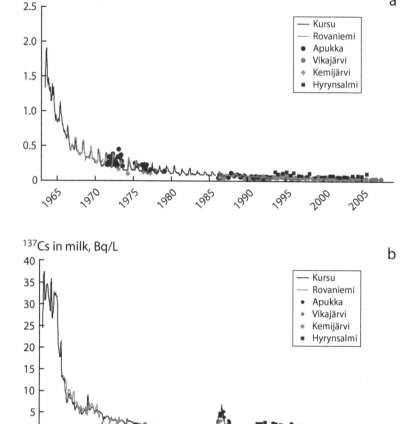

Figure 4.1 *Radioactivity concentrations of (a) ^{90}Sr and (b) ^{137}Cs in milk in Lapland since the early 1960s*

The 2009 AMAP radioactivity assessment confirmed that traces of radionuclide deposition from weapons testing and the Chernobyl accident could still be detected in the Arctic but that levels are now very low. However, their legacy continues to linger. Cold Arctic temperatures result in ^{137}Cs still being retained in the upper layers of thin nutrient-poor Arctic soils even 50 years after deposition. Meanwhile, ^{137}Cs that has been incorporated into boreal sub-Arctic trees is released again to the atmosphere during forest fires – a phenomenon that may become more common with global warming. One further point illustrates the long-term legacy of Cold War atmospheric testing and the futility of nuclear war. You will recall that one of the dose definitions we struggled through earlier in this chapter was called *collective effective dose equivalent commitment*. Well, the total collective effective dose equivalent commitment from all atmospheric nuclear explosions conducted up until the last one in 1980 was 30 million man-sieverts. At that time, only 12% had actually been delivered. The rest will reach humankind over millions of years.

This review has not considered underground tests. Provided they do not "vent", underground tests do not lead to the atmospheric transport of radionuclides. The long-term and poorly understood issue is the potential impact on groundwater. In addition to the testing of nuclear weapons, between 116 and 239 peaceful underground nuclear explosions (PUNES) have been reported to have taken place in the former Soviet Union, usually for mining, civil engineering purposes, stimulating oil and gas recovery and controlling runaway gas well fires. Seventeen of these PUNES occurred above or close to the Arctic Circle. In the United States, 26 PUNES were reported to have occurred between 1961 and 1973 under a programme called Operation Plowshare. The Peaceful Nuclear Explosions Treaty of 1976 set limits on how powerful PUNES could be, but the 1996 Comprehensive Test Ban Treaty has banned all such explosions.

Marine Discharge of Nuclear Wastes From Nuclear Reprocessing Plants at Sellafield in the United Kingdom and at Cap de la Hague in France

AMAP assessments found that the most widespread occurrence of radionuclides freely dispersed in the Arctic marine environment has not been from Cold War legacy sources in northern Russia. The main source has been from long-term waste discharge from nuclear reprocessing plants at Sellafield in the United Kingdom and at Cap de la Hague in France. They are called "planned" discharges, meaning they have resulted from the normal operation of these facilities and not from an accident or another malfunction. *Reprocessing* is the name given to activities that recover uranium and plutonium from spent nuclear fuel so they can be reused. Sellafield is located in north-western England. It is the only currently operating nuclear reprocessing facility in the United Kingdom and its normal operational practice has included the discharge of low-level radioactive waste into the Irish Sea. Sellafield discharges of the long-lived fission product technetium-99 (^{99}Tc) briefly peaked during the late 1970s to levels of more than 175 TBq per year. A terabecquerel (TBq) is 10^{12} becquerels. From then until 1993, discharges ran at about 5 TBq per year but then rapidly rose again after the introduction in 1994 of a new process designed to eliminate a large backlog of waste. Discharge of ^{99}Tc quickly reached levels of 190 TBq per year. Technetium-99 is highly soluble in seawater and it is therefore unaffected by sedimentation processes (in contrast to plutonium). Ocean currents have carried ^{99}Tc from Sellafield into the North Sea and northwards into the Barents Sea. The transport time from Sellafield to the Norwegian monitoring station at Hillesøy near Nordkapp is estimated to be about four years. By 2000, the annual average activity concentrations of ^{99}Tc in the seaweed *Fucus vesiculosus* at Hillesøy was between 300 and 320 Bq per kilogram of dry weight and ^{99}Tc activity could be detected as far away as the East Greenland Current.

The reprocessing facility at Cap de la Hague has also routinely discharged low-level radioactive wastes into the sea, but the main concern here was with iodine-129 (^{129}I). In the late 1990s, discharge of ^{129}I from Cap de la Hague was running up to 1.83 TBq per year. By 1999, 2,113 kilograms (Cap de la Hague) and 865 kilograms (Sellafield) of ^{129}I had been discharged into local sediments. To put this into perspective, it is one order of magnitude greater than the total ^{129}I inventory released to the environment by nuclear weapons testing. Sampling taken one month after the August 2000 sinking of the Russian submarine *Kursk* could not detect leakage from the vessel but confirmed the rising concentrations of ^{129}I derived from the European processing plants.

AMAP included information on the dispersion of ^{99}Tc in its reports to Arctic Council ministers, whose declarations at that time encouraged the United Kingdom and France to reduce Sellafield and Cap de la Hague discharges. In 2004, a new technology was introduced at Sellafield designed to extract ^{99}Tc and discharge levels quickly fell to 14 TBq per year. Since then, levels decreased and continue to fall in the Barents Sea and the Greenland Current region – in seawater and in *Fucus vesiculosus*. The 2009 AMAP radioactivity assessment reported that ^{129}I discharges from Cap de la Hague have fallen to between 1.0 and 1.5 TBq per year.

Local Arctic "Hot Spots" and Potential Sources of Radioactive Contamination

Ocean dumping of radioactive material into Arctic seas: I am going to begin this section with some memories of what was I think one of the most seminal meetings of my life. The tale began quietly enough with the second meeting of the AMAP task force in Toronto (Canada) in November 1992. It was the first time that any of us had met the new Russian delegate, Vitaly Kimstach. At the time, Vitaly was deputy chairman of Roscomhydromet, an enormous organisation responsible for most environmental monitoring in Russia. If we could in some way have added the cumulative influence of all the rest of us in environmental decision making, Vitaly would still have vastly outranked us! However, what really impressed everyone was his intelligence, knowledge and obvious ability to help bring Russia into the circumpolar monitoring plans of AMAP. Later, we learned about other attributes, including a wonderful sense of humour. We wondered for how long he would represent Russia in AMAP, but we never anticipated how things would actually come to pass. Great political upheavals were occurring at almost lightning speed. The Cold War crumbled and we all looked for a new future as the Soviet Union vanished with the emergence of the new Russian Federation. As part of the policy of glasnost (openness), Russian president Boris Yeltsin established a commission to report on the state of the Russian environment. Later known as the Yablokov Report, it included astonishing revelations concerning the Soviet legacy of radioactive waste and derelict radioactive equipment disposal in the Arctic. It was released in Russian early in October 1992.

Now we come to that remarkable meeting in February 1993. It was held in beautiful wooden buildings at Holmenkollen on the mountains overlooking Oslo and was organised by the International IAEA and the Norwegian government. We were isolated from the rest of the world by quite an impressive curtain of security arrangements. Vitaly had been a member of the Yablokov Commission and he led the Russian delegation. He had the unpleasant task of delivering technical details on the full extent of Soviet-era dumping of radioactive material into the Arctic seas. To give you an idea of the impact of glasnost, the delegates included naval engineers from Russia and NATO all sitting down together and examining naval information that must have been highly secret just weeks before! It marked a watershed moment in circumpolar environmental cooperation.

During one of the coffee breaks, Lars-Otto Reiersen and I were talking in the cloakroom about the implications of the meeting to AMAP. Vitaly walked in. He looked around at the otherwise empty room and then addressed Lars-Otto: "I hear you have a vacant position in the AMAP secretariat." "Yes", replied Lars-Otto. "Then, I have an excellent candidate for you," continued Vitaly. "It is me." Perhaps it is necessary to have grown up during the Cold War to appreciate the impact of Vitaly's proposal. It was not straightforward, but within a year, Vitaly and his family were living in Oslo and Vitaly began his remarkable contributions to the Arctic Messenger's story.

At the Holmenkollen meeting, we were given many of the details that lay behind the following facts (taken from the IAEA report of the meeting):

- Between 1965 and 1988, 16 marine nuclear reactors from seven former Soviet Union submarines and the icebreaker *Lenin*, each of which suffered some form of reactor accident, were dumped at five sites in the Kara Sea.
- Between 1960 and 1991, low-level liquid radioactive waste was discharged at sites in the White, Barents, and Kara seas.
- Between 1964 and 1991, low- and intermediate-level solid radioactive waste was dumped at sites in the Barents and Kara seas.
- Of the discarded marine reactors, six of the 16 still contained their spent nuclear fuel.
- Approximately 60% of the spent nuclear fuel from one of the three icebreaker reactors was disposed of in a reinforced concrete and stainless steel shell container.
- The vast majority of the low- and intermediate-level solid radioactive waste was disposed of in containers of unknown composition.
- The Kara Sea disposal sites for the 16 marine reactors and low- and intermediate-level solid radioactive waste varied in depth from 12 to 380 m. The icebreaker reactors and part of their spent nuclear fuel were reportedly disposed of in Tsivolka Fjord at an estimated depth of 50 m.

That February 1993 meeting at Holmenkollen marked the birth of the International Arctic Sea Assessment Programme (IASAP), organised by the IAEA with cooperation from the governments of Norway and the Russian Federation. It first assessed the risks to humans associated with radioactive waste dumped in the Kara and Barents seas and then examined the possibility of appropriate remedial actions. The conclusions of the assessment were published in 1997 and 1998 by the IAEA. The Russian information given in 1992 and at Holmenkollen in 1993 estimated that when the dumping occurred, it consisted of a total inventory of 37 PBq (a petabecquerel or PBq is 10^{15} Bq) being left in the ocean.

The study found that some of the low-level waste containers had probably leaked because elevated levels of some radionuclides in sediments could be detected within a few metres of the containers.

However, no measurable increase of radionuclides in the outer parts of the fjords or in the open Kara Sea was detected. To estimate risk posed by the dumped reactors and high-level waste, the study examined details concerning the construction and corrosion characteristics of their containment and developed several scenarios that could result in release of radionuclides to the ocean. They were:

- A slow release scenario occurring from the gradual corrosion of the barriers, waste containers and the fuel itself.
- Two catastrophic scenarios, causing an instant or accelerated release of the remaining radionuclide inventory.

The scenario release rates were then analysed using knowledge of the environmental behaviour of radionuclides to estimate radiation doses for three human population groups; for the world population; and for flora and fauna. The population groups consisted of an "extreme exposure" group of seafood consumers, an average north-east Russian population and a hypothetical group of security personnel patrolling the foreshore of the fjords close to the dumpsites. It was estimated that for all three scenarios, the maximum annual individual doses for members of the public were small (less than 1 mSv). Even the security patrols on the fjords were estimated to experience a radiation dose comparable to natural background radiation. The collective radiation doses for the slow release scenario for the world's population over the next 1,000 years were also estimated to be low, as were the radiation dose rates to a range of populations of marine animals from zooplankton to whales.

On the basis of their evaluation, the IASAP concluded that remedial measures, such as attempting to raise the dumped reactors, were not justified. At first, this may be a surprising recommendation, but it is far easier to understand when one realizes the potential hazards of causing a catastrophic failure of waste containment during any recovery operation. The report did recommend regular monitoring of the dumpsites.

The 1998 AMAP radioactivity team reached the same conclusions as those of the IASAP. In their 2009 assessment, the radioactivity group reviewed results of a survey conducted in 2002 and found that levels around the dumpsites did not show evidence of significant leakage to the environment. In fact, levels had generally fallen. The decision not to interfere with the ocean dumpsites does not mean they do not pose a long-term risk. It simply means it is safer at the moment to leave the material where it is. At present, the primary need related to the radioactive material dumped into the Arctic Ocean during the Soviet era is to ensure that regular monitoring continues.

Safe decommissioning and disposal of redundant Soviet-era nuclear submarines and related facilities: The studies of the dumping of obsolete nuclear submarines and of nuclear waste in coastal waters of the western Russian Arctic may in some ways have been something of an anticlimax. However, it raised international awareness of a related waste disposal issue and precipitated a major and little publicized period of technological cooperation that is still continuing today. The end of the Cold War eventually resulted in 198 former Soviet nuclear submarines being taken out of service – most of which were based around Murmansk. What has happened to them since 1995 and to Russian facilities associated with the management of the nuclear fuel cycle of these vessels, including the disposal of spent nuclear fuel?

This story begins with the creation in 1990 of the Nordic Environment Finance Corporation (NEFCO). It is an international finance institution that was established in 1990 by the five Nordic countries – Denmark, Finland, Iceland, Norway and Sweden – to help arrange for capital investments related to addressing environmental issues in the Nordic region. Its portfolio has now included environmental projects in Central and Eastern European countries, including Russia, Belarus and Ukraine. In 1994, NEFCO turned its attention to environmental issues adjacent to the Nordic countries and set up the Barents Region Environmental Programme to identify potential investment projects in the region. Overall organisation was coordinated by a steering group appointed by NEFCO and consisted of members from the Russian regional environmental authorities, the ministries of defence and of environment of the Russian Federation, Norway and Finland and the NEFCO and the AMAP secretariat. AMAP was charged with setting up two expert groups to do the work: one designed to address nonradioactive issues and the other to examine projects concerning radioactivity.

The involvement of AMAP in this work was a little problematic for Lars-Otto, Lars-Erik Liljelund and me (Lars-Erik and I were then vice chair and chair, respectively, of AMAP). We were only authorized to work on activities included in the work plans agreed on by the eight Arctic environment ministers at ministerial meetings. There was no procedure for changing our work plan between the two-year ministerial sessions. Getting involved in remediation activities associated with redundant military equipment in Russia was quite a significant step beyond the work we were expected to be undertaking. We came up with the procedural solution of asking the AMAP working group if it had any objection to the secretariat doing work that was supportive of Arctic environmental integrity, providing they had the person-power and budget to do it. No country objected and the secretariat was on its way. The cooperation between AMAP and NEFCO marked the beginning of a characteristic that has made AMAP so valuable to the Arctic Messenger and to Arctic governments. Not only has AMAP proved to be able to identify Arctic environmental issues, but it is also now sought after as a partner for the design and appraisal of practical activities undertaken to address such issues. Linking these two objectives has strengthened the ability of AMAP to keep its research and monitoring activities relevant to the needs of developing and implementing Arctic circumpolar environmental policy. This model later developed into the Arctic Contaminants Action Programme (ACAP), originally founded as an Arctic Council plan to address the most significant Arctic pollution sources identified through AMAP. ACAP became the Arctic Council's sixth permanent working group in 2006.

At the end of 1995, the AMAP secretariat forwarded proposals from the two expert groups to NEFCO. The following five issues were recommended for consideration by NEFCO as being priorities for practical remediation projects (not in any order of priority):

• Handling and transport of radioactive waste and spent nuclear fuel
• Regional storage for radioactive waste and spent nuclear fuel
• Development of alternative techniques for decommissioning nuclear submarines
• Nuclear safety at the Kola civilian nuclear power plant
• Risk and impact assessment associated with these proposed projects

An estimated budget was provided for each project. Other cooperative activities were and have been involved in working with Russia to secure its environmental management of radioactive wastes in Arctic areas. However, I believe that the initiative shown by NEFCO has been the most effective in catalyzing multi-stakeholder investment in addressing the issues. NEFCO was very successful in obtaining the co-support or interest of other international funding mechanisms, including the Global Partnership Commitment arising from the G8 meeting in Kananaskis (Canada) in 2002. As a result of these cooperative funding arrangements, 164 of the 198 Russian submarines taken out of service since 1994 had been decommissioned by 2008. Of the remaining 34 submarines, 11 were being decommissioned, leaving only nine remaining in the north-west Russian Arctic. The remaining submarines were located in far-eastern Russia. By the end of 2013, only one of the decommissioned submarines in the western Arctic remained.

The following will give you an idea of the broad scope of international cooperation involved in this effort up to 2008 (It is illustrative and not a summary):

- France provided the design of a new incinerator subsequently built at Zvezdochka near Arkhangelsk, without which decommissioning of the submarines would have been very difficult.
- Norway financed the decommissioning of four submarines, including Victor I, II and III submarine types.
- The United Kingdom financed the decommissioning of another three submarines.
- Norway and the United Kingdom shared the costs of dismantling another submarine.
- The United States has assisted Russia with the dismantling of more than 31 submarines, including at least one of the Typhoon class.
- Canada has funded the dismantling of one submarine and has joined other countries in funding the dismantling of others.
- Germany funded the building of storage facilities that enable the safe onshore storage of sections of dismantled submarines.

Many of these projects were considered to be demonstration projects. For each project, a class of submarine was selected that would provide blueprints for future work on that class. Cooperation has not been strictly limited to submarine demolition. For example, in 2008, the European Bank for Reconstruction signed a cooperation and implementation agreement with the Russian Rosatom's Centre for Nuclear and Radiation Safety. The projects include:

- Defuelling and decommissioning the *Lepse* floating base that stored spent nuclear fuel from nuclear icebreakers. It had long been identified by AMAP and others as being a major potential source of radioactivity to the Arctic environment. The 2012 annual review of NEFCO reported that all spent nuclear fuel had been removed and that the *Lepse* had been towed to the Nerpa shipyard – 40 kilometres from Murmansk. Financing of the decontamination and demolition of the *Lepse* (which will take place in a specially constructed building) is now being financed by the European Bank for Reconstruction and the Northern Environment Partnership Programme.
- Setting up facilities to manage the defuelling and dismantlement of a Papa class nuclear submarine, including the manufacture of defuelling equipment.

- Upgrading the radiation monitoring and radiation emergency response system for the Arkhangelsk region.

These examples of cooperative actions between the Russian Federation, other countries and international financial mechanisms by no means exhaust the list. Countries recognized the serious risks posed by the high density of potential sources of radioactive release in the Barents Region and pulled together to address the challenge. Nothing of this nature can happen without funding and a number of financial mechanisms have been created for this purpose, in addition to those I have described, such as the Arctic Council's Project Support Instrument. There is a great deal more to be done, but back in 1993, I would never have predicted that so much unsecured nuclear waste would be safely dealt with by 2008.

Radioisotope thermoelectric generators (RTGs): Thermoelectric generators powered by a radioisotope source were installed in remote Arctic locations in the former Soviet Union and Alaska to provide electrical power (usually for isolated marine navigation beacons). By the time of the 1998 AMAP assessment, the 10 remaining RTGs in Alaska were all located at a seismic monitoring station operated by the U.S. Air Force on Burnt Mountain. After 2001, they were replaced by solar-powered equipment. According to the AMAP 2009 assessment, the Russian (formerly Soviet) RTGs were located at that time on the Baltic coast (111); north European Arctic coast (1020); north-east Sea Route (422); and the far-eastern coast of Russia (217). The radioisotope heat sources in the U.S. and Russian RTGs were usually ^{90}Sr (strontium-90) in the form of a solid ceramic that is strong, fire resistant and has a low solubility in water. It has been calculated that the Russian RTGs produced an activity ranging from 0.7 to 15 PBq. This is a sufficiently high level of activity that in the event of direct exposure, death could occur in a relatively short period of time. Due to the remote location of the RTGs, little or no security arrangements protected them which made them potentially vulnerable to thieves who have tried to gain access to the generators or to other materials. One RTG was lost in the sea off Sakhalin Island in 1987 and two more were accidently dropped from a helicopter during transport in 2004. No loss of ^{90}Sr was detected as a result of these accidents.

Although the safety record of RTGs has been high, their vulnerability to interference and consequential risk of ^{90}Sr release was highlighted in the AMAP 1998 and 2002 reports. A similar collaborative recovery, disposal and replacement programme, as we have just examined with respect to obsolete nuclear submarines, has now been put in place. Therefore, the AMAP 2009 assessment was able to report extensive removal and decommissioning of RTGs from Arctic Russia, mainly through cooperation between Russia, Norway and the United States but also involving Finland and Canada.

Evaluation of contributions to environmental radioactivity in the Arctic from Russian nuclear facilities in Siberia (Mayak, Tomsk-7 and Krasnoyarsk-26):   Three large nuclear facilities have been operating in central Russia for more than 50 years. They are of interest relative to the Arctic because they have routinely discharged nuclear wastes into river systems that ultimately flow into the Arctic Ocean. The three facilities are Mayak, Tomsk-7 and Krasnoyarsk-26. Until the birth of the Russian Federation, it was difficult to obtain sufficient information about these facilities in order to evaluate their significance as sources of radioactivity to the Arctic Ocean. After the advent of glasnost, this situation changed. Largely through the work of a joint Russian-Norwegian expert group, it became possible for the 1998 and 2002 AMAP radioactivity assessments to examine these sites in an Arctic context.

The Mayak facility was originally built to produce weapons-grade plutonium and discharged intermediate-level waste directly into Lake Kyzyltash. The Techa River is the outflow from this lake, which eventually reaches the Ob River at Khanty-Mansiysk (via the Iset, Tobol and then the Irtysh rivers). The Ob ultimately flows into the Kara Sea. In the early years, the Asanov Swamp along the banks of the upper Techa River acted as a filter and accumulated large quantities of radionuclides, but as discharge rates later fell from Mayak, the swamp became a radionuclide source to the river. Between 1951 and 1961, a system of dams and canals was constructed in the upper part of the Techa River to retain much of the radioactivity, and after 1951, most waste was discharged into Lake Karachay, a small lake with no outlet. In 1997, it was reported that the total amount of radioactivity in the lake was about 4,400 PBq. After discharges to Lake Karachay were terminated, water levels dropped significantly. In 1997, the lake partially dried out during a summer drought. This resulted in about 22 TBq of radionuclides being redistributed by local winds and led to more than 1,800 square kilometres of territory receiving deposition of ^{90}Sr at levels greater than 3.7 kBq per square kilometre (with 1.0 kBq being equal to 1,000 becquerels).

The incident at Lake Karachay resulted in contamination of parts of the same geographical area that had occurred following an accident at Mayak in 1957. The latter is now known as the Kyshtym accident. At this time, a storage tank of high-level radioactive material, which was buried more than 8 metres underground, exploded and released about 74 PBq of radioactivity into the atmosphere at up to an altitude of 1.0 kilometre. The dominant long-lived radionuclide released was ^{90}Sr. For about 11 hours after the explosion, the radioactive cloud moved north-east for 300–350 kilometres, resulting in a deposition footprint of between 800 to 20,000 square kilometres (depending on contours of contamination criteria). If you do some more reading, you will find this area defined as the East Ural Radioactive Trace (EURT). Approximately 10,000 people received a collective dose of 1,300 man-Sv and were evacuated for 7–670 days, depending on where they lived. The effective dose in the most exposed group reached 0.5 Sv. It is possible that a significant proportion of the ^{90}Sr derived from the Kyshtym accident may have reached the Arctic Ocean, but the lack of contemporary measurements does not allow the accident source to be distinguished from other sources present at the time.

The Tomsk-7 facility is located on the River Tom, shortly before it flows into the Ob – well upstream of the Ob/Irtysh confluence at Khanty-Mansiysk. It was originally used for the production of weapons-grade material, but by 1998, the two remaining reactors were also being used to provide electricity for the town of Tomsk. Radioactive waste from Tomsk-7 had been discharged into reservoirs and injected into deep underground strata. The plutonium-producing reactors at Tomsk were shut down in 1998.

In 1993, a chemical explosion at the site occurred in a tank containing uranium nitrate that resulted in uncontrolled emissions to the atmosphere. A study by Porfiriev in 1996 indicated significant contamination over an area of about 189 square kilometres. A report on the event was published by the IAEA in 1998. Using two different methodologies, the IAEA estimated that the average value for the total activity of all material released was about 30 TBq. These figures are substantially higher than early estimates.

The Krasnoyarsk-26 facility is located on the Yenisei River, which also eventually flows into the Arctic Ocean. It was originally used for the production of weapons-grade plutonium and for the storage of radioactive wastes. More recently, it developed electrical power for the region. At the time of the 1998 AMAP assessment, it was estimated that between 26,000 and 37,000 PBq of radioactive waste had been injected into underground storage. I am not aware of any information of what this may mean to local groundwater contamination now or in the future.

Attempts to quantify how much of the radioactive discharges from the three Siberian facilities may have reached the Arctic Ocean have been inconclusive. This is partly because of the lack of monitoring data from the 1960s–1980s along the rivers from the nuclear facility to marine waters. The most likely signal of marine contamination attributable to these facilities would be from Mayak. In this case, it is clear from the 2009 AMAP assessment that ^{90}Sr activity progressively declines from Mayak along the Techa-Iset-Tobol-Irtysh-Ob river system. Upstream of the Irtysh River's confluence with the Tobol River, ^{90}Sr activity in the Irtysh River is essentially at background levels but is at an order of magnitude higher downstream of the confluence, which also suggests significant input from Mayak-related sources. A little upstream of the same confluence, ^{90}Sr activity in the Tobol River is about 200 times higher than in a typical Russian river (5–6 Bq per cubic metre). The influence of Mayak ^{90}Sr discharges can be detected downstream as far away as the confluence of the Irtysh and Ob rivers.

AMAP examined two studies published in the mid-2000s that attempted to use plutonium isotope ratios (^{240}Pu : ^{239}Pu) to investigate whether Mayak discharges had reached the Ob estuary and the Arctic Ocean. However, the results still do not settle the issue. If you are interested, take a look at the references by Lind and colleagues and Skipperud and colleagues listed in the bibliography. Whatever the historical situation may have been, monitoring indicates that the three large nuclear sites are not at present contributing significant levels of radionuclides to the Arctic Ocean via the Ob and Yenisei river systems.

The Thule aircraft accident: On January 21, 1968, a B52 strategic bomber from the U.S. Air Force with four nuclear bombs on board crashed onto the sea ice in North Star Bay just after taking off from the airbase at Thule in north-west Greenland. During the accident and subsequent (nonnuclear) explosion, some plutonium from the bombs was dispersed onto the ice. Most of this inventory was quickly recovered, but some of the ice-embedded material was lost into the water column when the ice melted later that summer. It was estimated following the cleanup that about 2.5–3.0 kilograms of plutonium were lost. Sediment core studies have shown that this corresponds to roughly the amount that lies in marine sediments around the crash site. The reason that the lost inventory has remained so close to the site of its original deposition is that plutonium adheres (or partitions) very strongly to sediment particles. However, the distribution of plutonium in sediment samples collected in 2003 was found to be very patchy and some lateral transport of up to 17 kilometres from the impact location has occurred. Despite this pattern, most of the lost inventory lies within a radius of 4 kilometres from the point of impact. The plutonium also appears to be relatively unavailable to biota because bivalves living buried in the sediment carry plutonium levels several times lower than are found in the sediments, although they are above background concentrations. The situation at Thule is not considered to present risks to human health even if local shellfish are consumed.

Komsomolets **submarine:** In April 1989, the Soviet nuclear submarine *Komsomolets* caught fire following an explosion and sank 180 kilometres south-west of Bear Island in the Norwegian Sea. The wreck lies at a depth of about 1,650 metres. The single nuclear reactor shut down safely during the accident and was estimated to contain an inventory at that time of about 2.8×10^{15} Bq of ^{90}Sr and 3.1×10^{15} Bq of ^{137}Cs, together with other long-lived fission and activation products. Now more than 25 years after the accident, ^{90}Sr and ^{137}Cs dominate the fission-product total inventory and it is estimated that most of the activation products will have decayed before corrosion leads to significant releases to the environment. Two torpedoes with mixed uranium/plutonium warheads were also on board and were estimated to contain approximately 1.6×10^{13} Bq of weapons-grade plutonium.

The 1998 AMAP assessment reviewed monitoring data up to 1997. Surveys in late 1993 that included Russian, Dutch and U.S. scientists detected cracks in the hull of the ship, a hole in the torpedo compartment and leakage from one of the warheads. These ruptures were subsequently sealed in 1994. Monitoring found that some leakage of ^{137}Cs was detectable through a reactor ventilation tube at an annual rate of less than 0.5 TBq and modelling suggested this rate of ^{137}Cs release would continue for the next 2,000 years. The nuclear warheads are not as well protected as is the submarine reactor and could therefore corrode more quickly. However, the released plutonium is likely to strongly bind to sediment and therefore most contamination will remain close to the wreck. At the wreck site, very little vertical mixing of the water column has been detected, and at the same depth, a slow horizontal deep-water current moves towards the north-east. To sum it all up, the *Komsomolets* is not considered to represent a significant hazard to humans now or in the future.

Natural Radioactivity and TENORM

According to the United Nations Scientific Committee on the Effects of Atomic Radiation (UNSCEAR) in 2000, the annual worldwide per caput effective dose of radiation from natural and man-made sources could be summarized as follows:

- *Natural background:* 2.4 mSv, with a typical range of 1 to 10 mSv but with some large populations exposed to 10 to 20 mSv
- *Diagnostic medical procedures:* 0.4 mSv, with a range of 0.04 to 1.0 mSv depending on level of health care
- *Atmospheric nuclear testing:* 0.005 mSv decreased in the Northern Hemisphere from 0.15 mSv in 1963. Levels are higher at high northern latitudes, where they vary considerably with diet.
- *Chernobyl accident:* 0.002 mSv in the Northern Hemisphere decreased from 0.04 in 1986
- *Nuclear power production:* 0.0002 mSv

You will notice that the large variation in these figures relates to diagnostic medical sources. The effective dose is much higher in countries with developed economies. Consequently, the U.S. National Research Council reports that the average effective dose in the United States is approximately 50% natural and 50% man-made, with the latter being made up of the sum of diagnostic radiological imaging and the use of radiation to treat disease (totalling 48%) and consumer products (2%).

For many people, it comes as a surprise to learn that most of us receive such a high proportion of our lifetime dose of radiation from natural sources, especially from radon gas that enters our increasingly sealed homes and workplaces. We will take a brief look at this fact because it helps put the anthropogenic sources into perspective. This is particularly useful when it comes to evaluating when specific anthropogenic sources may represent present or future environmental and human health risks on local, regional or larger geographic scales.

Here are the four main sources and global levels of annual effective dose from natural sources:

- *Cosmic radiation:* This varies with altitude but has a global average of 0.30 mSv. It is increased by air travel at a rate of about 0.4 mSv for every 100 hours of flying.
- *Terrestrial radiation:* This is highly variable depending on local and regional geology. Doses can be as high as 260 mSv in northern Iran or 90 mSv in Nigeria. In Canada, the estimated highest annual dose is

approximately 2.3 mSv measured in the Arctic (Northwest Territories). The global average is about 0.5 mSv a year.

- *Dose from inhalation:* Earth also contributes to our levels of exposure in a highly variable way that reflects local geology. Radon gas, which is produced by the uranium decay series, irradiates the lungs when inhaled. Radon naturally disperses as it enters the atmosphere from the ground, but concentration can build up in unventilated buildings. The global average annual effective dose of radon radiation is approximately 1.2 mSv. People who smoke tobacco can greatly increase the dose at "hot spots" in their lungs. Radon gas decays into a series of highly radioactive particulate metals (radon decay products) that can cling to the sticky hairs (trichomes) on tobacco leaves. The particles are not water soluble and make their way into tobacco. In the lungs, the radon decay products – lead-210 (half-life = 22.3 years) and polonium-210 (half-life = 138 days) – are retained in the bronchioles. If a person has been smoking for many years, the concentration of these radionuclides directly on tissues of the bronchioles can become very high and intense localized radiation doses can occur at the bronchiole "hot spots".

- *Dose from ingestion:* Several sources of natural radiation penetrate our bodies through food ingestion, drinking and breathing. If we ignore radon decay, potassium-40 is the main source of internal irradiation. The average global effective dose is approximately 0.3 mSv a year.

The acronym used for naturally occurring radionuclide material is NORM. As we have seen, the ultimate sources of NORM include the primordial radionuclides we met in the introduction to this chapter, including cosmogenic radionuclides, such as carbon-14 and tritium (^{3}H or hydrogen-3). Cosmogenic radionuclides rarely reach levels that pose risks to human health or the environment. However, NORM derived from potassium-40 (^{40}K) and from primordial isotopes of elements from the radioactive series originating from ^{238}U (uranium series), ^{235}U (actinium series) and ^{232}Th (thorium series) can in certain circumstances lead to significant radiation exposure. Each of these series includes many intermediate radionuclides until arriving at lead – their final stable and nonradioactive destination.

In industrialized societies, health authorities are very interested in knowing when people may be exposed to more NORM than is "normal". To explain this apparent contradiction, we will need to greet another acronym. This is TENORM, or technologically enhanced NORM. Since Earth's formation, geological processes have tended to concentrate potassium-40 (^{40}K) and uranium and thorium series nuclides in "favoured" minerals and geological formations that have limited contact with the biosphere. TENORM is produced when humankind disturbs these enriched geological formations of rocks, soils, water or other natural materials, bringing their radionuclides into contact with the biological environment. The types of human activity that can produce TENORM include uranium mining, phosphate and elemental phosphorus production, phosphate fertilizer production, coal ash generation, oil and gas production, drinking water treatment, metal mining and processing and geothermal energy production.

The radiological exposure to individuals from TENORM wastes occurs in three main categories:

- Associated with onsite disposal of wastes, such as mine tailings. This type of disposal can lead to groundwater contamination and to airborne releases of radioactive particulates and radon gas.
- From the use and/or disposal of these wastes. For example, solid mine waste may be used to provide fill around homes or as a constituent of soils. This can lead to a buildup of radon gas in homes, direct exposure to individuals located nearby, contamination of soil and the crops growing in that soil and groundwater contamination.
- NORM-contaminated materials may be used in construction materials, such as in concrete aggregate.

TENORM has not historically received a great deal of attention in a circumpolar Arctic context. However, the AMAP 2009 radioactivity assessment showed that for at least the last 40 years, TENORM has been an issue of considerable concern for the public in a number of discrete geographical areas and generally relates to the first of the listed categories. These were mainly areas where uranium mining had taken place, followed by the abandonment of the mine. In Canada, the focus of attention has been on three locations – all in the western part of the Northwest Territories. They are:

- Port Radium located on the east shore of Great Bear Lake.
- The transportation route from Port Radium across Great Bear Lake and up the Mackenzie River to Fort McMurray.
- The Rayrock Mine, located 145 kilometres north-west of Yellowknife.

The mine at Port Radium operated from 1932 to 1940 to produce radium salts and again from 1942 to 1960 to produce uranium for military use. Between 1964 and 1982, mining for silver took place at the site. AMAP reported that 6,200 tonnes of uranium were extracted over the life of the mine and about 910,000 tonnes of radioactive tailings waste were left at the site. Approximately 20% of this waste was deposited in depressions around the mine and the remainder dumped into Great Bear Lake. At the time when the mine was in operation, knowledge of radiation health effects, particularly with respect to low-level exposure and long-term effects, was in its infancy. Consequently, radiation protection standards were much lower than they are today.

From the 1980s, the local indigenous people of the Déline First Nation began to voice concerns about the state of the abandoned mine and to draw conclusions related to the incidence of cancers in their community to the mine. In 1999, the Canadian federal government began to investigate the site and between 2001 and 2004 worked with the Déline as part of the Canada/Déline Uranium Table (CDUT) to fully characterize the environmental conditions at the mine. This study also considered the potential for human exposure to radiation during the transport of uranium ore across Great Bear Lake and up the Mackenzie River to Fort McMurray. Local Déline indigenous people were employed in this work, but records do not show they worked at the mine.

The CDUT survey found that some parts of the exposed tailings contained uranium series radionuclides up to 37,000 Bq/kg. Gamma radiation levels at the site varied from natural background (100–150 nGy/hour) to 740 nGy/hour (a nGy, or nano Gy, is 10^{-9} gray). Water samples from Great Bear Lake in close vicinity to the mine showed slight elevation of trace metals, but fish from the lake carried no detectable body burdens.

The CDUT report paid particular attention to the potential exposure of Déline workers involved in the Northern Transportation Route. Oral histories referred to exposure to "yellow powder". This was originally assumed to be yellowcake (uranium concentrate). However, yellowcake was only produced at Port Radium from 1958 to 1960 and was shipped out in drums by air (not by the Northern Transportation Route). The yellow powder referred to in the oral reports probably referred to sulphur powder shipped to the mine site from 1950 to 1960 for use in the acid leach plant. A dose reconstruction was carried out for the CDUT to estimate historical radiation exposures to 35 Northern Transportation Route workers and their families. The average dose estimated for the ore transport workers was 76 mSv/year. The cumulative doses during the period of employment varied from 27 to 3,015 mSv. The study reported that based on the radiation doses calculated in the dose reconstruction, 1–2 cancer deaths would be expected amongst the 35 ore transport workers, in addition to the 9–10 cancer deaths that would "normally" be expected in a similar, nonexposed group of 35 people. Radiation doses to family members who lived near Port Radium or along the transportation route were estimated to be similar to background doses.

Following the publication of the CDUT report, the Canadian federal government implemented a remediation plan for the site that included sealing open entrances and placing soil caps on the tailings piles. It was finally concluded that the abandoned mine did not pose a concern under several occasional visit exposure scenarios for campers and fishermen. The Port Radium story is a sad lesson on the anxiety and concerns that can be caused when industrial activities provide little or no environmental and health information to local residents and when such sites are abandoned with inadequate or no remediation. In many respects, the Rayrock Mine was a carbon copy of this experience.

Rayrock operated for only a short period of time – from 1957 until 1959. During this period, 70,000 tonnes of ore were processed to provide 207 tonnes of uranium concentrate. Radioactive tailings deposited on land partly invaded three small lakes. Abandonment followed the all-too-common practice of the day in the Northwest Territories. It was not until several site assessments had been undertaken in the 1990s that the Canadian federal government stepped in to clean up and remediate the abandoned property in 1996–1997. This involved sealing mine entrances and ventilation shafts, moving tailings piles and covering them with soil and vegetation. The site is periodically monitored, and in 2010, Environment Canada reported that further maintenance of the tailings piles is required.

The Long and the Short of It

I have always liked this little phrase as a pithy way to wrap up a conversation. When my parents used it, I knew the time for discussion (and dissent) was over. It has been in use for quite a long time because I remember encountering it at school when reading *The Merry Wives of Windsor* and wondering if my parents picked it up from Shakespeare. You will meet it from now on whenever I want to wrap up a chapter and move on to another topic and also at the very end of the Arctic Messenger's story. What should we remember about radioactivity in the Arctic? Here is my selection:

- For the circumpolar terrestrial Arctic, the major source of radiation resulting from human activity remains atmospheric fallout primarily from the atmospheric testing of nuclear weapons and, to a much lesser extent, from the 1986 accident at the Chernobyl nuclear power station in Ukraine. Although deposition is now low, the Arctic terrestrial environment remains sensitive because its soils are thin. Its slow-growing and long-lived flora includes a high proportion of mosses and lichens that take their moisture directly from the atmosphere. Consequently, the deposited radionuclides (fallout) are easily transferred up the food chain by such herbivores as caribou and reindeer.
- Three simple lessons can be drawn from the experience of the negotiation of the atmospheric test ban treaties. Success was helped by there being a small number of potential parties, by the existence of a strong supporting scientific foundation and by the power of public opinion.
- The main source of marine radionuclide contamination to the Arctic has been waste discharge from nuclear reprocessing plants at Sellafield in the United Kingdom and at Cap de la Hague in France.
- Major incidents at Sellafield (formerly Windscale), Mayak, Three Mile Island, Tomsk, Chernobyl and Fukushima show that despite extensive national and international regulation, much progress remains to be made in order to ensure nuclear safety.
- Real and/or potential sources of radioactive contamination in the Arctic are associated with the high density of temporarily stored nuclear waste in Arctic and sub-Arctic Russia. Hidden beneath this observation lies a disturbing reality: Very few countries have developed and implemented plans for the permanent disposal of their nuclear wastes. In the long term, this may represent the greatest safety challenge for the nuclear industry.

5

Heroic Efforts

Any scientific activity in the Arctic is likely to be very expensive and even today can be hazardous. There are several ways one can go about it. The project may have modest requirements for new data and/or does not require a scientist to be in a remote part of the Arctic. In this case, there may be little difficulty in finding sufficient additional funds to do the job. For example, the existing capabilities of satellite-borne remote sensing may be used or perhaps the data gathering can be done within or a short distance from an Arctic community. However, if the project requires physically travelling away from Arctic communities, ships and aircraft are involved and costs skyrocket. If the geographical coverage of the project and/or the length of observations is extensive, the mobilization of national and perhaps international scientific cooperation will be needed for the science itself and for the logistics.

If the objective is to study how something changes over time and space (such as concentrations of a particular contaminant), a network of observing stations will be required. This strategy usually starts with some existing national networks and the task is to expand these precursors in ways that reflect the time and space demands of whatever is the object of study. The key attributes for success are ensuring that methods and equipment provide comparable data, the data are freely available and that arrangements will be sustainable for a long period of time. Much of the Arctic Messenger's story is derived from this type of monitoring (such as that organised by Arctic Monitoring and Assessment Programme [AMAP]). Recently, the Arctic Council launched an activity called the Sustaining Arctic Observing Networks (SAON), designed to enhance the circumpolar capability of this type of approach in the Arctic and to expand the focus beyond the types of data traditionally used for environmental monitoring. A different strategy is needed if the objective is to investigate how environmental processes function in the Arctic.

Such studies are multidisciplinary in nature and thus progress often requires integrated teams of specialists from a wide range of disciplines. The logistics required include ships, aircraft and remote sensing. This has been a common approach taken in ocean-atmosphere studies. It is very expensive and getting such activities organised is rather like planning a military campaign. Almost certainly, the project will be national or international in terms of scientific effort and logistics. The example par excellence has been the International Polar Years (IPYs), but there are a number of others – some of which were not directly focused on the Arctic but which had Arctic repercussions. I call them *heroic efforts*.

The idea for the first IPY came from Karl Weyprecht, an Austrian explorer and naval officer who took part in the Austro-Hungarian Polar Expedition of 1872–1874. From this experience, he decided that major advances in polar geophysical science required large-scale international cooperation. He foresaw the need for multidisciplinary teams that plan a programme of research, conduct the work and then collectively interpret the data. His ideas led to the creation in 1879 of the International Polar Commission, which consisted of Austria-Hungary, the Dominion of Canada, Denmark, Finland, France, Germany, the Netherlands, Norway, Russia, Sweden, the United Kingdom and the United States. It was this commission that organised the first IPY, which took place between 1882 and 1883. Twelve countries took part and 13 multinational expeditions set out for the Arctic and two for the Antarctic. This approach was quite an innovation. Previously, visits to the Arctic or Antarctic were primarily patriotic adventurous explorations or had national strategic goals. It was a dangerous business. Seventeen of the 24 Americans involved in one of the participating Arctic IPY-1 expeditions (led by Adolphus Greely) starved to death when a supply ship was missed. If you fly south from Alert along the east coast of Ellesmere Island, with the cooperation of the weather, the ruins of Fort Conger – where much of this sad story was played out – can still be seen. It is a desolate place. Nevertheless, the first IPY collected a huge amount of information over its 12 months of operation and it provided the foundation of our knowledge of Earth's magnetic field.

The next IPY took place 50 years later in 1932–1933. This time, 40 countries participated. As was the case with its predecessor, research was focused on some specific issues – this time, mainly related to atmospheric science, including geomagnetism and the nature of the jet stream.

The third IPY took place over the period of 1957–1958. It was actually given a different name: the International Geophysical Year (IGY). This title was a more accurate description of the research focus. Like its predecessors, it focused almost exclusively on geophysical phenomena. These included the nature of cosmic rays, the aurora, geomagnetism, ionospheric and atmospheric physics, meteorology, seismology, gravity, oceanography and solar activity.

Sixty-seven countries participated in the IGY, but more importantly (at ages twelve and thirteen), so did I! Every second night through the winter of 1957–1958, my school friends and I spent 30 minutes staring at the sky. We had been recruited as (involuntary) "volunteers" by our science schoolmaster to help with a project designed to find out the cause of the aurora borealis (northern lights). We were equipped with illustrations of the aurora and a duplicated sheet to record what we had seen. Each of us was assigned a period of the evening to do our duty for science. In order to escape light pollution from the town, only boys who lived in places without street lighting were recruited. The only excuse for not going out was a cloudy sky. We had to be very careful because we knew that three other neighbouring schools were taking part in the project and that our science master was checking up on us by randomly taking time slots himself. The aurora is not a common sight over Ashford (in south-east England). I so much wanted to see it during that year, but I had to wait until the 1970s before seeing this amazing spectacle for myself. Involvement of children in the IGY was an idea carried forward in IPY 2007–2008.

The achievements of the IGY included the discovery of the Van Allen radiation belts and of the mid-ocean submarine ridges, which helped support the then-emerging theory of plate tectonics. And it led to the modern explanation of the aurora!

Fifty years later, in 2007–2008, came the most recent IPY. I was still working for the Canadian government during the very early planning stages. At that time, it was far from being a *fait accompli*. Marty Bergmann (then with Fisheries and Oceans Canada) and I were given the task of preparing the first presentation on the objectives and possible costs to the Canadian cabinet. The presentation was much changed by the time it reached the cabinet room, but feedback told us it would be a hard sell. I retired soon after this, but Marty moved on to become director of Canada's Polar Continental Shelf Program, where he continued to be involved in IPY planning. We kept in touch, and from Marty, I learned that much of the credit for securing Canada's commitment to IPY was due to the work of Peter Harrison. In those years, Peter was roving between several Canadian environmental ministries as a deputy minister. Marty's death in a plane accident at Resolute Bay in August 2011 was a great loss to Canada's Arctic scientific community.

The scientific organisation of the 2007–2008 IPY was quite different from that of its predecessors. Gone was the singular focus on geophysics. Instead, more than 200 projects were undertaken that supported one or more of the six themes. They are documentation of the current status of the environment in polar regions; understanding environmental and social change and projecting future change; linking polar environmental processes with other regional and global processes; investigating the frontiers of science in the polar regions; using the unique features of the polar regions to study terrestrial and cosmic geophysics; and investigation of the cultural diversity, resilience and sustainability of circumpolar human societies. Interdisciplinary observational strategies were then developed to address the six research themes. A number of the most recent advances in knowledge included in this story came from the 2007–2008 IPY.

More than 60 countries participated in the 2007–2008 IPY. It will go into the history books as setting a new standard for how an international heroic effort can be organised and executed. Much attention was given to showing the general public why the research themes were and are important. More than 50 of the projects were related to education, outreach and policy implications. One such project created the Association of Polar Early Career Scientists (APECS) that has become an ongoing legacy of IPY. It continues to promote interaction and cooperation between students and young and established researchers. Organisations such as AMAP now routinely invite representatives of APECS to contribute to and participate in their meetings.

The scientific literature is still producing new insights into polar processes based on the fourth IPY – five years after the completion of the programme. Once the scientific community had begun to digest the information gleaned from the 2007–2008 IPY, thoughts began to turn towards wondering how a much longer mobilization of scientific effort and logistics could be arranged and coordinated. The first idea to come forward was called the International Polar Decade. This essentially consisted of a proposal to run a 10-year-long IPY. However, this concept was probably neither practical nor sustainable and did not enjoy a great deal of support from the powerful agencies that fund major international research activities. A new idea has since emerged, known as the International Polar Initiative. This seeks to build on the IPY's success of creating a framework to identify a set of key science issues for polar regions and of coordinating scientific and infrastructural resources to apply to these issues. It envisages that at least initially, the activities will be built on existing resources. An implementation plan would be developed by a number of new bodies and working groups. The plan would guide research and studies on the identified issues until the advent of the next IPY (suggested for 2032–2033). It remains to be seen whether this idea sees the light of day. Key factors will probably include whether existing national and international Arctic science and monitoring programmes, their coordinating organisations and their funding mechanisms believe the new proposal adds significant benefits to existing arrangements.

Many different organisations have and continue to organise heroic efforts to study specific science issues in the Arctic but with a smaller scope than the 2007–2008 IPY. Here is a glimpse of one such organisation. The Arctic Ocean Science Board was set up in 1984, with participation from research and government institutions in 16 countries. Its accomplishments have included:

- The Greenland Sea Project that was co-sponsored by the International Council for the Exploration of the Sea (ICES), 1987–1993. Coordinated national surveys involving marine surveys led to totally new levels of understanding of ocean and climate interaction in Greenlandic and Nordic seas.
- The International Arctic Polynya Programme. Recurrent polynyas (ice-free areas surrounded by ice) play unique roles in Arctic oceanography. This programme was initiated in 1989 and ran for more than 15 years before it finally evolved into a self-sustaining project called Polynyas in a Changing Arctic Environment (PACE).

The Arctic Ocean Science Board merged with the International Arctic Science Committee (IASC) in 2009. The IASC has convened International Conferences on Arctic Research and Planning (ICARP) in 1995 and 2005. ICARP III is planned for 2015 in Japan. Its objectives will include identifying key Arctic science needs and discussing how these may be addressed using national programmes and resources. The IASC annual bulletin (available online at www.iasc.info/files/PRINT/ Bulletin2013.pdf) provides an overview of some of the collaborative Arctic science mechanisms. The IASC has often been the forum where an idea first germinated before being brought to fruition through extensive international collaboration in other organisations (particularly those controlling infrastructure and other resources).

Distinctions between the two strategies described at the start of this chapter on organising Arctic studies ("monitoring" and "process-related research") are often blurred. Of course, science is science. Knowledge derived from long-term monitoring to address questions on environmental changes over time and space and know-ledge on the processes involved in those changes are intimately linked. This is one reason why AMAP carefully avoids using this categorization. At the end of the day, we must ensure the long-term sustainability of circumpolar monitoring (such as that provided by AMAP) and of heroic efforts focused on overriding environmental process issues in the Arctic.

6

Acidification and Arctic Haze

Casca: "I durst not laugh for fear of opening my lips and receiving the bad air."

William Shakespeare, Julius Caesar

Before we start, here is another simple revision of our school chemistry. Water molecules exist in equilibrium with hydrogen ions H^+ and hydroxide ions OH^-.

$$H_2O <-> H^+ + OH^-$$

A solution is acidic if the H^+ ions are in excess and is basic (or alkaline) if the OH^- ions are in excess. The pH scale is used to measure the acidity of a substance. The scale ranges from 0 to 14. A pH value of 7 is neutral. A value of less than 7 is acidic and a value greater than 7 is basic. It is a logarithmic scale. Therefore, pH 5 is 10 times more acidic than pH 6 and 100 times more acidic than the neutral pH 7. Normal rain has a pH of about 5.6 due to an interaction with carbon dioxide, which results in the formation of carbonic acid.

With this technicality over, the acidification story can begin. In Part 1, we briefly noted that in the 1960s, terrestrial and freshwater ecologists began to describe widespread deterioration of forests and freshwater ecosystems in northern Europe, eastern Canada, and northeastern United States. The cause was quickly identified. Sulphur and nitrogen oxides released from combustion sources were reacting with water in the atmosphere to form acids of nitrogen and sulphur that increased the acidity of rain and snow. It was probably the first time the general public heard the phrase *acid rain*.

It is worth pausing for a moment to appreciate several of the then-unique features of this situation. First, the sources of the offending sulphur and nitrogen were ubiquitous and included nonferrous metal smelting and the combustion of hydrocarbons, ranging in scope from industrial scale activities (including electrical power generation) to automobile exhaust and also at that time from domestic heating. Coal was a particularly significant source. When I was a child in the south-east of England in the 1950s, I remember the dense, suffocating winter smog that hung over London whenever my parents would take me to the city. Thousands of people died at this time from respiratory complications in the large industrial cities during long winter periods of atmospheric inversions that trapped the industrial and domestic heating smoke in the lowest few hundred metres of the atmosphere.

One of the abatement measures taken was to simply increase the height of smokestacks. It improved local air quality but also increased the export of pollution to other areas. This leads us to another "unique" feature of the acid rain situation. The most impacted areas showing ecosystem stress from acidification were generally (but not always) hundreds of kilometres distant from potential sources. It was realized that two factors were responsible for the observed pattern of acidification. The first is pretty obvious: It is the direction of prevailing wind. Therefore, much of the cargo of acidifying substances generated, for example, in the industrial heartlands of the United Kingdom and Germany was exported in the atmosphere by south-westerly winds towards the Nordic countries. The second factor was less obvious but was easy to understand. Not all regional environments are equally vulnerable to acid rain. For example, soils and lakes sitting on a bedrock that is chemically basic (the opposite of acid) will buffer the incoming acid rain, resulting in few ecosystem effects. However, if the bedrock is acidic (such as one of the many types of granite, schist and gneiss), the overlying terrestrial and freshwater systems will already be under a significant degree of natural acid stress. There will be little buffering capacity to deal with anthropogenic acid rain. Consequently, the most acid-sensitive terrestrial and freshwater ecosystems were generally those situated on the Precambrian shields in Fennoscandia, Scotland and north-eastern North America. The crystal clear waters of a loch in Scotland are an artefact of their natural acidity and low nutrient concentrations (mainly nitrogen and phosphorus) that severely restrict plankton growth.

Arctic haze is the name given to a visibility-reducing dirty brown haze that has been noticed by Inuit, whalers and Arctic explorers since the nineteenth century and since the 1950s by pilots flying polar routes. Known as Poo-jokt by the Inuit, the phenomenon is mainly seasonal, occurring with a peak in late winter and spring. By the late 1960s and early 1970s, atmospheric studies showed that Arctic haze is a visible manifestation of the long-range transport in the atmosphere of a mixture of anthropogenic pollutants. These pollutants come from the midlatitude industrial combustion of hydrocarbon-based fuels and are mostly made up of small particulate matter (including aerosols) and acidifying pollutants (anthropogenic sulphur and nitrogen oxides). The pollutants migrate towards the Arctic in north-flowing air streams in the lower 5 kilometres and especially in the lower 2 kilometres of the atmosphere. The lower-altitude streams mainly appear earlier in the winter and originate from more northern mid-latitudes than the higher-altitude streams that appear later and have travelled from more southerly sources. Once in the Arctic, they build up in the high-pressure, stable Arctic air masses. These acidifying substances (mainly sulphate and, to a lesser extent, nitrate) can therefore be deposited into the developing Arctic snow pack, where they remain until released into the Arctic terrestrial and freshwater environments at the time of spring snowmelt.

For the first time, European and North American politicians were forced to recognize that countries were producing atmospheric air pollution that crosses national boundaries. Consequently, an impacted regional environment could be many hundreds of kilometres distant from several different source regions that together were cumulatively responsible for the observed environmental degradation. The only prospect for dealing with this complex erosion of environmental quality was the development of legally binding national and international environmental controls on emissions. This conclusion was not easily accepted in some countries. For example, at the close of the administration of President Jimmy Carter in 1980, it appeared that the United States was poised to take action on acid rain. However, influential lobby groups vigorously opposed it. Indeed, it was not until the administration of George H. W. Bush that concrete actions were introduced into the domestic Clean Air Act. These were largely the same "denial" groups that will consistently appear in the Arctic Messenger's story concerning persistent organic pollutants, stratospheric ozone depletion and climate warming. The reference in the bibliography by Naomi Oreskes and Erik Conway provides a good overview of how effective these groups can be when they intervene in the political decision-making processes in North America.

As outlined in Part 2 of this story, countries decided to use the United Nations Economic Commission for Europe (UNECE) to organise cooperative regulation of the substances responsible for acid rain.

The UNECE includes not only Europe but also Russia and all the other countries of the former Soviet Union, together with Canada and the United States. It therefore embraces most of the heavily industrialized countries of the Northern Hemisphere, with the exception of China and India. Under the UNECE, a brand-new treaty was established in 1979, called the Convention on Long-Range Transboundary Air Pollution (CLRTAP), to enable countries to work cooperatively to reduce acid rain. At the time, it was a unique and bold enterprise. Europe was divided between East and West and the Cold War showed no symptoms of the thaw that was to dramatically arrive in the late 1980s. Cooperation between East and West was not something that happened easily. The CLRTAP was (and still is) a framework convention. It basically sets out some general aims and principles. On this mutually agreed foundation, it enables parties to "attach" legally binding protocols that target specific pollution issues as they emerge.

Over the years since 1979, several protocols to address acidification have been negotiated under the CLRTAP, with the latest and most ambitious being the 1999 Gothenburg Protocol to Abate Acidification, Eutrophication and Ground-Level Ozone. As a general rule, the early protocols have evolved from a flat-rate emission reduction philosophy to a science-based approach in the Gothenburg Protocol. Under the Gothenburg approach, on the basis of the acid rain sensitivity of the environment that will ultimately be the expected area of deposition, geographically gridded emission ceilings are established for the source regions. This description is mine and is not totally accurate, but I think you will understand how it works.

The scientific basis for the development and operation of the Gothenburg Protocol comes from the European Monitoring and Evaluation Programme (EMEP). It feeds the integrated emissions, atmospheric transportation and deposition assessments that lie at the heart of the protocol's effects-based emissions-reduction strategy. Another core scientific body is the Working Group on Effects (WGE), which looks at the sensitivity of the receptors and monitors the impacts. Two key concepts that link it all together are known as *critical loads* and *critical levels*. Critical loads are defined as a "quantified estimate of an exposure to one or more pollutants below which significant harmful effects on specified sensitive elements of the environment are not expected to occur". Critical levels are defined as "concentrations of pollutants in the atmosphere above which direct adverse effects on receptors (such as freshwater or forest ecosystem elements) may occur". Areas where critical loads may be exceeded can be identified by comparing geographically gridded critical load maps with modelled geographically gridded deposition data for present or theoretical emission scenarios.

In 2012, the CLRTAP Gothenburg Protocol was amended to provide more ambitious national emission reduction commitments to be achieved by 2020. The revision included new sets of emission limit values for key stationary and mobile sources of acidifying substances. It also introduced emission ceilings for fine particulate matter. This was good news for climate scientists, as we will see when we look at the impact of black carbon (a short-lived climate forcer) on Arctic climate warming. Regional policies, such as the directives of the European Union, have also become increasingly important for achieving emission reductions.

But what does all this mean for the Arctic? At the time when the Rovaniemi Declaration of 1991 established the Arctic Environmental Protection Strategy and Arctic Monitoring and Assessment Programme (AMAP), little monitoring data for the Arctic were available. It was clear that in some Arctic areas, emissions of acidifying substances and heavy metals from several smelters in Russia, such as in the Kola Peninsula at Nikel and Monchegorsk and further to the east at Norilsk, were responsible for local or regional terrestrial and freshwater ecosystem damage. I first visited the Kola Peninsula in the Barents Region at the very end of the Soviet era. Downwind of each smelter was a devastated landscape of lifeless forests and lakes that extended in some areas across borders into Norway and Finland. However, the spatial extent of effects of these within Arctic sources had not been quantified and no comprehensive attempt had been made to evaluate any impacts from sources outside the Arctic. Therefore, in 1991, our question relating to the Arctic could not be answered and AMAP set about establishing acidification monitoring arrangements across the affected subregion.

The first AMAP acidification assessment was coordinated by Juha Kämäri and published in 1998. It was found that some ecosystem effects attributable to low deposition acidification from distant sources could be detected in the European mainland Arctic, primarily due to the vulnerability of this region that results from its underlying geology. No such effects attributable to low deposition from distant sources could be detected in eastern Siberia or Arctic North America, but in both these regions, very little data were available. The report documented the temporal extent of the effects to vegetation and freshwater ecosystems from within Arctic sources, particularly from the nonferrous metal smelters at Nikel, Zapolyarny and Monchegorsk on the Kola Peninsula and at Norilsk in the Taymyr region. In these regions, the visible effects were largely attributed to the direct impact of sulphur dioxide and to heavy metals (co-emitted from the smelters) accumulating in soils and water bodies.

The second AMAP assessment of acidification in the Arctic was published in 2006. This report found that 1990 emissions data suggested that critical levels for soils were exceeded at that time over large areas. However, by the time the 2006 assessment had been completed, emissions of sulphur dioxide from sources within the Arctic had decreased by about 23% since 1992, most of which was attributable to declines of as much as 82% from the Kola Peninsula smelters. The reductions were much lower at Norilsk. These declines were in part due to new emission control procedures but also to a decline in Russian industrial activity following the collapse of the Soviet Union. It was concluded that despite the dramatic local effects caused by the Arctic emission sources, most of the acidifying substances reaching the Arctic as a whole are arriving by long-range atmospheric transport from lower-latitude sources, mostly in Europe and North America and increasingly from Asia. Forest fires appear to be a growing source of Arctic air pollution, which includes acidifying substances and the climate-forcing agent black carbon. This source is likely to increase in significance as our global climate continues to warm. However, circumpolar monitoring data from such remote stations as Alert (Ellesmere Island, Canada) and Zeppelin (Svalbard) indicate that the overall background levels of sulphate and sulphur dioxide in Arctic air are decreasing in summer and winter, indicating that national and international controls are being effective. At the same time, the amount of Arctic haze–inducing substances arriving over Alaska and the western Canadian Arctic appears to have been increasing since the late 1990s and is indicative of the increasing importance for the Arctic of emissions from Eastern Asia.

Little information was available for the Chukotka region of Russia, but it lies within a prevailing air stream originating from industrial areas, particularly those involving coal-fired power generation in China and India. It is possible that this area is being exposed to increasing levels of acidifying substances. Elsewhere, air streams from North America, Europe and Asia that are known or believed to be carrying sulphur and nitrogen are not doing so at levels likely to cause widespread Arctic soil or freshwater acidification now or in the near future. In Arctic Canada, the combined maximum sulphur and nitrogen deposition (north of 60° north) is below the critical loads for soil acidity. It is not known at this time whether this situation will show local or regional perturbation resulting from the large-scale development of tar sands in northern Alberta.

Surprisingly, acidified soils around the Kola Peninsula smelters appear to be restricted – more or less – to the zones of visible vegetation loss. This is not an area of high rain or snowfall. Vegetation studies along the Barents Sea coast of Norway show that most of the sulphur in the leaves of small tundra plants arrived in association with dust particles and not directly from sulphur dioxide. Up to 80% of the sulphur entering the air at the Kola sources may be quickly deposited in this way. Other supporting studies have shown that acidic precipitation generally falls within 30–40 kilometres of the smelters. Outside this zone, it appears that the neutralizing activity of base cations from the local geology and from alkaline fly ash co-emitted from the source smelters are sufficient to prevent soil acidification reaching levels that may result in vegetation damage.

Ecologists recognized a series of concentric zones of impact (elongated by prevailing winds) close to the Kola smelters. Beside the smelters is the "industrial barrens" zone – in which almost all vegetation has disappeared – followed by zones of decreasing severity of effect. At the less impacted zones, the observed change remains significant ecologically because it is manifested through changes in plant community structure, especially with respect to lichens. Lichens are particularly sensitive to sulphur dioxide and preindustrial lichen-dominated heaths have essentially disappeared from areas anywhere close to the smelters.

Arctic freshwater bodies also vary in their ability to resist the effects of acidifying inputs primarily due to differences in how well they are endowed with a natural buffering capacity from such constituents as bicarbonates and organic acids. Northern Norway has a particularly high proportion of acid-sensitive lakes due to natural underlying catchment geology. It is therefore not surprising that around 1990, critical loads for northern Europe were being exceeded over large areas, especially where natural, local and long-range (from Kola smelters) acidification converged. However, the 2006 AMAP report noted improvements in freshwater chemistry in the Barents Region and predicted that continued implementation of the Gothenburg Protocol should greatly reduce the extent to which critical loads are exceeded in northern European freshwaters.

Long-term monitoring data have revealed that short-term pulses of highly polluted air are particularly significant. For example, in the 1990s, up to 30% of the total sulphate deposited at a remote station in Finland arrived over a period of only five days. This is ecologically

important because many plants, such as Scots pine and mature mountain birch trees, are particularly sensitive to sudden high-concentration
pulses of atmospheric sulphur dioxide, especially during the growing
season. Freshwater ecosystems are also very sensitive to pulses of highly
acidic water. In the Arctic, these conditions are most likely to occur in
spring, when the entire winter's deposition of acidifying substances can
be released through snowmelt over just a few days. In the Kola Peninsula region, where metal emissions are at least as toxic to freshwater
and terrestrial ecosystems as is acidification, the high-concentration
pulse associated with spring melt delivers metals and acids to the
ecosystem at the same time. Studies in Finland and Norway suggest
that the strength of the acidity peak in northern lakes during the period
of snowmelt is more important to invertebrate survival than is the
annual or seasonal acidity (pH).

What does the future hold? There are some indications that the
volume of emissions from Russian sources may be increasing. Providing
that this does not emerge as an upward trend, we should be able to
expect that increased Arctic environmental degradation resulting from
emission sources located within the Arctic (such as the Kola smelters) is
unlikely to increase in geographical extent. Full implementation of the
CLRTAP Gothenburg Protocol (including the amendments agreed to by
the CLRTAP Executive Body in 2012) makes it unlikely that European
and North American sources will result in widespread acidification
impacts in the Arctic. Two large questions hang over this prediction:
Will the growing East Asian sources increase in significance to the
extent that they result in widespread acidification and will climate
change alter the physical and chemical dynamics of the acidification
process in Arctic environments? The increasing oil and gas exploration
in the Arctic (due partly to increasing accessibility of these resources as
a consequence of melting sea ice) will also likely cause increasing local
air pollution problems.

After completing its 2006 acidification assessment, AMAP concluded that in the future, this subject should be an element of the
wider issue of air pollution (including black carbon). This reflects
the fact that the more recent air pollution agreements (such as the
Gothenburg Protocol) now use this approach rather than dealing with
specific substances, such as tropospheric ozone, sulphur, nitrogen and
small particulates (for example, the short-lived climate forcer black
carbon). This enables better attention to be paid to the physical and
chemical processes that occur between the various air pollution
constituents.

The Long and the Short of It

At present, the Arctic continues to experience discrete pockets of intense local environmental degradation due to local Arctic sources of acidifying emissions. At the same time, there is little evidence to suggest that emissions from outside the Arctic are causing significant Arctic impacts. However, data are essentially unavailable to enable an assessment to be made for the Eastern Asian Arctic.

Despite this conclusion, the acidification story is very important in an Arctic context because it:

- Showed how decisions to take remedial action to alleviate air pollution impacts close to sources (by building taller smokestacks) exacerbated the problem in distant regions, such as the Arctic.
- Exposed national and regional governments to the first powerful demonstration of how widespread anthropogenic pollutant emissions in one geographic region can – through the transport of pollutants in the atmosphere – lead to widespread significant environmental impacts in a different and distant region. It is a phenomenon that turned out to be a harbinger of other and more painful consequences of long-range transboundary air pollution for the Arctic. We will learn the details from the Arctic Messenger when we consider persistent organic pollutants and mercury.
- Demonstrated that governments were willing to work cooperatively to address the problem by creating the CLRTAP and by subsequently adhering to the emission-reduction regimes of its protocols.
- Created a forerunner to several actions taken later to address other pollution issues of universal importance to the Arctic.
- Demonstrated the power of lobby groups that opposed the need for action and which successfully held up meaningful controls on acidifying substances in the United States for at least 10 years.

Thus, the Arctic Messenger can leave this chapter in a cautiously optimistic frame of mind – unfortunately, a rare condition in this book! Quite possibly, the Arctic's escape from the widespread acidification seen in north-western Europe is in large part thanks to the Convention on Long-Range Transboundary Air Pollution (CLRTAP). Governments can do it when they try. Now we just have to make sure the CLRTAP is allowed to continue to flourish while at the same time we keep a vigilant eye on emissions from Eastern Asia as well as local point sources.

7

Stratospheric Ozone Depletion

By three methods we may learn wisdom: First by reflection, which is noblest;
Second by imitation, which is easiest;
And third by experience, which is bitterest.

Confucius, The Analects

When scientists in North America, Europe and the Soviet Union were beginning to detect and understand the causes of freshwater and terrestrial acidification, another potentially more serious problem (especially for the Arctic) was slowly being recognized: The stratospheric ozone layer that protects us all from the harmful effects of solar ultraviolet light was thinning.

Before we go further into this part of the Arctic Messenger's story, we need a basic understanding of the behaviour and nature of oxygen and ozone in the stratosphere. Oxygen can exist in three forms. Most commonly, it occurs as a molecule made up of two oxygen atoms (O_2). However, it can also occur alone as a single atom (atomic oxygen) or as a molecule of three oxygen atoms (O_3). This is ozone. In the stratosphere, highly energetic shortwave ultraviolet radiation from the sun can break O_2 molecules apart into lone oxygen atoms. When one of these free oxygen atoms bumps into an intact O_2 molecule, it can join up with the molecule to form ozone. However, a cascade of chemical and physical processes involving solar radiation and a number of naturally occurring compounds containing nitrogen, hydrogen and chlorine also continually break down stratospheric ozone. Therefore, the amount of stratospheric ozone present at any moment in time is the result of a dynamic process of production and removal.

Why Is Stratospheric Ozone Important?

Stratospheric ozone absorbs solar ultraviolet radiation (UV), which warms the stratosphere. More importantly, this absorption acts as a sort of filter. The ozone allows only a relatively small proportion of the highly energetic UV radiation to reach the troposphere below. It is within the troposphere that we and every other living organism play out our lives. Oxygen was absent from the earliest atmosphere, but that changed with the evolution of the first photosynthetic bacteria about 3.5 billion years ago. Since then, life on Earth has evolved in ways that depend on the protective stratospheric ozone filter.

The UV radiation arriving from the sun is classified according to its wavelength and, therefore, its energy. UV-A has the lowest wavelength, is the least energetic and is the least biologically active form. It is only partially filtered out by the ozone layer and thus it is the most common form to reach Earth's surface. UV-C has the shortest wavelength, is the most energetic and is potentially the most biologically active form. Putting it dramatically, UV-C is capable of ripping apart the chemical bonds that hold biological chemicals together. Fortunately for us, it is very rapidly mopped up in the atmosphere by molecular oxygen and nitrogen gas. We rely on well water in our home. Before it reaches our taps, however, the water passes by an enclosed UV-C lamp that kills any bacteria that may have been lurking in the well.

That leaves us with the intermediate form UV-B, whose properties lie between the extremes of UV-A and UV-C. It is partially absorbed by stratospheric ozone, but the residual amount that reaches Earth's surface is sufficiently energetic to cause effects in plants, animals (including humans) and even materials. (Think about what happens to unprotected plastics and fibreglass when left for some time in the sun.) The stratospheric level with the highest concentrations of UV-B lies at an average altitude of about 25 kilometres and is known as the *ozone layer*. UV-induced ozone production is on average highest over the tropics, but this inventory is then partially moved by circulatory patterns within the stratosphere towards mid- and high latitudes. Therefore, the normal situation is for lower levels of ozone over the tropics and higher levels over the poles.

We will look at the biological effects of UV-B exposure later. For the moment, we will just remember that life is no stranger to UV-B radiation and has evolved protective mechanisms to deal with its impacts. However, the effectiveness of this protection depends on UV-B irradiance being roughly within the limits that have existed during which life has evolved these mechanisms.

It is time for another little digression. Sometimes, people become confused because of an apparent contradiction. They read in the newspaper that high levels of ozone in the air are a major contributor to poor summer air quality, but on the next page, they read that scientists are worried about ozone depletion. The answer is quite simple. The ozone that is responsible for protecting us from UV radiation is in the upper atmosphere (stratosphere). The "excessive" ozone implicated in poor air quality is in the lower atmosphere (troposphere). Here, it is formed when hydrocarbons and nitrogen oxides (derived, for example, from automobile exhaust and other fossil fuel combustion sources) interact with sunlight. Ozone is a highly reactive molecule that injures biological tissues (plant and animal) – in particular, the respiratory tract of people. It is also damaging to such manufactured materials as rubber and nylon.

Ozone-Depleting Substances and the Thinning of the Ozone Layer

By the time Arctic Monitoring and Assessment Programme (AMAP) was established in 1991, it was well known that the ozone layer had dramatically thinned (been depleted) over the previous decade and a half. The cause had been identified and international actions to arrest the depletion had also begun under the Montreal Protocol, a brand-new treaty we will learn more about a little later. In support of the protocol, the World Meteorological Organization (WMO) and the United Nations Environment Programme (UNEP) produce periodic assessments of ozone depletion and on its associated environmental effects. These assessments, together with two assessments from AMAP, have provided most of the material for this chapter. The two AMAP reports were prepared by teams led by Elizabeth (Betsy) Weatherhead. The first was included in chapter 11 of the 1998 *Assessment Report: Arctic Pollution Issues*. The second appeared as chapter 5 in the 2005 *Arctic Climate Impact Assessment*.

Concern for the integrity of the ozone layer began in the late 1960s. This was in relation to the possible environmental impacts of the engines of the space shuttle in the stratosphere and of a planned fleet of civil supersonic transports (SSTs). In 1970, Dutch chemist Paul Crutzen showed that the nitrogen oxides NO and NO_2 (collectively abbreviated as NOx) formed by soil microorganisms could reach the stratosphere. There, they react with free oxygen atoms, thus slowing the creation of ozone (O_3) and also decomposing ozone into nitrogen dioxide (NO_2) and oxygen gas (O_2). What is more, this reaction did not involve the loss of NO and NO_2, which were functioning as catalysts. Harold Johnston then published a paper in *Science* in 1971 in which he specifically linked the potential impact of NOx erosion of the ozone layer to the planned operation of a fleet of SSTs. By this time, plans to develop a fleet of SSTs in the United States had been abandoned, but the proposal had awakened scientific interest in the health of the ozone layer.

James Lovelock is a British scientist whom most people recall as the founder of the Gaia hypothesis. Early in his career in the mid-1950s, he designed an instrument called the *electron capture detector* that could be used to detect very small amounts of an organic gas in the atmosphere. In papers published in 1970 and 1973, he showed that concentrations of chlorofluorocarbon (CFC) gases were very high and ubiquitous in the atmosphere. For example, CFC-11 was present at concentrations of 60 parts per million over Ireland and they were even found in Antarctica.

CFCs were first produced commercially in the 1930s. By the 1950s, they were widely used around the world as a propellant in spray bottles; as a coolant in refrigerators, freezers and air-conditioning units; and in the production of plastic foams and polystyrene. One of the reasons for their popularity was they were chemically inert and therefore resisted degradation. This is why they gradually accumulated in the atmosphere to the concentrations observed by Lovelock.

The next significant steps came in 1974. First, Richard Stolarski and Ralph Cicerone outlined how chlorine atoms could attack ozone catalytically if they ever reached the stratosphere. Then, in a paper published in *Nature*, Mario Molina and Frank Rowland described how this could and almost certainly was taking place. Their laboratory study showed that CFCs could spill over from the troposphere into the stratosphere, where, in the presence of UV radiation, they would be broken apart and their chlorine released into the stratosphere. Stratospheric ozone would then be attacked by the free chlorine atoms, resulting in the release of oxygen gas (O_2) and chlorine monoxide (ClO). As chlorine monoxide is very unstable, its chlorine would be quickly released and be available to attack another ozone molecule. Today, it is believed that a single chlorine atom released in this way could react with about 100,000 ozone molecules in something like a runaway chain reaction.

Molina and Rowland estimated that the continued use of CFCs would substantially thin the ozone layer, which would remain depleted for many decades due to the huge inventory of CFCs in the troposphere reported by Lovelock. Two decades later, in 1995, Mario Molina, Frank

Sherwood Rowland and Paul Crutzen shared the Nobel Prize for chemistry "for their work in atmospheric chemistry, particularly concerning the formation and decomposition of ozone". It is now known that the chemistry of ozone depletion is quite complex in the details, but it still conforms to the general explanation collectively worked out by Molina, Rowland and Crutzen.

Further studies suggested that within about 60 years, the ozone layer would be depleted by about 7% (soon shown to be a major underestimate). These developments precipitated a fierce reaction from segments of the chemical industry and associated lobby groups, largely in the United States. One of the main arguments they posited was that volcanic eruptions were the main source of chlorine in the stratosphere. Rowland essentially debunked this idea by pointing out that most volcanoes emit huge amounts of water vapour, along with other substances, such as chlorine. Consequently, most such substances are washed out in the troposphere. He showed that when El Chichón erupted in 1982, hydrogen chloride increased in the stratosphere by only 10%. The much larger eruption of Mount Piatubo in 1991 led to an even smaller increase. Meanwhile, chlorine levels had continued to steadily rise between the two eruptions. Clearly, volcanoes were not the source of rising levels of stratospheric chlorine. Undeterred, the lobbyists fought on. Books by Dotto and Schiff and by Oreskes and Conway listed in the bibliography provide narratives of how all this unfolded. Just for the record, in 1994, the WMO estimated that less than 18% of the stratospheric chlorine then being measured in the atmosphere came from natural sources. These are mainly forest fires and an unusual gas released by some marine plankton.

By the way, the sulphate aerosols injected into the stratosphere by these two volcanoes enhanced ozone destruction, an effect that increased depletion for several years. The strength of this response depends on the concentrations of chlorine and bromine in the stratosphere. Therefore, once these levels go down, the perturbations caused by such volcanoes weaken.

Oreskes and Conway also record how the public quickly began to make up its own mind about the use of CFCs and began to abandon products that used CFC propellants in spray cans. By 1979, the first regulations controlling CFC use as a propellant gas in spray cans came into effect in the United States. They were quickly followed by similar actions in Canada, Norway and Sweden. However, other uses of CFCs remained unregulated, and after a brief falloff, their global production began to rapidly grow in the early 1980s.

The next stage in this story began in 1985, when Joseph Farman and colleagues published what is now a classic paper. They had discovered the Antarctic ozone hole. Since the late 1950s, Farman had been measuring stratospheric ozone above Halley Bay in Antarctica (76° S and 27° W) using a ground-based instrument. In 1982, he noticed a massive springtime ozone reduction of about 40% in comparison with data from the 1960s. He knew that NASA satellites had not reported this apparent anomaly and he assumed he had an instrument problem. However, a new instrument in the following spring detected another sharp drop in ozone levels. When he looked at his old records, he found that the spring anomaly had been occurring since the mid-1970s, and even before then, a more modest decline had been taking place. After another instrument located on the Argentine Islands (65° S and 64° W) to the north-west also detected the steep decline, Farman and his colleagues published their paper in the journal *Nature*.

NASA quickly found that its satellite-borne instruments had been detecting the ozone decline, but its computerized data analytical programme contained cutoff values to remove any data that did not appear to be credible. This is a common feature for modern instrumentation that is capable of deluging scientists with too much information, but on this occasion, the baby had been thrown out with the bathwater. Once NASA corrected its analyses and the Nimbus-7 satellite survey was published, the world was in for a shock. The corrected satellite data revealed a springtime Antarctic "hole" in the stratospheric ozone layer. It was so large that it covered a geographical area larger than that of the United States. We now know that during the two to three months of the Antarctic spring/early summer, about 50% of the total amount of ozone in the stratosphere vanishes. Over the two decades following the publication of Farman's paper, the spring Antarctic ozone hole grew by up to 27 million square kilometres.

It is time for another quick digression. If you are tempted to do more reading about ozone depletion, you will quickly encounter Dobson units which are used to measure the total amount of ozone in the atmosphere above a point on Earth's surface. One Dobson unit is equivalent to a layer of pure ozone 0.01 millimetres thick at standard temperature and pressure.

Back to the Farman paper and the NASA reanalysis. They caused quite a scientific commotion. There could now be no doubt that stratospheric ozone depletion was real and already far more extensive than formerly estimated. Why was it so intense above Antarctica? The answer came from an atmospheric chemist and it provided the basis for tidying up our fundamental understanding of what is going on when CFCs and other oxygen-depleting substances meet in the stratosphere. Susan Solomon suspected the involvement of polar stratospheric clouds. In winter, the Antarctic stratosphere cools down and descends close to the surface. At the same time, strong westerly winds circle the pole as a vortex that contains and traps the polar atmosphere. As temperatures in the lower stratosphere plunge to below -78°C, polar stratospheric clouds are formed. They consist of minute ice particles, along with nitric and sulphuric acid. They can only develop at extremely low temperatures and are much more common in the Antarctic than the Arctic. Solomon suggested that the ice particles create a surface or substrate on which the ozone molecules and chlorine from the decay of CFCs can efficiently interact. In the absence of UV radiation during the polar winter, an inventory of CFCs can build up on the ice surfaces. Supportive information quickly arrived.

First, it was demonstrated in laboratory studies that ice could form a substrate as proposed by Solomon. Meanwhile, measurements of chlorine monoxide in the stratosphere were found to be 100 times larger than could be expected without the icy particle substrate being involved. Furthermore, airborne instruments showed that whenever chlorine monoxide levels are high in the stratosphere, ozone levels are low and vice versa. This is exactly as would be predicted if UV-induced breakup of chlorine-containing chemicals is responsible for ozone depletion.

International political action was remarkably swift. Clearly, any regulation of CFCs would only be effective if it could be implemented globally. In 1985, the Vienna Convention for the Protection of the Ozone Layer (VCPOL) was negotiated. The convention entered into force in 1988 and has now been ratified by 197 countries. It is another example of what is known as a framework or umbrella convention. It sets out such elements as aims and objectives as well as methods of operation, such as decision making and procedures whereby concrete actions can be undertaken. The "workhorses" of framework conventions are their accompanying protocols. They spell out any legally binding obligations to implement the objectives of the convention and describe in detail the responsibilities of the participating countries (parties). For the Vienna Convention, the "workhorse" is the Montreal Protocol on Sub-stances That Deplete the Ozone Layer. It was agreed on in September 1987 and entered into force in January 1989. At the time of writing, it has been ratified by all 197 countries that are parties to the Vienna Convention. In its original form, the protocol stipulated that the production and consumption of compounds that deplete ozone in the stratosphere (CFCs, halons, carbon tetrachloride and methyl chloroform) were to be phased out by 2000. (It was later accelerated to 1995.) Halons are unreactive gaseous compounds of carbon and a halogen (five elements occupying Group VIIA of the Periodic Table). They were used as fire suppressants, which included chlorine and bromine, while methyl chloroform was used as a food and grain fumigant. Since 1989, the protocol has been strengthened through seven revisions, mostly in response to new scientific information, and has taken on board actions on substances that in 1989 were not understood to be capable of stratospheric ozone depletion. These include hydrochlorofluorocarbons (HCFCs), methyl bromide and hydrobromofluorocarbons (HBFCs).

The catalyst that propelled the birth of the Montreal Protocol was no doubt the continuing stream of scientific reports that emphasized the urgent need for international controls on ozone-depleting substances. The most influential was the Ozone Trends Panel organised by NASA. Its report released in 1988 concluded that ozone depletion was not unique to the Antarctic but was occurring globally. The panel reported that between 1969 and 1986, stratospheric ozone had declined in winter by 6.25%, 4.7% and 2.3% at corresponding latitudes of between 53° N and 64° N, 40° N and 52° N and 30° N and 39° N. At the same time, it was shown that after lasting for eight weeks each spring, the Antarctic hole slowly migrates as it dissipates. Its migration leads towards more populated areas, including southern Chile, southern Argentina and the Falkland Islands. In other words, unless we lived in the tropics, the ozone layer was thinning above our heads.

Ozone-Depleting Substances in the Atmosphere Since January 1989

Warning: For the next few pages, we are going to be faced with some difficult-to-pronounce chemical names that look as if they were designed for an elocution exam. Their acronyms are not much better. If you find this difficult, do not despair. You will be able to follow the storyline by just remembering that CFCs and methyl chloroform were first-generation depleting substances. Most (but not all) of the remainder were second-generation substances intended to replace them.

January 1989 was the year when the restriction on the first batch of ozone-depleting substances (the CFCs and methyl chloroform) under the Montreal Protocol entered into force. The reductions in use of CFCs was originally achieved by replacing them with substances that still contained chlorine or other halogens but which had shorter lives in the atmosphere. Notably, these were hydrochlorofluorocarbons (HCFCs), methyl bromide and hydrobromofluorocarbons (HBFCs). However, these replacements were still a source of chlorine or bromine to the stratosphere. Consequently, they were scheduled for elimination in subsequent revisions (termed *adjustments*) to the protocol.

The actions against CFCs and methyl chloroform were none too soon. As we have seen, a huge reservoir of ozone-depleting substances had built up in the troposphere and stratosphere because most of them were very resistant to any form of degradation. Therefore, CFC concentrations in the atmosphere continued to increase well into the 1990s despite the fact that due to the protocol, emissions were falling. It was as if we had turned down the taps when filling up the bath but had not turned them off. The tub continued to fill. By 2005, the tide was slowly beginning to turn. Tropospheric chlorine had decreased to roughly 92% from its peak value reached between 1994 and 1995. Most of the decline came from the relatively rapid loss of methyl chloroform because this substance has a comparatively short atmospheric life of between six months and 25 years. In 2010, the WMO reported that the total tropospheric abundance of chlorine from ozone-depleting substances had declined by 2008 to 3.4 parts per billion (ppb) from its peak of 3.7 ppb. The rate of decline was lower than anticipated probably because of leakage of CFCs from old equipment and from a more rapid increase in use of HCFCs than expected, particularly in developing countries.

The 2010 WMO *Ozone Report* showed that declines in CFCs were responsible for much of the decrease in total tropospheric chlorine over the previous few years. They are now expected to decline throughout the present century. The decline of methyl chloroform levels made a smaller contribution to the decrease in total chlorine than in earlier WMO assessments because it had already been largely removed from the atmosphere. Bromine from halons stopped increasing in the troposphere during 2005–2008. Methyl bromide levels continued to decline during the same period due to reductions in production, consumption and emission. Some uses are not controlled by the Montreal Protocol.

These trends are very encouraging. You will remember that ozone-depleting substances gain access to the ozone layer by wafting up from their tropospheric reservoir. There can be no doubt that the Montreal Protocol is working. However, the WMO 2010 assessment was not entirely good news. We have already noted that HCFC use has been growing more quickly than expected and this is reflected in troposheric levels. For example, the most abundant HCFC (HCFC-22) increased more than 50% faster between 2007 and 2008 than between 2003 and 2004. Similarly, tropospheric abundances and emissions of hydrofluorocarbons (HFCs; substitutes for CFCs and HCFCs) continue to increase, with significant sources being from Europe, Asia and North America.

HFCs present a thorny regulatory problem. They do not erode the ozone layer, but their growth in use is a direct result of their being chosen as an apparently benign replacement for the ozone-eating CFCs and HCFCs. However, they are not benign. They are very potent greenhouse gases. We will learn more when we reach the "short-lived climate forcing" topic in the climate chapter. Attempts have been made to control them under the Montreal Protocol, but so far, these attempts have been thwarted by those who feel that such action should only be considered under the Framework Convention on Climate Change. The present "action" arm of this agreement is the Kyoto Protocol, which has so far been able to achieve very little.

The part of the Arctic Messenger's story concerning HCFCs and HFCs strikes a theme that will reappear again in the chapter on persistent organic pollutants. Having created a global problem with the original ozone-depleting substances, we have managed to replace them with substances that in many ways are just as hazardous. We have a knack for choosing the third method for gaining wisdom as described by Confucius in the *Analects* quote at the beginning of this chapter.

Perhaps the most encouraging parts of the WMO 2010 report concern the stratosphere. After all, this is where the damage is done. Total chlorine is declining in the troposphere and stratosphere from peak values reached in the 1990s. Stratospheric bromine is no longer increasing. The declines in the sum of stratospheric chlorine are largest in mid-latitudes and smallest in Antarctica.

However, in March 2014, Johannes Laube and colleagues published a paper in *Nature Geoscience* that reminded us that it is unwise to be complacent. Their study reported for the first time the presence of three CFCs and one HCFC from samples collected in air over Tasmania (from 1978 to 2012) and from firn snow in Greenland (collected in 2008). The four substances first appeared in the 1960s and two of them continue to accumulate in the atmosphere. They are, of course, unmentioned in the Montreal Protocol, but the paper estimated a combined total emission of 74,000 tonnes. The source of all four substances is unknown, but the study demonstrates the importance of monitoring, without which we can never have the whole story.

Projected Recovery of the Stratospheric Ozone Layer

In the mid-1990s, global average stratospheric ozone levels stabilized, but they have not yet shown any substantial increase. Here is the situation from data analyses available from the WMO in 2010. For the period 2006–2009, the average ozone values were:

Latitude 90° N to 90° S: 3.5% below 1964–1980 averages
Latitude 60° N to 60° S: 2.5% below 1964–1980 averages

For the same period but using a different method of recording, the average mean total ozone values were:

Latitude 35° N to 60° N: 3.5% below 1964–1980 averages
Latitude 35° S to 60° S: 6.0% below 1964–1980 averages

Therefore, substantial ozone depletion affects almost all of Europe, Russia, Australia, New Zealand and North America and a part of South Africa and South America. However, smaller decreases in stratospheric ozone have been observed in mid-latitude regions of the world and no statistically significant depletion has been detected over the tropics. Where I live in southern Canada, the ozone layer has thinned by an average of about 7% since the 1980s.

In the Antarctic, the ozone hole continues to appear every year between late August and early October. In spring 2000, it reached a record geographical extent by occupying 28.3 million square kilometres. At this time of year, some vertical profiles from stations located near the South Pole show almost complete ozone destruction in the lower stratosphere.

In the Arctic, ozone depletions have also been seen over the last 20 years. However, the Arctic stratosphere is usually not cold enough for a sufficiently long time to allow the widespread formation of polar stratospheric clouds that are essential for the dramatic nature of seasonal ozone depletion above the Antarctic. This is because the high altitude winds that circle the Arctic (the polar vortex) and which trap cold air over the North Pole are not as effective at doing so as their counterparts in Antarctica. Consequently, in the 1998 AMAP assessment, Elizabeth Weatherhead described Arctic ozone depletion as resembling the type of Swiss cheese that is riddled with bubbles. Overall, average ozone levels in the Arctic spring remain comparable to 1990 levels. We will return to the nature of Arctic depletion events later in the present chapter and again when we look at climate change in the Arctic.

If the Montreal Protocol has arrested erosion of the ozone layer, when can we expect a full recovery to pre-1970 values? It will not be soon, but the short answer is that by the end of the present century, the layer should have recovered. This is assuming that the protocol continues to be implemented and is allowed to respond positively to any further scientific findings on the behaviour of ozone-depleting substances.

There are several reasons that make a general statement such as this very difficult. The first is rather obvious: It depends on latitude. The second is that ozone-depleting substances are greenhouse gases. In addition, a healthy ozone layer absorbs ultraviolet radiation. Our depleted ozone layer is allowing more of this highly energetic radiation to reach Earth's surface and to warm the troposphere. In other words, ozone-depleting substances are directly and indirectly contributing to global warming. It is ironic that by taking action on ozone-depleting substances, the Montreal Protocol has probably achieved much more concrete action to slow global warming than has the Framework Convention on Climate Change and its accompanying Kyoto Protocol!

Global warming is relevant to projecting timetables for ozone recovery because it impacts the rate of this recovery. This is where things become rather complicated. As the troposphere warms, the stratosphere cools. The WMO has reported that between 1980 and 1995, the global mean lower stratosphere cooled by 1–2°C and the upper stratosphere by 4–6°C. At the same time, modelling studies and new observations suggest that the stratospheric circulation (known as the *Brewer-Dobson circulation*) is increasing. It is believed that such acceleration will increase the poleward transport of ozone away from the tropics (where most ozone is produced), resulting in decreased column ozone levels in the tropics and increased levels elsewhere.

Global warming can also lead to tropospheric changes that are expected to enhance stratospheric ozone levels. For example, troposphere warming accelerates the processes of ozone formation and affects its transport into the stratosphere. In addition, increased tropospheric levels of methane (CH_4) will result in more methane transported to the stratosphere. Here, methane interacts with the chlorine that is destroying ozone and converts it into an inactive form.

Therefore, although the WMO 2010 report projected that global ozone levels will increase roughly as the levels of depleting substances decline, the increase is expected to be accelerated by the greenhouse gas–induced cooling of the upper stratosphere and changes in stratospheric circulation. This suggests that a return to pre-1980 ozone levels at Northern Hemisphere mid-latitudes could occur by (or before) 2050 and even rise above 1980 levels by the latter part of the present century. The return of the lower-stratospheric ozone layer over the Arctic to pre-1980 levels is also expected to be accelerated by the strengthening of the stratospheric circulation and by stratospheric cooling. Ozone dynamics in the Antarctic are thought to be less sensitive to global warming because changes in stratospheric circulation are expected to be less pronounced. The return of ozone levels to pre-1980 levels over southern mid-latitudes is not anticipated until the second half of the present century because of the slower recovery expected over Antarctica.

The Biological Effects of UV Radiation and Trends in UV Exposure

Despite the fact that the Arctic and Antarctic have experienced the most intense episodes of increased UV irradiance resulting from ozone depletion, there has been (with a few exceptions) a dearth of information on what this means to human and environmental health in these two polar regions. Therefore, in this section, we will have to resort to observations made largely at other latitudes and from experimental studies.

We have already noted that human health and environmental scientists are worried about the thinning of the stratospheric ozone layer. This is because before thinning occurred, the layer absorbed much (but not all) of the incoming solar UV-B and some of the UV-A radiation. Therefore, most of the UV that is encountered by plants, animals and near-surface aquatic life is UV-A. In humans, excessive UV-A exposure first results in congested blood capillaries in our skin, called *erythema* (sunburn), and localized oxidative stress. The result is a pigment-darkening effect (tanning) caused by redistribution of the pigment melanin that is stored in skin cells, called *melanocytes*. Overexposure to UV-A is associated with a number of human medical outcomes, including cataract formation, premature aging and toughening of the skin and immune system suppression.

UV-B is more likely to produce erythema than UV-A. It also results in an effect on the melanocytes, but this time, it involves the production of new melanin. The main cause for concern is that UV-B is sufficiently energetic that it can cause photochemical damage to DNA. When DNA absorbs UV-B radiation, some of the bonds that maintain the shape of the DNA molecule are broken and the shape can change. If this happens, it can interfere with the way in which protein-building enzymes inter-pret the DNA code. DNA has always been exposed to a certain amount of UV-B radiation. Consequently, it has evolved mechanisms to recognize and repair such damage. However, when some damage is not repaired, aberrant reading of the DNA code may result in serious health effects, including skin cancer.

A small degree of UV exposure is needed by humans to support the synthesis of vitamin D. This vitamin is important for the growth, development and maintenance of bone. Vitamin D deficiency increases the risk of osteoporosis, bone fractures and the condition known as rickets in children. Low levels of vitamin D are also thought to increase the risk of several internal cancers, some autoimmune diseases and cardiovascular diseases, including hypertension. People with dark skin need more UV exposure than fair-skinned people to synthesize the same amount of vitamin D. As we age, the efficiency of this process declines. It is not yet clear whether oral vitamin D supplementation provides similar benefits.

Erythemal UV exposure is a term used to measure UV irradiance (the amount of energy arriving at a surface) that takes into account the potential of UV-A and UV-B to cause biological damage. However, it has been very difficult to demonstrate a direct correlation between the amount of exposure to UV-B radiation and a particular biological effect. Factors contributing to this difficulty include varying degrees of suscep-tibility amongst different species and even between populations from the same species sampled from different geographical regions. They also include interactions with other stressors, such as nutritional status, or the presence of biologically active chemicals. For example, in humans, a wide range of substances, including birth control pills, cyclamates, tetracycline antibiotics and coal tar distillates used in some antidan-druff shampoos and some cosmetics, can exacerbate the photochemical effects of UV-B radiation. Similarly, aquatic organisms in areas of crude oil and heavy metal pollution (including cadmium, selenium and copper) are more sensitive to the effects of UV-B.

UNEP produces periodic ozone assessments that focus on the human and environmental effects of ozone depletion. These reports are therefore complementary to the physical/chemical assessments pro-duced by the International Meteorological Organization (IMO). According to the 2010 UNEP assessment, peak values of erythemal UV radiation in mid-northern latitudes may have tripled by 2065 if it were not for the actions taken under the Montreal Protocol. Levels of this magnitude would have carried the risk of substantial human health and environmental impacts, which hopefully we will not now experience. As we have seen, implementation of the Montreal Protocol has stabilized the accumulation of ozone-depleting substances in the upper atmosphere. Consequently, stratospheric ozone levels are no longer declining. Therefore, Northern Hemisphere mid-latitudes are at present experiencing UV-B irradiances only 5% greater than in 1980. Of course, peak values are much higher at high and polar latitudes, where ozone depletion has been larger. We will return to projections on the levels of UV irradiance for the rest of the present century at the end of this chapter.

Here is a closer look at some of the human health and environmental effects that have been attributed to chronic UV-B radiation exposure. The most well-established evidence concerns the incidence of several types of skin cancer.

There are two broad categories of cancer involved: cutaneous malignant melanomas and nonmelanoma skin cancers (basal cell carcinoma and squamous cell carcinoma). The latter two are the most common forms of skin cancer. A crusty area that develops on exposed skin, known as an *actinic* or *solar keratisis*, appears to often be a precursor to squamous cell carcinoma. Fair-skinned people spending much of their lives in sunny locations are the most susceptible, but treatments are usually effective and mortality is low. Cutaneous melanoma is less common but is much more dangerous.

According to the UNEP 2010 assessment, the incidence of all three of these tumours has risen significantly over the past five decades, particularly in people with fair skin. For cutaneous malignant melanoma, there has been an average increase of 1–3% per year. The annual incidence of cutaneous malignant melanoma also varies geographically. For example, the incidence in Europe ranges from between five and 24 per 100,000 to more than 70 per 100,000 in higher ambient UV radiation regions of Australia and New Zealand. In Australia, melanoma is now the third-most reported cancer in men and women. The number of cases of melanoma in dark-skinned people is fewer than in fair-skinned people. It has historically been uncommon in people under age 20, but in 2007, Lange and colleagues reported an increase of 2.9% per year between 1973 and 2003 in the United States. How much of this may be linked to societal trends, such as sunbathing, is unclear.

In addition to the temporal and geographical coincidence of these epidemiological trends with corresponding trends of ozone depletion and increased UV irradiance, there are patterns of skin cancer occurrence on the body that relate to cumulative UV exposure and lifestyle. For example, head and neck tumours are common in elderly people. In younger people, the highest rates of cutaneous malignant carcinoma occur on the trunk for males and on the limbs for females.

In the Arctic, skin cancer rates have generally been low, presumably due to a combination of the protective skin pigmentation of indigenous peoples, the low average UV exposure and the heavy clothing usually worn. However, the 2005 ACIA report quoted studies showing that females of Danish descent in Greenland have been showing a higher incidence of melanoma and nonmelanoma cancers.

It is not surprising to learn that our eyes are vulnerable to UV radiation. There is now strong epidemiological and experimental evidence for an association between exposure to UV-B and the incidence of age-related cortical cataracts. There is also some evidence of association with other forms of cataract and for a condition called *pterygium*. Here, however, the case is less strong. Pterygium is an inflammatory invasive growth on the conjunctiva and cornea that can impair vision.

Suppression of the immune system is the last impact of UV exposure on humans that we will look at briefly. Many animal studies have shown that during an infection, there are critical stages when UV-B exposure can increase the severity of symptoms and duration of the disease. For example, solar UV radiation can trigger the reactivation of the latent herpes simplex virus. This effect has some very practical generic implications for health care because it is now known that UV exposure before (and possibly immediately after) inoculation can reduce the strength of the immune response. Consequently, the effectiveness of the vaccination can also be reduced.

The recovery of the ozone layer is expected to lead to an overall reduction in the levels of UV radiation arriving at Earth's surface. Compared to 1980, UV-B irradiance towards the end of the twenty-first century is tentatively projected to be lower at mid- to high latitudes by between 5% and 20%, respectively, and higher by 2–3% in the low latitudes. The expectation of lower levels at mid-latitudes reflects the impact of climate warming we have just reviewed. If we do indeed find ourselves in a situation of 20% less UV-B irradiance at mid-latitude in the Northern Hemisphere in the future than in 1980, health scientists and practitioners could find themselves worrying about the possibility of widespread vitamin D deficiency.

Environmental effects have been more difficult to evaluate than those related to human health, partly because of the difficulty in collecting information but also because seasonal variation in UV irradiance complicates the detection of long-term impacts. What follows is a selective summary of what is known.

It has been rare to have information directly related to Arctic UV impacts. However, a recent meta-analysis of several studies examined the response of mosses and flowering plants from the Arctic and Antarctic to varying UV-B irradiance. The authors concluded that plants responded to increased UV-B by reducing leaf growth and area by about 1% for each 3% of increased UV-B irradiance. By extrapolation, this suggests that in such areas as the southern tip of South America, where a 20% increase in the summertime UV-B has occurred, plant growth decreased by up to 6% between 1979 and 2009. This estimate is roughly as would be expected from observations elsewhere and from experimental studies. If you would like to know more, take a look at the reference in the bibliography by Newsham and Robinson. There is good evidence to support the theory that the reduction in plant growth occurs because the plants are redirecting their energy to UV radiative protection mechanisms, such as the production of UV screening resins, antioxidants and for DNA repair. For example, the increased production of UV-absorbing compounds in an Antarctic leafy liverwort was calculated to require 2% of the carbon photosynthesized by each plant. This is not a huge proportion, but remember that these plants are not living in a friendly environment. Perhaps a better indication of the cumulative cost of metabolic protective measures comes from studies on a Patagonian shrub: *Grindelia chiloensis*. When the plants were artificially screened from natural ambient UV-B radiation, they responded by growing taller, putting on more biomass and increasing their leaf area. This was correlated with a reduction of 10% in the energy used in the synthesis of UV-B protective resin.

Less vegetative biomass can be expected to have impacts on the food chain. The 2005 Arctic Climate Impact Assessment (ACIA) describes how this can even result in changes to the dynamics between sub-Arctic birch trees and caterpillars of the autumn moth, especially in combination with climate warming. The moth compensates for the reduced size and nutritive value of birch trees exposed to higher UV by eating up to three times as much as normal. This causes significant forest damage. The situation is exacerbated by winter climate warming, which is increasing the overwinter survival of moth eggs. There is very little information relating to trophic impacts on larger herbivores.

The first limiting factor on the impact of UV radiation on aquatic life is how far it can penetrate below the surface. In clear oceanic and lake waters, UV-B can penetrate tens of metres, but in cloudy waters, it can be almost entirely absorbed in the upper centimetres. This roughly corresponds with the upper mixed-layer euphotic zone, within which photosynthetic phytoplankton are responsible for most oceanic and much lake primary production. The UNEP 2010 ozone assessment showed that phytoplankton, fish eggs and larvae, zooplankton and other animals from higher trophic levels that inhabit the euphotic mixed layer can be adversely impacted by UV radiation. For example, UV radiation penetrates well into the euphotic zone in the Gulf of Maine, a key nursery area for the embryos and larvae of the Atlantic cod. In experimental studies, significant mortality of cod embryos and a significant decrease in the length of young cod occurred following exposure to UV radiation equivalent to the 10-metre depth level in the Gulf of Maine. This represented a level of UV exposure that is common in many fish-spawning areas.

Calcium carbonate efficiently absorbs UV radiation. In the climate chapter, we will spend some time looking at how increased levels of carbon dioxide in the atmosphere are leading to acidification of the oceans. At the same time, we will see how increasing acidity makes it difficult for many marine organisms to incorporate calcium carbonate into their shells and other skeletal structures. For example, in experimental studies, Gao and colleagues found that increased acidity depressed calcification rates in the marine coccolithophore *Emiliania huxleyi*. This resulted in a thinning of the calcium carbonate–containing scales (coccoliths) that surround the organism. Therefore, the effectiveness of the coccoliths to protect the organism from UV radiation is diminished. This could be a significant finding. Planktonic coccolithophores are responsible for much of the photosynthetic primary production in the world's oceans. However, remember that most UV radiation is absorbed within the first metre from the ocean surface.

Some shallow water fish species have shown increased incidences of skin conditions, including cancers, that are normally associated with high UV exposure. However, adult fish appear to be generally quite resilient – perhaps because they show behavioural adaptations, such as simply swimming to deeper water. Last year, in Cyprus, we visited some large aquaculture fish tanks constructed in Ptolemaic times (about 300 BCE). The walls of the tanks had deep but blind tunnels that provided the fish with a midday refuge. Nothing new under the sun?

Once again, climate change complicates assessing future aquatic exposure to UV radiation. For example, the turbidity of ocean coastal waters is much influenced by the amount of coloured dissolved organic matter being added by terrestrial runoff. This depends on the amount and nature of precipitation. Regional changes in patterns of rainfall predicted in climate models will therefore affect the intensity profile of UV in adjacent marine waters. In the Arctic, species that have evolved to live in an environment in which very little UV penetrated the sea ice are now experiencing ice-free summers.

This is the end of our short review of ozone depletion. Despite the apparent effectiveness of the Montreal Protocol and the emerging scenario of ozone layer recovery, we may have some unpleasant interludes on our way to (or beyond) pre-1980 ozone levels, especially in the Arctic. However, these will be much easier to consider when we have taken a look at polar atmospheric circulation and climate change.

Finally, from the perspective of the Arctic Messenger, I think the situation is not entirely satisfactory. Yes, ozone levels have been stabilized and the massive depletions seen in the Antarctic may have been avoided in the Arctic. However, 25 years ago, leaders of Arctic indigenous peoples' organisations knew that ozone depletion was more pronounced at both poles and were asking about the implications to their health and their natural environments. I expect the reader will have noticed that most of the answers to their questions today come not from Arctic data but from extrapolations and inferences from studies at mid-latitudes. Sometimes, this approach works. Sometimes, it does not.

The Long and the Short of It

The ozone story is probably the best good-news story in this book. Here are some points to remember:

- It demonstrated that when faced with a global environmental problem, governments are able to work together to find and implement a concrete solution: the Montreal Protocol.
- The protocol legally binds parties to ban identified ozone-depleting substances according to agreed time schedules.
- Perhaps most importantly, the protocol contains mechanisms to allow it to keep track of and to assess new scientific information on previously recognized and newly recognized depleting substances and to add new substances to those subject to elimination or control.
- The Montreal Protocol has worked well. Levels of ozone-depleting substances have been decreasing after reaching a peak in the 1990s and ozone levels are no longer decreasing. It is expected that

stratospheric levels of ozone at mid-latitudes in the Northern Hemisphere will return to 1980 levels before mid-century. Recovery will take somewhat longer in Southern Hemisphere mid-latitudes. Recovery is expected to be slower in the Arctic and especially in the Antarctic, with significant spring ozone depletion continuing into the second half of the present century.

• The success of the Montreal Protocol in controlling ozone depletion has meant that increases in UV-B radiation have been relatively low away from regions impacted by the Antarctic ozone hole.

• Without the Montreal Protocol, levels of UV-B radiation would have reached levels in mid-latitudes probably capable of causing serious consequences for the environment and for human health.

However, there are unsatisfactory aspects:

• By the end of 1975, the scientific explanation as to how CFCs would deplete the stratospheric ozone layer had been described, but this was denied by a large-scale lobbying initiative.

• Although the United States banned the use of CFCs as spray propellants in 1979 (quickly followed by a few other countries), not much happened. Global CFC emissions rose and the intense lobbyist campaign continued to deny a link between anthropogenic chemicals and ozone loss.

• It was not until 1985, when NASA confirmed the existence and huge extent of the Antarctic ozone hole, that governments began to seriously address ozone depletion as a global problem that required much more than just a ban on CFCs, such as spray can propellants. Ten valuable years had been lost.

• The tardiness of governments to react quickly to warnings from environmental scientists will be a recurring theme, as will the influence and power of lobbyists to hinder action.

• Another recurring theme is our continued inability to learn from our mistakes. We have a gift for replacing one set of chemicals with others that either cause the same problems as their predecessors or get up to other mischief. In other words, we have difficulty in reaching any of the three methods of attaining wisdom, as outlined in the quote by Confucius at the start of this chapter.

• Climate warming is aiding recovery of the ozone layer. This is expected to result in ozone levels that are substantially higher in the Arctic than in 1980. If so, the Arctic could be receiving 20% less UV irradiance than in 1980, thus raising concern that Arctic human populations could face vitamin D deficiency. Therefore, we cannot regard the climatic enhancement of ozone level recovery as a favourable development, especially when we consider the collateral negative impacts of global climate change.

- There are still unexpected surprises, such as the discovery in 2014 of three HCFs and one HCFC that appear to have been undetected in the atmosphere since the 1960s.

8

Persistent Organic Pollutants and Heavy Metals (Including Mercury)

And another spirit is at hand,
conceived in the restless city,
growing from the rubbish yards of houses and from smoking factories,
a spirit swelling with confidence,
multi-coloured all-embracing
garish quick-nosed ever pacing.

Dáithí Ó hÓgáin, *"Putting Out the Hag"* from
Footsteps From Another World

This chapter tells a tragic story caused by humankind's hubris, "swelling with confidence" in its attitude towards its environment. Here, the Arctic Messenger will tell us how a class of chemicals called *persistent organic pollutants* (POPs) and some metals (especially mercury) have accumulated through the Arctic ecosystem to top predators and to indigenous peoples. At the heart of the story is the conspiracy of a number of independent properties of these substances that target the Arctic ecosystem. What makes the tragedy so raw is that the potential for this polar conspiracy could have been recognized long before the evidence was stumbled on in the Arctic. As we will see, even more disturbing is the evidence that for POPs, world governments have still only understood and acted on a small portion of the situation described by the Arctic Messenger.

Before we begin, I have a warning. The simplest approach to this chapter would be to describe what we know today about POPs and heavy metals in the Arctic and global environments and to end with a review of the present state of international actions to deal with them. This would have been easy for the reader and for me. But it would have lost the historical perspective of how we came to understand the environmental and human health characteristics of these chemicals over the last 30 years. Therefore, I want the story to be told as it really happened, with science and policy evolving together. When told in this way, the Arctic Messenger's story reveals lessons on the effectiveness of inter-action between science and national and international policy. It is most effective when organisational measures are set up to force interaction between the science and policy camps and when pathways exist that can channel the resulting scientific information into regional and global political processes to deal with the issues exposed. At the same time, we will see how difficult it is for governments to take actions that involve short-term economic pain.

The consequence of this historical approach is that you, the reader, will not enjoy a smooth passage. You will find yourself flipping from atmospheric physics to human health and POPs to international control actions to the environmental chemistry of mercury to environ-mental toxicology and back to atmospheric and oceanographic physics – all within a few pages. It will be a rocky ride, but along the way, I think you will experience some of the emotions and disappointments of those of us who took this journey.

Early History: Beginning to Understand the Behaviour of POPs and Mercury in the Arctic

In the early 1980s, I was working for the Canadian government and managing the Beaufort Environmental Monitoring Programme (BEMP). Across the floor in a different directorate was Garth Bangay. We encoun-tered Garth earlier during the formation of the Arctic Environmental Protection Strategy and the Arctic Council. Garth had a knack for attracting hot potatoes, and at that time, it was the cleanup of aban-doned Distant Early Warning (DEW) Line stations. Sixty-three of these linked radar stations were built from Alaska to southern Baffin Island to warn North America of any approaching military force that would be taking the polar route. These stations were later abandoned with one pulse in the late 1960s and another in the late 1980s and early 1990s. This marked the end of the DEW Line. The upgraded replacement (the North Warning System) involved comparatively few stations. The aban-donment of those first stations was very simple. The military simply packed its personal items and any classified material and departed, leaving behind the buildings, electrical equipment (containing polychlor-inated biphenyls [PCBs]) and barrels of fuel and other fluids outside.

Northern Canada appears in coffee-table books as a vast pristine and pure environment. However, for Canadian governments, it has always represented a promise of huge nonrenewable resource development. It is a philosophy that has never been more intense than it is today. In the mid-1970s, Arctic indigenous communities became more and more uneasy about the potential impact of "southern" activities in their homelands. At the same time, they noticed changes in the health of their families and of the animals on which their diet depends. They were frustrated that whenever a southern activity was completed in the North, the standard abandonment plan was to leave everything behind – be it toxic tailings ponds or barrels of unidentified fluids. This situation was also frustrating for those of us who tried unsuccessfully to ensure that new development activities (usually involving public subsidy in some form) came with realistic abandonment plans that were secure and independently funded – only to be brushed aside with the rubric of "national economic growth" and "jobs for Canadians". Then, with bankruptcy frequently shutting down such operations a few years later, pollution treatment facilities at the abandoned sites would often be required "in perpetuity" at public expense. I suppose we should have been partly happy. In earlier times, everything would have been left to rot. The abandoned DEW Line stations were the last straw that exhausted the patience of Arctic indigenous communities.

It was obvious to Garth that the greatest potential hazard at the abandoned DEW Line sites came from the PCBs in electrical fluids and in the paints and sealants (that contained PCBs to make them less brittle at a cold temperature). It was well known that some of the fluids had leaked. To know if these fluids could cause a dietary hazard for indigenous peoples and to gather funding for cleanup, he needed to know how far away from a station the contamination had spread and exactly what it had spread into. In 1985, he set up an "interagency working group on contaminants in native diets". At the same time, he funded some sampling around a few of the sites. After overcoming an analytical problem at one of the Environment Canada laboratories, he found that PCBs were showing up everywhere he looked, not just at the DEW Line stations. However, PCBs were not the only problem. To help Garth, I used some of my programme funding for Michael Wong to make a literature search of all scientific publications that reported on contaminants in wild foods harvested in northern Canada. Michael's report also became available in 1985. It was, to use modern slang, a "game changer". The data were sparse and different methodologies made comparisons difficult. Nevertheless, there was no escape from the obvious conclusion: PCBs were widespread in wild aquatic foods of indigenous peoples in the Canadian Arctic, as were a number of other man-made organic substances that had almost certainly never been used at the DEW Line sites. In addition, several metals were showing up that needed investigation, especially lead, cadmium and mercury. It was later found that the PCB signature for a DEW Line station could extend to about 20 km. Beyond that, something else was going on.

We quickly noticed that except for lead, cadmium and mercury, all the substances that were appearing at surprising and unexplained concentrations were organochlorines. These are almost exclusively man-made substances that are structurally built around rings. Each ring is made up of six carbon atoms, but one or more rings can be linked together. Each carbon atom can hold either a hydrogen atom or a halogen – in this case, chlorine. At that time, we had evidence that the organochlorine chemicals shown in Table 8.1 were present in the Arctic, but the list would later grow and eventually include substances that are not organochlorines.

The general public had known about some of the environmental issues associated with a number of these substances since Rachel Carson published *Silent Spring* in 1962. In this very famous book, she linked the dramatic decline of bird populations in the United States to the intensive use of pesticides, especially DDT. Her work launched the environmental movement in the United States. At the University of Stockholm just two years later, Sören Jensen was trying to measure DDT in human blood using electron capture gas chromatography. He was constantly noticing another group of substances he could not identify. The substance group was in his blood, in his hair and that of his wife and young daughter and even in the soil outside. He was more baffled when he found the group in wildlife samples that were dated before the use of

Table 8.1 *Organochlorines found in the Canadian Arctic environment in the late 1980s*

Pesticides	aldrin, chlordane, dichlorodiphenyltrichloroethane (DDT), dieldrin, endrin, heptachlor, hexachlorocyclohexane (HCH), mirex and toxaphene
Industrial Chemicals	hexachlorobenzene (HCB) (also a pesticide and a by-product) and polychlorinated biphenyls (PCBs) (also a by-product and a breakdown product from other POPs)
By-Products	polychlorinated dibenzo-p-dioxins (dioxins) and polychlorinated dibenzofurans (furans). Dioxins and furans can be produced naturally in very small quantities from, for example, forest fires.

organochlorine pesticides. In 1966, he solved the problem. The mysterious substances were PCBs. It was rapidly confirmed that they were ubiquitous in the environment. We must thank Sören Jensen for exposing just how widespread PCB contamination had become.

It is rarely recognized that big steps in science are often made possible by the availability of a new type of technology. In the ozone chapter, we learned how James Lovelock designed the electron capture detector, which was capable of measuring very small concentrations of trace gases. This advance enabled chemists to measure such substances as DDT, PCBs and freons at picogram levels of concentration. Therefore, we should probably also be thanking James Lovelock for opening this door to modern environmental chemistry.

PCBs were first synthesized in Germany in 1881 and were soon put to a wide spectrum of uses, which included hydraulic oils, flame retardants, pesticides, coolants, paints and sealants. Concerns about their toxicology grew from the 1930s, and by the early 1970s, most European and North American countries banned all uses except when they were constrained in closed systems. By the end of the 1970s, most of these same countries had abandoned production and import, but use of PCB-containing fluids in existing equipment, such as transformers, was still allowed.

Returning to the Arctic situation, where were these chemicals coming from and were the levels significant?

Garth organised a small monitoring programme to help provide answers, and by late 1988, he had some preliminary data on his desk. Michael Wong's report and Garth's first data set were also "game changers" for me. Reorganisation in 1988 took me into Garth's directorate and we set about solving the puzzle until he took off for more lofty responsibilities and even hotter potatoes a few years later. Russel Shearer joined my group at about this time, and together, we gathered a wonderful team over the next 15 years, with Kathy Adare, Jennifer Baizana, Siu-Ling Han, Jill Jensen, Sarah Kalhok, Marilyn Maki, Janine Murray, Carol Reynolds, Simon Smith, Jason Stow and Jill Watkins.

At the end of February 1989, Garth set up a science evaluation meeting in Ottawa to examine his data set. More than 40 specialists from North America, Northern Europe and the Soviet Union participated. Very few direct air or water measurements were available to reveal the pathways taken by the contaminants to reach the Canadian Arctic. It was known that radionuclides, acidifying substances and some metals from industrial activities were being transported in the atmosphere from sources outside the Arctic. There was just not enough data to be confident with making the same conclusion for organochlorines. However, only PCBs of the 13 or so organochlorines found in the Canadian Arctic had ever been used there in any quantity and one (toxaphene) had never even been registered for use in Canada. The source could not be local. Freshwater and marine ecosystems were contaminated. The only natural transport mechanism that would be common to both was the atmospheric pathway from distant sources, so we began to take a special interest in this possibility.

The critical component of this early work was a small survey of levels of several contaminants in the diet, blood and breast milk of Inuit from the small community of Broughton Island, off the east coast of Baffin Island. The data were collected over two periods: in 1985 and between 1987 and 1988. It showed that all the local Inuit foods tested contained PCBs and mercury, with the highest levels being in narwhal and beluga blubber (the thick fatty layer that lies in the innermost layer of the skin of marine mammals) and in polar bear fat. Blood PCBs exceeded Health Canada's "tolerable" levels in 63% of children and in 39% of women of childbearing age. Only four breast milk samples were available. Three were comparable to levels in southern Canada, but the fourth exceeded Health Canada's "tolerable" level. Blood mercury levels exceeded Health Canada's guidelines in three persons – all males. Clearly, the indicated level of contamination could not be ignored.

The first clue to unravelling this mystery came from the Arctic substances themselves. They all share four basic characteristics:

1. They are very resistant to biological and chemical degradation because the chlorine-carbon bonds are very strong. As a general rule, the more chlorine (or other halogen) substitutions that exist on the molecule, the more long lived the substance. Any degradation that does happen is usually temperature dependent, grinding to a halt at polar temperatures.

2. They bioaccumulate and biomagnify. Animal physiological processes typically evolved to get rid of unwanted substances by making them water soluble. However, organochlorines are said to be lipophilic. This means they are concentrated in fats and have little solubility in water. As a result, when an animal takes in an organochlorine molecule, it is unlikely to excrete it. Therefore, the animal accumulates more and more of the substance until it eventually dies (most probably by being eaten). Biomagnification simply reflects that a food web consists of a lower level of primary producers that accumulate the organochlorines from the nonliving environment and whose

organochlorine "harvest" is progressively squeezed into fewer and fewer individuals as they pass up the food web. The top predator accumulates the grand total of all the organochlorine intake of all the plants and animals in the trophic levels below that have contributed to its meals. A coastal or lake food web could see plant plankton (phytoplankton) passing their cumulative burden initially to herbivorous zooplankton, then through carnivorous zooplankton to small fish to large fish to the gull, osprey, otter or seal. The cost the gull pays for being well fed is to potentially end up with a body burden about 25 million times higher than the concentration in coast or lake waters in which our gull takes its daily paddle. Of course, it is not a problem if there are no contaminants that are capable of being biomagnified in the water. Figure 8.1 provides a simplified view of the Arctic marine food web with its opportunities for different degrees of biomagnification.

3. Organochlorines are toxic, but this toxicity takes several forms. Each substance is associated with a number of acute effects that are fairly simple to detect. However, hidden below this is an array of other outcomes associated with long-term (chronic) exposure even at much lower doses. We will look at these later, but they come in a large variety of forms, ranging from increased incidences of tumours to reproductive tract and fertility abnormalities to behavioural effects. It is an old tenet of toxicology dating from Paracelsus (a Swiss

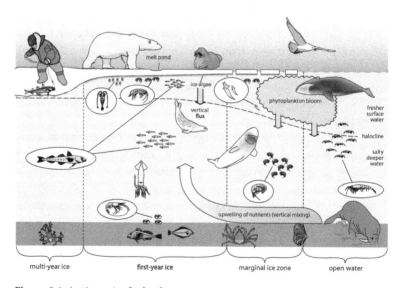

Figure 8.1 *Arctic marine food web*

physician and polymath who lived between 1493 and 1541) that the dose makes the poison. If you combine this thought with the organochlorine biomagnification characteristic, you will begin to glimpse where this tale is heading. We will return to the adage from Paracelsus when we notice that a fair portion of the chronic toxicity of organochlorines and related substances may be because they are interfering with hormones and that the effect may depend more on when an animal is exposed rather than on the dose itself.

One final point about toxicity. We noticed earlier that organochlorines do not "want" to degrade. This is not quite true. Several do undergo some transformation (especially in biological systems), but unfortunately, this often results in substances (residues) with similar or higher toxic properties that resist any further degradation.

4. Organochlorines are said to be semivolatile. Some of us would explain this property in terms of its vapour pressure, while a more exacting chemical engineer would use the term *fugacity*. However, the Arctic Messenger wants us all to understand what is going on, so we will explain it this way. Any substance – whether it be a solid or a fluid – will have a tendency for its molecules to escape as a gas. This escape activity or volatility largely depends on temperature. A semivolatile substance is one where the balance of the behaviour to evaporate or to condense takes place over the range of environmental atmospheric temperatures we find in nature. If it is warm, the substance will tend to evaporate. If it is cold, it condenses.

Something else was intriguing. Metals are natural, but gaseous mercury and lead can travel in the atmosphere and another form of methylmercury can biomagnify.

We found it remarkably easy to come up with a working hypothesis to explain the puzzle when we used the overseas expertise from Garth's 1989 evaluation and from discussions with Richard Addison, Len Barrie, Dennis Gregor, Lyle Lockhart, Derek Muir, Don Mackay and Ross Norstrom. Readers have probably worked it out already. Organochlorines released into the environment in low or mid-latitudes evaporate into the atmosphere, where they stay until the air reaches a cold enough temperature for them to condense. This is regulated by their vapour pressure (or fugacity) characteristics. On a warm day, up they go again until another cold spell leads to condensation. In this way, they progressively hop towards the Arctic (the grasshopper effect), where they are trapped by the cold temperatures (cold trapping). The ability to resist degradation robs the Arctic of a mechanism to get rid of them, while their efficient biomagnification leads to amplification in Arctic top predators and people who use these animals as their main source of food. The end result can be contaminant exposure at levels considered to be of concern in top predators (including people) even when the ambient concentrations in water and air around the animal are close to detection limits.

We will pause here for another digression. I remarked earlier that the widespread accumulation of organochlorines from mid-latitude sources in the Arctic should not have come as a surprise. Richard Addison was reporting organochlorines in beluga whales and ringed seals from the Canadian Arctic as early as 1973 and 1974. He considered his Arctic study area as a reference area for his studies on seals on Sable Island (300 kilometres south-west of Halifax, Nova Scotia). Meanwhile, Brynjulf Ottar was describing the concept of "the grasshopper effect" (Figure 8.2) and "cold trapping" by successive volatilization and condensation as early as 1981 and Don Mackay was working out the physical dynamics of contaminant movement in and out of the gas phase at about the same time. I should take the opportunity to introduce these two interesting scientists. Brynjulf Ottar was the first director of the Norwegian Institute for Air Research (NILU) and was a pioneer in studies to understand long-range atmospheric transport, particularly with regard to acid rain. He was clever and charming. Russel and I were fortunate to meet him once at separate meetings shortly before his death. For me, it was at a Comité

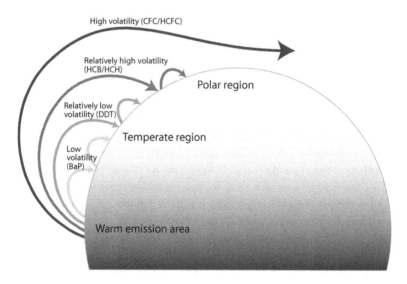

Figure 8.2 *Schematic representation of the "grasshopper effect"*

Arctique meeting in Oslo, and in about 30 minutes, he gave me enough insight on how to organise international action on organochlorines to last the rest of my working life. Don Mackay must have few rivals in environmental chemistry. Now past his mid-seventies, he is still turning out original research like a graduate student. However, despite his amazing scientific output, I think his greatest legacy is the outstanding brood of young environmental chemists and physicists who have sprung from his care and who are now carrying on his science and mentoring tradition with yet another generation. He is quite the patriarch!

It is time to return to the cold trapping idea and its connections to biomagnification. The hypothesis could be tested, but it would require a hemispheric monitoring and research effort to do it. Timing was serendipitous. This was when Finland proposed that Arctic countries should cooperate to protect the Arctic environment and when Norway championed the idea that it could not be done without a circumpolar monitoring programme. In another fortunate coincidence, in 1990, the Canadian federal government announced the Green Plan, a well-funded environmental programme directed towards sustainable development. One of its components was a monitoring programme exclusively directed towards contaminants in the Arctic in relation to indigenous diets: the Northern Contaminants Programme (NCP). The Green Plan did not survive for long, but our NCP is still alive and is directed since I retired at the end of 2004 by Russel Shearer. Our stable funding base made it possible to not only build an adequate monitoring effort in Canada but to also strongly participate in circumpolar monitoring emerging under Arctic Monitoring and Assessment Programme (AMAP). It was this combination of effort by scientists from around the entire circumpolar Arctic that enabled the conversion of the working hypothesis into an accepted concept.

While we were beginning to focus on the conspiracy of the airborne pathway with the power of biomagnification in Arctic Canada, research programmes organised under the Oslo and Paris conventions and by the Helsinki Convention had decided that contamination levels of organochlorines measured in the North and Baltic seas could only be explained if there was a significant atmospheric pathway as well as river discharge and direct discharge into the ocean. At that time, these seas were facing a dramatic rise in mortality of seals caused by distemper (morbillivirus). However, the severity of the epidemic seemed to be associated with PCB levels and immunodeficiency in the seal populations. Seals in waters with much lower PCB concentrations off western Ireland and Scotland were far less impacted by the infection. Meanwhile, in Canada and the United

States, scientists working on the rehabilitation of the Great Lakes, which had suffered years of abuse from chemical dumping, found that even when levels of pollution input were drastically reduced, the pollutant levels in the lakes failed to fall at the expected rates. There had to be another organochlorine source for the lakes, and of course, it soon revealed itself to be atmospheric transport. Monitoring in these different regions was well ahead of Canadian efforts in the Arctic, but as yet, Europe had not paid a great deal of attention to its far north. The emerging realization of the role of long-range atmospheric transport of these substances raised serious implications.

The startling lesson being learned was that for substances with the characteristics of organochlorines and mercury, you do not have the ability to control the quality of your local or regional environment. Instead, you are at the mercy of emissions far away.

By mid-1990, an evaluation of all the information on Arctic contaminants we could find was completed. It was eventually published in 1992 as five papers in a single edition of the journal *Science of the Total Environment*, volume 122 (1–2). The paper which reviewed how these substances were reaching the Arctic, was quite noncommittal about whether the atmospheric, marine or freshwater pathway dominated. However, in the "Arctic Marine Ecosystems Contamination" paper in the same publication, evidence in support of the cold trapping hypothesis was discussed. I think this noncommitted stance was partly because the best data then available were for hexachlorocyclohexane (HCH). It was the only one of the Arctic organochlorines we were monitoring at the time that possesses a significant degree of water solubility; indeed, it was later found that because of this, a sizeable portion of the HCH in the Arctic could arrive by marine transport. We now know that any relatively soluble chemical should behave in the same way as HCH (such as toxaphene and endosulfan) and that a sizeable portion of perfluorinated sulphonates (a substance we will look at later) also arrives in the Arctic by oceanographic transport. Look at the discussion by Derek Muir and Rainer Lohmann (2013) if you would like to learn more.

Early History: Stimulating International Action to Control the Entry of POPs and Heavy Metals to the Environment

Despite the caution of the bench scientists, Garth, Russel and I were convinced of the veracity of the grasshopper/cold trapping hypothesis and were treating it as proven. We were familiar with the work of Jozef

Pacyna and Michael Oehme (published in 1988) that supported the atmospheric transportation hypothesis. We also knew that organochlorine levels in seals in waters around Svalbard were similar to those of seals in Arctic Canada. With his usual tenacity, Garth thought we had enough justification to seek international controls on organochlorines. Of course, we did not have anything like the required information, but that did not matter to Garth. I took a look at existing agreements related to chemicals and the environment. Nothing existed that would commit countries to control these substances. There was not even an international collaboration to discuss the possibility that a combination of long-range transport and biomagnification could cause significant effects at locations far from sources. The only partially relevant agreements were the Basel and Rotterdam conventions that dealt with waste disposal and international trade of toxic substances. Garth was not discouraged one little bit, and in spring 1990, I was writing to the United Nations Environment Programme (UNEP), the World Health Organization (WHO) and the Convention on Long-Range Transboundary Air Pollution (CLRTAP) to investigate the possibility of an appropriate multilateral environmental agreement (MEA) to deal with POPs. In the case of CLRTAP, I suggested the creation of a POPs protocol.

UNEP and WHO did not respond, but the CLRTAP opened its door slightly. We encountered this convention earlier when we looked at acid rain in the Arctic. Because it was set up to deal primarily with acid rain, its delegates were very familiar with the concept of long-range atmospheric transport being responsible for environmental effects occurring at localities distant from sources. As it is a framework convention, it enables parties to "attach" legally binding protocols that target specific pollution issues as they emerge. I was invited to make a presentation to a subsidiary body in August 1990 in Geneva. Hans Martin, who was a science manager with the Canadian Atmospheric Environment Service, was a mentor for me at that time and I often turned to him for advice. He had a long history of working on the acid rain CLRTAP protocols. The presentation went quite well and it was decided to forward the proposal to the Executive Body (the convention's decision-making body).

The August 1990 meeting and most of the subsequent ones under the CLRTAP took place in the old League of Nations buildings in magnificent grounds just above Lake Geneva. I quickly found allies there, mainly from the Nordic countries, Germany, the Netherlands and the United Kingdom. Most importantly, I met Lars Lindau from Sweden, who was a veteran at the CLRTAP, and Lars Nordberg, who headed the secretariat for the convention. Lars Lindau already had a mandate to conduct an evaluation of POPs for a different organisation under the UNECE, but it was not directly linked to a convention that could introduce chemical controls. Organochlorines are a class of POPs. As history later showed, the Swedes were wise to use this term and Canada swiftly embraced it. The change made no difference to our earlier evaluation of what was happening in the Arctic, and if you go back a few pages, you can always switch the word *organochlorines* for the acronym *POPs*. However, you cannot do it the other way around, as some POPs are not organochlorines. The only technical difference resulting from adopting POPs was that we could now deal with substances on the basis of their environmental behaviour rather than on their chemical structure. Later on, we will see why this was such a good idea. Lars Lindau and I quickly decided to combine our efforts. At our first meeting in Stockholm, he introduced me to Sören Jensen (who we met earlier discovering that PCBs were everywhere in the environment) and Bo Jansson, who over the years has continued to teach me about the behaviour of POPs. In December of the same year, the Executive Body agreed to establish a task force on POPs that Lars Lindau and I would co-chair. We were charged with developing the scientific case for a legally binding POPs protocol under the CLRTAP. A task force co-chaired by Czechoslovakia and Germany was given a similar mandate for heavy metals. We decided to produce two reports for POPs: an "Overview of Persistent Organic Pollutants in the Environment" and a "State of Knowledge Report on POPs". The latter included a summary of the former, together with socioeconomic and technological information relevant to potential international controls.

Another quick little digression about working with UN organisations. They have a reputation for endless formality and bureaucracy. It is true, but the criticism is partly unfair because it comes from internal mechanisms designed to protect the national sovereign interests of all countries. However, for a while, lucky organisations have a secretariat headed by a resourceful individual who can respect impartiality while also cutting straight through the tangled webs of diplomacy. Their teams take up their talents, but when they leave, the versatility vanishes so quickly. Such was the case with Lars Nordberg, who even managed to enliven our lives with a keen sense of humour. The first CLRTAP POPs meeting was also the first UN session I chaired. Within an hour, we were in chaos when Lars saved the day by whispering in my ear: "Perhaps it is time for a little coffee break"!

The task force on POPs and the task force on heavy metals were now faced with gathering and assessing as much information on environmental contamination in the Northern Hemisphere by POPs and heavy metals as we could get our hands on. We knew we would expose huge data gaps, but if we were swift, perhaps some of them could be addressed in time to be included in our reports to the CLRTAP. It was a perfect opportunity for the emerging AMAP. Fortunately for us in Canada, our new NCP funding base enabled us to not only find out what was going on in our Arctic but to also participate in international partnerships through AMAP to design and carry out circumpolar research and monitoring that could feed into the CLRTAP process.

Before returning to the POPs story, I want to take a few paragraphs to describe how we designed the NCP. A little earlier, I mentioned the BEMP. It was a large interdisciplinary project watching over a then-unique oil and gas exploration effort in the Beaufort Sea. It was unique because industry was working in an offshore pack ice environment from artificial islands in winter and drill ships in summer. By today's standards, the BEMP was very well funded. An environmental assessment and review had essentially concluded that hydrocarbon exploration and production could be safely undertaken in the Beaufort offshore, providing that the enormous list of environmental issues identified was comprehensively monitored. How could this monitoring be done? One possibility was to find a way in which all the issues could be compressed. While mulling over this possibility, I was asked to review a funding proposal to use a process called *adaptive environmental assessment and management* (AEAM) that had been developed by C. S. Holling and C. J. Walters at the University of British Columbia and at the International Institute of Applied Systems Analysis (IIASA) in Vienna. The proposal was not related to the Beaufort and it was not funded, but I was fascinated by their methods. I arranged to meet the proponents – Alan Birdsall, Robert Everitt and Nicholas Sonntag, together with Bill Brackel of Environment Canada – and we began to develop a funding and management regime for the BEMP using the AEAM methodology.

We held a number of workshops that involved the hunter-and-trapper associations of all the indigenous communities on the Beaufort coast, representatives of industry operations (such as drilling engineers, drill ship and icebreaker captains and spill countermeasures people), wildlife biologists, toxicologists and oceanographers. I will call these people "the team". Work began by compiling a list of valued ecosystem components (VECs), which were those entities society (particularly indigenous society) would be very upset about if they were damaged. Next, the team created a long list of all the identified potential or purported impacts (at this stage, not necessarily linked to VECs). The team then built a conceptual model of the Beaufort Sea ecosystem and superimposed on it a similar model of the industrial activity. The next step was to create a series of impact hypotheses. These linked specific industrial activities through items on the potential impact issues list step-by-step through the ecosystem model to the purported unwanted effect on the "vulnerable" VEC or VECs. When the team evaluated each hypothesis in this way, there were three possible outcomes: (1) We could decide that more information was needed, requiring modification to our research or monitoring for that hypothesis; (2) we could judge the hypothesis valid, which would require a change in the relevant industrial activity; or (3) we could judge the hypothesis to be invalid and deem that the purported impact issues required no further attention.

Because of the inclusiveness of the process and of its unambiguous end points, the BEMP began to enjoy the support of indigenous communities, together with the operational people in industry, and we were able to justify the whole programme to our political masters in Ottawa. Scientists were a little slower to warm to this form of decision making, but in less than a year, most became advocates. The result was that we now had a manageable list of specific research and monitoring topics that all "stakeholders" had agreed on. Furthermore, we had a management framework that involved the same participants evaluating future results and modifying our research and monitoring activities.

By about 1987, the Organization of Petroleum Exporting Countries (OPEC)–inspired global shortage of oil vanished. Consequently, offshore oil and gas exploration in the Canadian Beaufort Sea was abandoned for nearly 20 years. The BEMP ground to a halt, but the experience had a major impact on how we later approached the management of research and monitoring. As we pondered how to organise the NCP, I realized that we needed a mechanism that would involve, be embraced by and support cross-fertilization with northern indigenous peoples, scientists and government agencies with responsibility for human health and the environment. It would have to set tangible policy-relevant objectives and deliver on them or it would lose the confidence of Northerners and would not escape the long knives whenever budgets would become lean.

Naively, I thought that perhaps the "AEAM" methodology we had modified for the BEMP could be pressed into service and I asked Russel to give it a trial. We quickly found that it would not work, primarily because we could not find a surrogate for the BEMP's industrial scenario. Instead, we progressively built our own system. At its core was our decision to say that the strategic objectives of the programme were to:

1. Identify how contaminants reached the Canadian Arctic and where they came from.
2. Measure the contaminant levels in air, snow, water, soil, plants, fish, wildlife and people of the Canadian Arctic.
3. Assess the effects of the discovered contaminant levels on the health of Arctic ecosystems, including human health.
4. Provide information on contaminants and nutrition to people who live in the Arctic and eat traditional foods.
5. Promote agreement and cooperation with other countries and the circumpolar community to establish international controls of the contaminants of concern.

We needed clear and unambiguous answers to the first three objectives in order to understand and inform Northerners about the hazards and risks they faced (fourth objective) and to take action to halt the entry of the offending substances into the environment (fifth object-ive). We then invited scientists and indigenous peoples' organisations to send in proposals that would fit under one or more of the strategic objectives. A committee of science managers and representatives of Canadian indigenous peoples' organisations then collectively made funding decisions based on the prospects of a proposal to address one or more of the strategic objectives. The indigenous organisations under-took activities related to the last three of the objectives.

About nine months later, the researchers were required to report on progress and to attend an interdisciplinary workshop in which the significance of results was discussed. We then sent out a new request for proposals – the wording of which reflected the workshop discussion in the context of the strategic objectives. The process was reiterated on an annual cycle. As time went on, we prepared "blueprints" that gave greater guidance to scientists on exactly what sort of information we were looking for and these were periodically modified, as some issues would be resolved and other needs identified. At intervals of about six years, major assessment reports were published, but in the meantime, information from the research and monitoring was flowing outwards to Arctic peoples, to regulatory agencies concerned with wildlife and human health, to AMAP and to activities working to control the con-taminants that worried us. Over the years, this initial framework has evolved considerably, but the core intent remains the same. It was a simple structure, but it is remarkable how few monitoring and research efforts do not set out such a framework. As with the BEMP, another fact probably helped enforce our management regime: It is always useful to be holding the purse strings.

Before we continue, it is time for a quick pause to clarify two terms. Think about *hazard* as referring to an inherent property of a substance or activity to cause harm. For example, drinking too much alcohol will lead to liver damage and ultimately kill me. However, the key phrase in this sentence is "too much" – in other words, the size of the dose. The word *risk* expresses the idea of the probability of an unwanted biological/health impact (such as liver damage) resulting from a given dose of a hazardous substance (such as alcohol).

The Symbiotic Relationship Between Emerging Science and International Negotiations for Controls on POPs and Heavy Metals

Returning to our topic, we had reached the stage where the CLRTAP had asked its task forces on POPs and heavy metals to provide the convention with state of knowledge reports. On the basis of these reports, the convention would decide whether to launch negotiations for international control on the release of these substances to the environment. The fortuitous creation of AMAP provided a unique opportunity to identify what circumpolar information would be vital to support the CLRTAP process and to coordinate the collection and synthesis of such information. The first part of this task (collecting high-quality information in the Arctic) took much effort and organisation, and as a result, AMAP was not able to directly contribute to the information synthesis before the CLRTAP 1994 deadline. However, it did so indirectly because the scientists who were taking part in AMAP's setting up of circumpolar monitoring capabilities now found themselves and their work feeding directly into the CLRTAP POP and heavy metal syntheses. Several delegates to the AMAP working group (including me) also sat on the CLRTAP task forces on heavy metals and POPs.

In early 1994, we sat down at a meeting in The Hague and began to prepare the CLRTAP state of knowledge report on POPs. It was at this stage that I first met Ramon Guardans from Spain. Ramon is a remarkable polymath. In addition to leading the drafting of several key sections of the report, he subsequently launched a personal crusade: to raise awareness of POPs in the Mediterranean hinterland in Europe and North Africa and to increase the monitoring capacity of POPs in this region. He was successful on both counts and has continued to be very active at the interface between science and international pollution agreements, particularly with the later global Stockholm Convention we will soon explore.

What critical circumpolar information was collected and synthesized in the report? You could organise the answer according to three fundamental questions.

1. Are POPs being moved far from their sources in the atmosphere? Evidence supporting the long-range transport hypothesis to explain distribution of POPs was looking more and more convincing. It included a string of measurements of air concentrations of dioxins and pesticides made by Michael Oehme from Birkenes in southern Norway to Svalbard. The sampling frequency conducted over two years enabled episodes of contaminated air to be detected and for the prior history of those "contaminated" air parcels to be tracked backwards for about four days. This technique is known as *back trajectory analysis*. It indicated that most of the POP pulses or transport episodes to the Nordic Arctic came from Eastern Europe and the (then) Soviet Union. Mats Tysklind and his colleagues were able to use this technique to also follow the transport to southern Sweden of dioxins and furans in air from sources in Central and Western Europe. Meanwhile, studies in southern Ontario, Canada, were showing that concentrations of toxaphene, DDT, HCH, chlordane, endosulfan and dieldrin could be linked to their use in the southern United States, Mexico and the Caribbean.

At the same time, Frank Wania and Don Mackay had just about tied up the fundamental physics of the grasshopper/cold trapping hypothesis, which now carried the more impressive names of "global fractionation" and "cold condensation". There was no denying that it explained the patterns of concentrations of different POPs across Canada in Arctic air and snow. These included measurements made on an ice island drifting westwards off the northern coast of Axel Heiberg Island, on the Agassiz Ice Cap and at Alert on the northern tip of Ellesmere Island. It would be difficult to find three more remote locations, but all our (then-known) POPs were found there. Terry Bidleman and colleagues reported the ice island results and Dennis Gregor was responsible for the Agassiz measurements. Over several years, Dennis dug from the surface down to the fully consolidated ice. This enabled the volatilization dynamics of different pollutants to be measured as the snow was gradually compacted into ice. That information was vital to enable us to interpret the pollutant concentration history recorded in ice cores. One member of the team was from India and he had a reputation across the Arctic for his curries. I landed there once in a ski-equipped Twin

Otter en route from the meteorological station at Eureka to Alert. It was a strange sensation: Sitting on a barrel only 900 kilometres from the pole – at an altitude of 1,730 metres – with who knows how much ice under your bottom and eating this incredible curry!

The measurements made at Alert used a high-volume air sampler to measure polycyclic aromatic hydrocarbons (PAHs) and POPs at a weekly sampling frequency. The particle phase of the pollutants was trapped on a glass fibre filter and two polyurethane foam (PUF) plugs collected the vapour phase. The sampling record from Alert is continuing without interruption to this day. Between 1992 and 1993, the NCP arranged for four identical samplers to be deployed: one in the Yukon, one at Cape Dorset, one (through AMAP) at Dunai Island near the delta of the Lena River in Siberia and one (also through AMAP) in the western Russian Arctic. Norway had compatible samplers operating at Ny-Åle-sund in Svalbard and the United States installed similar equipment at Barrow in Alaska. Not all these stations are still operating, but for many years, we had a circumpolar ring sampling POPs at coordinated weekly intervals. The true wealth of information that was to be delivered by the continuous air samplers did not emerge until after the 1994 CLRTAP deadline. However, the AMAP contributing scientists had gleaned enough from applying back trajectory analysis to the circumpolar sampler records to confidently say in the CLRTAP state of knowledge report that the environmental loadings of most POPs reaching the Arctic were travelling in the atmosphere from sources in mid-latitudes – usually in the form of discrete episodes. In western Arctic Canada, pulses could come from as far away as Southeast Asia. In some regions, riverine transport is also important, but even here, it is thought that much of the riverine loading is from those POPs then being studied having been atmospherically transported to the river's watershed.

Atmospheric transport is swift and measured in days. The report explained that the oceans are important for the movement of the smaller number of POPs that have a relatively high water solubility (such as HCH), but their movement in this medium is on the scale of a year or so. Over time, the importance of the oceans will probably turn out to be their huge capacity to act as a storage reservoir for atmospherically deposited POPs and therefore a potential source when atmospheric concentrations go down and rising sea temperatures tip the balance towards evaporation. You have probably already realized from this last sentence why scientists working with POPs are casting a worried eye on climate warming.

2. Are POPs being accumulated in biota (the biological part of the ecosystem) far from source regions? POPs are thought to have the potential to cause impacts on biota at concentrations expressed often in parts per billion or even parts per trillion. This means that environmental chemists are frequently operating at close to the detection limits of their equipment. Great care must be taken in sampling and analytical procedures. Frequent interlaboratory comparisons are needed to ensure continuity and comparability between laboratories when investigating changes in levels in biological material between regions (spatial) or over time (temporal). One of AMAP's first tasks was to establish procedures to help compare data from different laboratories. Even today, achieving data comparability from different organisations remains a challenge, but back in the 1980s and early 1990s, this difficulty severely limited the ability to assess circumpolar data. Consequently, the 1994 POPs state of knowledge report was able to show that most POPs could be found in biota throughout the UNECE region, including North America, Europe and the former Soviet Union, but it was cautious about making comparisons. Nevertheless, it explained that although levels were generally highest in industrial catchment areas (such as the Great Lakes and the mouth of the River Rhine), another factor was at play. The longer the food chain, the higher the degree of biomagnification and the higher the pollutant body burdens in fatty tissues. Some top trophic Arctic predators carried similar concentrations as were being measured in the most heavily polluted industrial areas.

3. Should there be concern for deleterious impacts associated with POPs levels mainly attributable to atmospheric transport? The task force report began its answer by pointing out that it is very difficult to prove or disprove that a POP in the environment is responsible for disease or a change in terrestrial, aquatic or human populations and that exposure would always be to a mixture of similar POPs anyway. In practice, a weight of evidence rationale is used. This includes comparing "health" outcomes from dose response relationships in controlled laboratory studies with observations from wildlife or from human populations that are environmentally exposed to high levels of the pollutant. It showed that the main concern was not related to acute toxicity. For most POPs, this was not high at concentrations found in the environment. Instead, the concern was with exposure to comparatively low doses received either over a lifetime or at a particularly vulnerable period in the animal's life (usually at the embryonic stage or very soon after birth). There was a common family of health outcomes from laboratory studies of certain POPs (at that time, mainly PCBs, dioxins

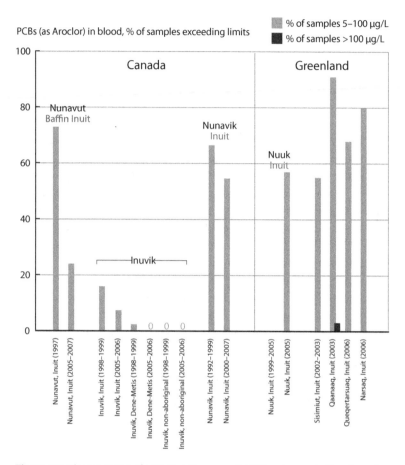

Figure 8.3 *The percentage of mothers and women of childbearing age in Canada and Greenland that exceed guideline limits for PCBs (measured as Aroclor) in blood*

and DDT) and of effects in wildlife populations exposed to high levels of the same substances. What is more, the report showed that because of biomagnification, it is possible in the special circumstances of Arctic ecosystems for POPs to reach concentrations in the fatty and oily tissues of top predators that are in the same general range as those that produced effects in the laboratory studies. Humans whose diet is at the end of a long food chain (this usually means an aquatic one) are the top predators of that food chain. Figure 8.3 provides a glimpse of the potential significance of biomagnification of POPs to people in the Arctic whose traditional diet includes a large proportion of top marine predators. The graph (from the AMAP 2009 *Assessment of Human Health in the Arctic*) shows the percentage of mothers and women of childbearing age

in Canada and Greenland that exceed guideline limit
ured as Aroclor[1]) in blood. The guidelines are those
indicate a "level of concern" (>5 µg/l) and a "level of
(Health Canada, 1986). Notice that in some locat
available, in the 1990s and since 2000, levels of
declined (such as Baffin Island and Inuvik). This is prou
combination of dietary change and a decline in environmental levc
response to recent regulatory controls on these substances.

We will delay a close examination of these potential health outcomes until the end of the POPs story, when we can take a look at the
most up-to-date information. However, the following are examples of the
types of human and wildlife outcomes that had been associated with
exposure to POPs in the laboratory and in wildlife and human populations at the time of the 1994 CLRTAP reports: immune system dysfunction; reproductive impairment caused by functional, physiological
(including hormonal), anatomical and behavioural changes; developmental abnormalities mainly appearing from the time of fertilization to early
infancy; carcinogenesis; and behavioural abnormalities (Table 8.3). In
1986, Rogan and colleagues published a prescient study involving more
than 900 children in North Carolina (United States) that suggested that
transplacental transfer of PCBs and DDT from mother to unborn child
depressed muscle reflexes in the newborn. This was really the beginning
of a new perspective on the toxicity of POPs that took a giant step forward
with the work of Joseph and Sandra Jacobson and their colleagues.
Beginning in 1992, they published a series of papers showing lasting
neurological and behavioural effects in children in Michigan (United
States) whose blood was carrying high burdens of POPs at the time of
their mothers' pregnancy. The first publication was just in time for
inclusion in the CLRTAP state of knowledge report. It was to have a great
impact on decision making under the CLRTAP and subsequent global
negotiations for control of POPs. It was also at about this time that Theo
Colborn was bringing together all the diverse information that pointed
towards a number of POPs being able to disrupt the way in which
hormones of the endocrine system coordinate the development of
animals (including humans) from conception to old age. Possibly many
of the health effects listed here involve hormone disruption.

I want to pause again for another little digression – this time about
biomagnification. It is key to understanding the behaviour of POPs and

---

[1] PCBs were marketed as mixtures of many different types known as congeners.
The most common mixture was called Aroclor.

...mercury in the Arctic and it all depends on the structure of the Arctic food web (trophic relationships). One of the neat tricks that biologists keep in their "toolboxes" to investigate trophic relationships is stable isotope analysis. This is how it works: As described in the introduction to radioactivity, the atoms of any given element will always have the same number of protons in the nucleus, but there can be small differences in the number of neutrons. Of these different forms (isotopes), only one is usually the most common in nature, but very small quantities of the other isotopes will be found. For example, most carbon is ^{12}C, but approximately 1% is ^{13}C. Similarly, for nitrogen, ^{14}N is much more common than ^{15}N. The word *stable* means we are thinking only about isotopes that do not transform into another element. Although stable isotopes share identical chemical behaviour, their mass varies according to the number of neutrons. It turns out that biological and geochemical processes sometimes have a slight favour of one isotope over another, resulting in a change in the ratio of occurrence of the two isotopes before and after the biological or geochemical process. The stable isotope content is described as delta values (δ), meaning parts per 1,000 (‰) differences from an international standard. You can use a mass spectrometer to calculate the ratio of the two isotopes in a sample.

The metabolism of animals tends to enrich the ^{15}N content of the body relative to the ^{14}N content. When animals in the lowest trophic level are eaten by animals in the next level, they pass along their $\delta^{15}N$ enrichment. The story proceeds from one level to the next until you finally arrive at, for example, the polar bear or ourselves. In marine systems, there is roughly a 3.2‰ enrichment of ^{15}N from one trophic level to the next. The ratios of ^{13}C and ^{14}C vary between different species of phytoplankton and some plants. This signature is passed on to herbivores that eat them and thus differences in $\delta^{13}C$ between different animals suggest dietary differences. The main value of this approach is that it provides a way of quantifying the trophic underpinning of biomagnification. It has also often solved fascinating mysteries in data sets where, for example, a population of walrus was found to have organochlorine concentrations much higher than usual. Walrus feed on clams and are therefore not too high in the food chain and generally do not carry worrisome body burdens of POPs. This group was carrying burdens more akin to polar bears. Stable isotopes solved the puzzle. They had added two more trophic levels to their food web by enjoying the taste of fish-eating seals.

Returning to the POPs state of knowledge report, about half the document was devoted to showing that international action to ban or to severely restrict the use of POPs was achievable. It occupied so much space because these substances had so successfully insinuated themselves into our lives and into the global economy. Furthermore, dioxins and furans as unintentional by-products from combustion presented major abatement challenges, especially in relation to the destruction of obsolete organic chemicals, such as other pesticides or plastics. Two people from the United Kingdom played important roles at this stage. First, John Murlis succeeded in interesting his UK government colleagues in international controls on POPs. His support was crucial during some particularly difficult moments in Geneva. When John left for academia, John Rae began to appear at our meetings and took on the difficult role of developing a logic path that would unambiguously identify substances to be controlled by a new protocol at that time and in the future. The logic path he developed became the foundation of the process for adding new substances not only for the CLRTAP protocol but also for the later global Stockholm Convention we will soon be examining. It was while John Rae was developing this work that I first began to receive phone calls from various libertarian lobby groups in the United States. I had never encountered these organisations before, but I found it very intimidating. The callers were always extremely polite and they told me they were lawyers interested in the proposals to control various chemicals and that they were recording our conversations. Their interest waxed and waned several times during the CLRTAP negotiations and later when the protocol became operational. I suspected that the questions were related to the evaluation of hazard and risk in environmental legislation in the United States, but not being aware of that legislation in any detail, I felt very vulnerable.

By the time we had finished the CLRTAP report in 1994, we felt we had a robust and practical case for international action on POPs. But what about heavy metals?

Up until now, I have said little about heavy metals and of the work of the CLRTAP task force on these substances. This is partly because I was much less involved in the metals work (than with POPs), partly because (with the exception of mercury) metals were never such a major human circumpolar health issue in the Arctic in comparison to POPs and partly because the most severe effects of metals in the Arctic were close to local sources of pollution (again with the exception of mercury). It quickly became evident from the NCP and AMAP that we needed to pay close attention to three metals: lead (Pb), cadmium (Cd) and mercury (Hg). All three are natural substances with known acute toxic characteristics but whose distribution in the environment can be greatly influenced by human activities. Therefore, the main task from an Arctic perspective was to measure levels in different environmental compartments, to assess levels against potential adverse effects and to look for patterns that may reflect transportation pathways.

In the early 1990s, the largest sources causing metal contamination in the Arctic from within the Arctic itself were thought to come from the nickel-copper smelting industry located in the far north of Russia in the Kola Peninsula, from the Urals and from Norilsk. Here, large swathes of denuded forest spread downwind of the smelters, reflecting the combined deposition of metals and acids. High local sources were also associated with mining operations in Greenland (Black Angel) and Canada (Nanisivik beside Strathcona Sound on Baffin Island) and at the Polaris mine on Cornwallis Island, where a lake (in the case of Polaris) or a deep fjord with restricted vertical water circulation (such as Black Angel) was used as a dumpsite for mine discharge. In fresh and marine waters, Arctic levels of lead, cadmium and mercury away from the local sources were generally quite low.

Lead usually strongly partitions to soils and sediments and is therefore not easily absorbed by plants and animals. The main atmospheric source came from leaded gasoline and levels were already declining by the early 1990s. Lead normally accumulates in bone. However, its main toxicological impacts are its interference with red blood cell formation and with the neurological development of children. Exposure in the Arctic was generally low, but there were significant exceptions. Local elevated levels were found in some indigenous communities and appeared to be related to the use of lead shot during hunting.

Cadmium partitions into the dissolved phase and is therefore easily taken up into the biotic environment. In mammals, cadmium accumulates (but does not biomagnify) in the kidneys and liver. Here, it is mopped up by a class of proteins called *metallothioneins*, which, however, can be overwhelmed by very high cadmium levels. If this occurs, kidney damage results, the metabolism of calcium is upset, and the skeleton decalcifies. In some populations of caribou and willow and rock ptarmigan in the Yukon and northern Norway, cadmium levels in kidneys were in the range where effects would be expected. However, it proved difficult to detect signs of impact and the cadmium source was believed to be related to local geology. A similar situation was found with the Finlayson caribou herd in the Yukon. This raised concern for the indigenous people who harvest this herd, particularly because their cadmium levels are already high due to cigarette smoking (the normal source of most excessive human exposure to cadmium).

Mercury is the heavy metal of most concern in the circumpolar Arctic. Throughout geological time, it has been naturally released to air and water via the erosion and weathering of minerals, such as cinnabar, and volcanic processes. From these reservoirs, it has recycled through other environmental compartments back into sedimentary rocks. Humankind has for many centuries added to this input before the industrial era, notably associated with use of mercury in gold and silver mining. However, it was not until the mid-nineteenth century that anthropogenic emissions from sources, such as the burning of fossil fuels (especially coal), metal smelting, waste disposal and waste incineration, led to significant widespread increases of mercury to the environment. This occurred when the combined natural and anthropogenic inputs exceeded the capacity of the natural cycle to move mercury back into rock, leading to a growing "pool" of mercury in air, water, biota and unconsolidated sediment.

The main form of mercury entering the air is gaseous elemental mercury (Hg^0) that has an atmospheric residence time of about one to two years – more than enough to ensure rapid long-range transport around the globe. Elemental mercury is generally unavailable to biota. In the atmosphere, it can be transformed into forms of ionic mercury (such as mercuric chloride) that may also occur as a primary emission. Within 100–1,000 kilometres from a source, ionic mercury is usually deposited onto the surface, where it is strongly bound to sediments and organic matter. This deposited mercury can then be reemitted into the atmosphere as secondary emissions, which therefore include mercury that originated from natural and anthropogenic sources. There have been many attempts to apportion the relative significance of sources to the environment, but the one I summarize here is based on the 2013 UNEP *Global Mercury Assessment*. This estimates that about 30% comes from annual anthropogenic emissions of mercury to air, another 10% comes from natural geological sources, and the rest (60%) is from reemissions of previously released mercury that has built up over decades and centuries in surface soils and oceans.

Most of the anthropogenic mercury found in the Arctic has originated from emission sources in mid-latitudes. Historically, this was from industrial Eurasia and North America, but the 2013 UNEP *Global Mercury Assessment* concludes that about 40% of all global anthropogenic atmospheric emissions now comes from East and Southeast Asia. China alone is thought to be responsible for about one-third of the global total. The same report also showed that artisanal and small-scale gold mining in South America and Sub-Saharan Africa is responsible for a greater proportion of global emissions than was previously believed. Back trajectory analysis (a technique we met in the context of transport of POPs) has shown pulses of mercury being imported to the Arctic from these areas. Although reductions in emissions since the 1950s and 1970s have been achieved in the European Union, countries of the former Soviet Union and North America, they are being offset by increased emissions from elsewhere, particularly Southeast and East Asia. This is consistent with the results from a number of regional and global air-monitoring programmes that allow us to detect regional trends in atmospheric concentrations. For example, from the report, we can see that in parts of Western Europe, atmospheric levels have been declining by 1.4–1.8% per year over the period 1996–2011, presumably reflecting reduced regional emissions. In North America, the Mercury Deposition Network and the Canadian Air and Precipitation Monitoring Network suggest a decrease of 2.2%–17.4% in rural levels between 1995 and 2005. Mercury levels in Mediterranean seawater have shown a decrease between 1990 and 2004 and a similar trend has been seen in the North Atlantic. Taken together, it appears that air and surface oceanic levels of mercury have been falling, while the mercury emissions in North America and Europe have declined from their peak values of the mid-twentieth century. In the Pacific Ocean, levels are increasing, presumably in response to the increasing emissions from East and Southeast Asia.

The net result of humankind's "overloading" of the natural cycling mechanism of mercury has been the accumulation of increasing concentrations in surface soils, aquatic sediments and the oceans. In the oceans, the mercury loading in the upper 100 metres has doubled, while at lower depths, it has increased by 10–25%. Elemental inorganic mercury (which makes up most of the emissions) is not very available to most biota, but under the right conditions, microorganisms are able to convert inorganic mercury into methylmercury, which then biomagnifies through the food web. It is thought that methylmercury remains in upper ocean layers for about 11 years before removal by uptake into the marine food web and by photochemical processes. The most significant biomagnification is found in freshwater and marine trophic pathways, and once again, predators at the top of the food chain accumulate the heaviest body burdens. For example, the concentration difference between water and a beluga whale in the Arctic can be 10 orders of magnitude. It is the amount of methylmercury in animal or human tissue that is of interest to a toxicologist. In the Arctic, methylation mainly takes place in freshwater lake sediments and wetlands depleted of oxygen (anoxic) but which have a supply of organic matter and in

Mercury species, %

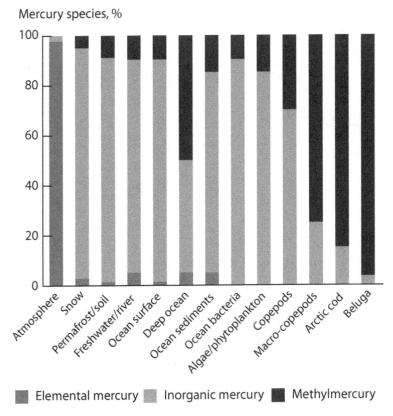

Figure 8.4 *Changes in the proportion of the different forms (species) of mercury (Hg) in the Arctic food web*

sediments of shelf regions of the Arctic Ocean. Publications by Elsie Sunderland provide a good introduction to the literature. Figure 8.4 illustrates how the proportion of the three mercury species changes with progression along the Arctic food web. Note that in top predators, such as the beluga whale, almost all the mercury present is in the form of methylmercury (MeHg).

What can we say about mercury levels in the Arctic?

This is not an easy question to answer. Measurements of mercury in lake sediments show levels that are about two to three times greater than before the onset of mid-latitude industrialization in the mid-1800s. Once again, long-range atmospheric (and to a lesser degree river and marine) transport has been dumping the wastes of industrialization in the Arctic for more than the last 150 years. Levels have risen in top predators by at least 10 times over the same period, with evidence that the increase could

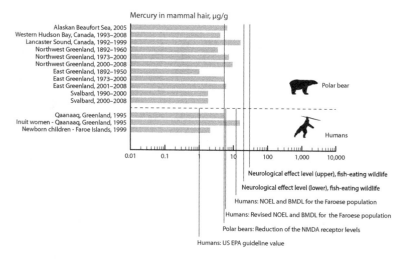

Figure 8.5 *Mercury concentrations in the hair of polar bears in relation to various effects and guideline levels*

be considerably greater. The progression of this increase can be seen by sampling hard tissues of animals that lived during this time period, such as human, beluga whale and ringed seal teeth, polar bear hair and falcon feathers. Many scientists have worked hard to trace the spatial and temporal Arctic accumulation of mercury resulting from mid-latitude industrialization. I would like to especially mention Rune Dietz, Eric Born and Frank Riget. Rune is a quiet and uncompromisingly meticulous researcher who enjoys great respect from his coworkers. I have worked with him periodically since 1981, when we were involved in marine studies between Baffin Bay and Greenland. In a study published in 2011, Rune and his colleagues described results from 117 north-west Greenland polar bear hair samples taken between 1892 and 2008. In addition to modern material, they analysed older material obtained from dated museum specimens. Mercury concentrations showed yearly significant increases of 1.6–1.7% from 1892 to 2008. The concentrations in the 2006 and 2008 sample sets were 23- to 27-fold higher, respectively, than the baseline level from 1300 AD in the same region. Figure 8.5 provides a summary of this work and shows how Hg concentrations have substantially increased in polar bears over the last 100 years in Greenland and north-east Arctic Canada. The figure also indicates the potential health significance of these levels by comparing them to several established effects and guideline levels. Note that the x-axis is logarithmic.

AMAP has prepared two assessments dealing with mercury (1998 and 2011). The AMAP 2011 *Mercury Assessment* is very comprehensive and was completed just in time to provide a welcome stimulus for completion of negotiations for the Minamata Convention. The assessment found evidence that in some parts of the High Arctic, atmospheric mercury levels have shown a recent decline, presumably in response to reduced emissions in Europe. If so, this trend could soon be reversed by increasing emissions from East Asia. In other areas, either no decline or no statistically valid trends can be seen. Part of the difficulty here is that short-term variability from natural processes can be greater than variation between years. More than 50% of the mercury delivered to the Arctic arrives through atmospheric deposition, mainly during the polar spring, which is also when atmospheric mercury depletion events (AMDEs) occur. This phenomenon was first reported in 1997 by Bill Schroeder and his colleagues at an AMAP meeting in Tromso and published in *Nature* the following year. They found that just after the polar sunrise occurs in the spring at Alert on Ellesmere Island, the concentration of gaseous atmospheric elemental mercury drops dramatically. Other polar monitoring stations quickly reported the same observations (in the Arctic and Antarctic). The greatest depletion is correlated with a corresponding reduction in surface air concentrations of ozone. What is going on? The chemical and physical processes involved require sunlight in order to perform their tricks. The gist of a rather technical explanation is that over the winter, sea spray results in the injection of bromine into the atmosphere, which, when the sun appears in spring, reacts with ozone (explaining its depletion at this time). A string of short-lived intermediate reactive compounds are then formed involving bromine, which lead to the conversion of elemental gaseous mercury into a reactive ionic form that quickly deposits onto the snowpack. Mercury concentrations in the snowpack can increase by 100 times during AMDEs, but it is thought that up to about 75% may be reemitted back into the atmosphere as elemental mercury within a few days. One of the unanswered questions is what happens to the remaining 25% because it is potentially available for conversion into biologically available methylmercury. Uncertainty on the significance of AMDEs to the overall budget of mercury gain and loss to the Arctic ecosystem is at present a key research issue. For more information on AMDEs, take a look at the 2008 review by Alexandra Steffen and colleagues.

The AMAP 2011 assessment team looked at more than 80 data sets from localities at higher latitudes than 60° north to examine trends of mercury levels in marine mammals, marine fish, marine invertebrates, seabirds, freshwater fish and land mammals. They reported "a recent increase in 16% of data sets, a recent decrease in 5% of data sets and no change in or fluctuating trends in the remaining 79% of data sets". As expected, biomagnification has led to a general linear trend of increasing concentrations with increasing trophic level within any given food web. However, an interesting west to east trend was seen with levels. Most of the species showing increasing levels from freshwater and marine environments were collected from Arctic Canada. Here, marine species were showing levels increasing by 5% or more per year. In contrast, all species collected north of Europe showed either no trend or decreasing levels. The Faroe Islands, Iceland and Greenland were transitional between these two patterns. Even polar bears in Svalbard appeared to be carrying lower levels. A study published in 2012 showed that mercury levels in Svalbard polar bear teeth have dropped over the last 40 years but with no changes over the same period in the stable isotope ratios of nitrogen ($\delta^{15}N$) and carbon ($\delta^{13}C$). You will remember from our digression on food web dynamics that this suggests that it is unlikely that the trend to lower mercury concentrations reflects changes in their diet. Most probably, it represents decreasing atmospheric deposition to eastern North Atlantic seawater and to lower regional emissions from Europe.

The basic question that matters to most of us is whether these levels are likely to have an impact on Arctic wildlife and the people living in the Arctic. Mercury has been known to have toxic properties for many years. The English term *mad as a hatter* shows that even in the eighteenth and nineteenth centuries, it was understood that long-term chronic exposure could lead to serious neurological and behavioural abnormalities. This colloquial phrase comes from the dementia suffered by the hatmakers at the time, who used mercury to treat the felt used in hat production. In more modern times, these properties gained notoriety in Minamata, Japan, where from 1932 to 1968, industrial wastewater was discharged into Minamata Bay and the Shiranui Sea. Aggressive biomagnification of methylmercury took place, leading to very high concentrations in local shellfish and fish. People who consumed these species began to suffer from a wide range of neurological and muscle problems, including difficulties with vision, hearing, speech, numbness, muscle weakness and, in more severe cases, insanity, coma and death. The syndrome became known as *Minamata disease* and it was eventually realized that these outcomes were caused by exposure to methylmercury through diet. What was especially disconcerting was the realization that children were particularly vulnerable and could acquire a dose leading to significant effects while still in the womb.

In the 1970s, Canadians in Ontario also became tragically familiar with methylmercury through very similar circumstances. In 1962, a chloralkali industrial plant began to discharge wastes into the Wabigoon and English rivers, but it was not until 1970 that commercial fishing was closed due to the discovery of widespread mercury contamination. Meanwhile, identical symptoms to those found in Minamata with the same sad consequences were being detected in communities of indigenous peoples who lived and fished on the Wabigoon and English river systems, including Grassy Narrows. In 1970, direct disposal was curtailed from the industrial plant, but emissions into the atmosphere continued until 1975.

In the years since the tragedies of Minamata and the Ontario Wabigoon and English rivers, toxicologists and health authorities have established toxicity threshold values for mercury that enable the potential for health effects to be assessed. Laboratory and field studies mainly on fish and fish-eating mammals are consistent with one another. Chronic mercury exposure at levels much lower than those that characterized the Minamata and Ontario incidents can result in a wide range of neurological outcomes in birds, fish and mammals. This is because methylmercury is able to cross the blood-brain barrier. The mechanism of disruption of neurological processes is thought to be partly related to the high affinity of methylmercury to protein thiols and the activity of N-Methyl-D-aspartic acid (NMDA), an amino acid derivative concerned with many neurological functions. Other outcomes include kidney and liver lesions, endocrine dysfunction and reproductive abnormalities, but there is considerable variety in effects and susceptibility. In part, this is related to how different species take up and excrete methylmercury. In land mammals and polar bears, the kidney is the main storage organ, but in birds and marine mammals, it is the liver. In addition to neurological effects, methylmercury is associated with abnormal spawning behaviour in predatory freshwater fish.

The 2011 AMAP assessment team summarized the available data to show that in some areas, organs and tissues of Arctic predators at the top of the food web are carrying levels of mercury that exceed thresholds for biological effects. This includes polar bears, beluga whales, pilot whales, hooded seals, several seabird species (ivory gull, glaucous gull and black guillemot), northern pike, lake trout and landlocked Arctic char. Arctic marine fish rarely show levels exceeding thresholds, but this is not the case with polar bears and toothed whales. For example, the 2001–2008 data set (mentioned earlier) of hair taken from East Greenland polar bears exceeded the proposed neurochemical effect level of 5.4 µg Hg/g dry weight in 93.5% of the cases. Toothed whales appear to be especially vulnerable (partly because they do not have the polar bear and bird option of moving mercury into hair or feathers) and they showed high concentrations of mercury in brain tissue and associated signs of neurochemical effects.

We should have a short digression here about the metal selenium. Since the 1970s, it has been known that there is usually a strong positive correlation between selenium and mercury concentrations in many species, especially marine mammals. It is believed that these species are partly protected from mercury toxicity by selenium that is also taken in with their diet. In a number of fish-eating birds and mammals, it has been found that when methylmercury levels are low, most of the mercury in the liver is in this form. However, as methylmercury levels rise, an increasing proportion in the liver is in an inorganic (nonbiologically available) form as a selenium-mercury complex. This suggests that perhaps selenium is involved in a demethylation process. There is also evidence that selenium acts as an antioxidant to protect against oxidative stress due to mercury exposure. Selenium is present in very high concentrations in certain tissues of many Arctic marine species, such as the livers of bearded and ringed seals and in muktuk (the fatty layer just below the skin of whales) – all of which feature prominently in the traditional Inuit diet.

In this brief review of the POPs and heavy metal contamination, I have tried, more or less, to match developments in knowledge with parallel "diplomatic" work to put in place international controls on the offending substances. You will have noticed, however, that this scheme has broken down with the mercury story, where, with the exception of human exposure and impact, we have brought ourselves close to being up to date.

In 1994, when the CLRTAP heavy metals report was being written, we had evidence that once again, some of the most exposed human populations to mercury were in the Arctic or sub-Arctic, where fish and/or marine mammals form much of the diet. Canada uses a blood-mercury guideline of 20 micrograms per litre of mercury that mothers and women of childbearing age should not exceed. In the United States, the guideline is set at 5.8 micrograms per litre. In the 1990s, these values were being commonly exceeded in coastal Arctic communities. For example, at this time, 52–76% of Inuit mothers and women of childbearing age in Nunavik (Canada) and 68% of Inuit mothers in Nunavut (Canada) exceeded the U.S. guideline (Figure 8.6). The AMAP 2011 *Assessment of Mercury in the Arctic* showed that there are indications of some

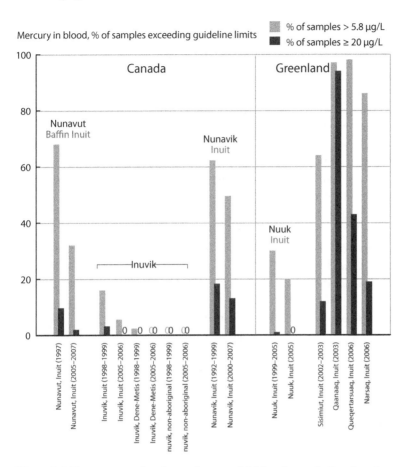

Figure 8.6 *The percentage of mothers and women of childbearing age in Canada and Greenland that exceed guideline limits for mercury in blood*

reduction in blood levels (particularly on Baffin Island), but these concentrations remain high. Do these levels pose a risk to human health?

We did not know the answer, but some important clues were emerging. We have already seen that in mammals, mercury can cross the placenta into the foetus. Extrapolating from knowledge of acute exposure effects, it was thought that children may be particularly vulnerable to upsets in their neurological development while in the womb. In 1992, Philippe Grandjean and Pál Weihe and their colleagues published a paper that woke us all up. It was a study from the Faroe Islands of umbilical cord blood sampled from more than 1,000 placentas taken from mothers at the time when they gave birth. Surprisingly high levels of mercury were found. Maternal dietary information included in the study clearly related the levels to a diet dominated by fish and marine mammals. This was not good news and it featured significantly in the heavy metals information compiled by AMAP for the CLRTAP 1994 report. I remember the resounding impact it made on the Executive Body to the convention when it considered what actions to take. Grandjean, Weihe and colleagues then began a benchmark study of a cohort of Faroese children who had been exposed prenatally to known concentrations of mercury through placental transfer from their mothers. The lasting legacy of this work will be seen a little later.

Now that we have digested a brief outline of what was known in the mid-1990s about POPs and heavy metals pollution in the Arctic, we can return to the history of international controls on these substances. At the end of 1994, Lars Lindau and I presented the POPs state of knowledge report to the Executive Body of the CLRTAP. At the same time, the heavy metals task force presented its report. To our great relief, it was decided that a legally binding POPs protocol should be developed under the convention. However, these things usually take two or three years to complete and the convention's negotiating body (called the Working Group on Strategies and Review) had only the calendar year of 1997 available. It was decided to disband the POPs and heavy metals task forces and replace them with a POPs preparatory working group and a heavy metals preparatory working group. The new working groups were instructed to prepare draft protocols but were to refrain from negotiating any intractable issues. When these arose, alternative text would be prepared and explanations provided. Deiter Jorst from Germany led the metals working group and I chaired the one on POPs. The first meeting of the POPs preparatory working group was held in March 1995 and was attended by John Buccini, who was en route from his final session as chair of the Organisation for Economic Co-Operation and Development (OECD) Chemicals Group in Paris. He was immediately intrigued by the POPs story and we will meet him again a little later.

To cut a long story short, our two draft protocols were accepted in December 1996 by the convention as providing the basis for final negotiations that were then conducted by Lars Björkbom. I first met Lars about four years earlier during the negotiations to set up the Arctic Environmental Protection Strategy and AMAP. At the time, he was Sweden's ambassador for the environment. He was a highly respected and veteran diplomat with an amazing memory. During brief moments of relaxation in Geneva, he would entertain us with stories. They included his experiences at the 1972 United Nations Conference on the Human Environment held in Stockholm. It was the first major UN conference devoted to relationships between human activity and the environment and laid the political groundwork for the birth one year later of the UNEP.

In June 1997, the first AMAP state of the Arctic environment report was delivered at a ministerial meeting in Norway. Its chapters on POPs and heavy metals therefore became available at an opportune time. In December of that year, the Executive Body to the Convention considered and approved the text for two new protocols: one on POPs and the other on heavy metals. In May 1998, both protocols were signed by parties of the convention at a special conference held in Aarhus, Denmark. Before I was caught up in this work, I thought that international negotiations were played out in the conference room. They are not. They take place on the phone between meetings, in the conference corridors, during meals and in little groups outside the main room – often working and writing throughout the night. It is here that most of the "give and take" of compromise is played out. The conference room is the "Heathrow Airport", where, all being well, it all comes together or vanishes – never to be seen again.

The POPs protocol originally took control actions on 16 substances (11 pesticides, two industrial chemicals and three by-products/contaminants). It banned (for signatories to the protocol) the production and use of aldrin, chlordane, chlordecone, dieldrin, endrin, hexabromobiphenyl, mirex and toxaphene and set in motion defined steps that led to the eventual elimination of the production and use of DDT, heptachlor, hexaclorobenzene and PCBs (at that time still in use in Russia and Eastern Europe). The use of DDT, HCH (including lindane) and PCBs was severely restricted and conditions were prescribed for the disposal of banned products. It also included a schedule for the reduction of emissions of dioxins, furans, PAHs and HCB and set out procedures for waste incineration.

The Heavy Metals Protocol targeted cadmium, lead and mercury. Parties were required to reduce their emissions for these three metals below their levels in 1990 (or an alternative year between 1985 and 1995). The protocol aimed to cut emissions from industrial sources (iron and steel industry and nonferrous metal industry), combustion processes (power generation and road transport) and waste incineration. It laid down limit values for emissions from stationary sources and suggested best available techniques (BATs) for these sources, such as special filters or scrubbers for combustion sources or mercury-free processes. It also required parties to phase out leaded petrol. The protocol introduced measures to lower heavy metal emissions from other products, such as mercury in batteries, and proposed the introduction of management measures for other mercury-containing products, such as electrical components (thermostats, switches), measuring devices (thermometers, manometers, barometers), fluorescent lamps, dental amalgam, pesticides and paint.

The two CLRTAP protocols have continued to evolve as new monitoring information on environmental levels and effects on POPs and heavy metals has become available. For POPs, this has been through the addition of seven new substances controlled by the protocol (hexachlorobutadiene, octabromodiphenyl ether, pentachlorobenzene, pentabromodiphenyl ether, perfluorooctane sulfonates, polychlorinated naphthalenes and short-chain chlorinated paraffins). At the same time, the obligations for DDT, heptachlor, hexachlorobenzene and PCBs, together with the emission limit values for waste incineration, have been tightened. For heavy metals, the CLRTAP Executive Body agreed in December 2012 to introduce more stringent emission limit values for emissions to the atmosphere of particulate matter (the significance of which we will note when we look at black carbon and climate) and of the metals cadmium, lead and mercury. The emission source categories for the three heavy metals were also extended to the production of silico- and ferromanganese alloys, thus expanding the scope of industrial activities for which emission limits are established. These measures are to be achieved by the use of best available techniques.

Back to POPs again. The CLRTAP was not an ideal instrument under which to deal with these substances. The convention is nested within the United Nations Economic Commission for Europe (UNECE), which, as mentioned previously, has a geographic scope that is restricted to Europe, all the states of the former Soviet Union, Canada and the United States. In other words, it does not have a global reach, although with the exception of China and India, it does have the potential to include a substantial proportion of the world's manufacturing capacity and unintentional emissions. However, it was a wonderful start, and we must remember that back in 1990, no other international organisation with legal teeth was interested.

From the hindsight of history, the big achievement of the protocol was to pave the way for a global approach to control these substances. The advent of the 1992 Earth Summit and its associated Agenda 21 provided UNEP with new tools to achieve international cooperation for the management of such chemicals as POPs. The most important was the establishment of UNEP Chemicals, a new organisational unit based in Geneva.

The next few lines illustrate the pedantic machinery of how things are done in UN circles, but it also shows how the machine can be a racing car in the right circumstances. At the last preparatory meeting for the 1992 Earth Summit, Iceland succeeded in introducing the need for global action on POPs into Agenda 21. However, the text appeared in chapter 21 ("Oceans") rather than in chapter 19 ("Toxic Chemicals"). At the March 1995 Reykjavik preparatory meeting for the 1995 Washington Conference to establish the Global Plan of Action for the Protection of the Marine Environment from Land-Based Activities, Iceland delivered a science synthesis document derived from the CLRTAP POPs state of knowledge report and from the POPs section of an interim assessment report produced by AMAP. Just two months later, in May 1995, the Governing Council of UNEP invited the Inter-Organization Programme on the Sound Management of Chemicals (IOMC), the International Programme on Chemical Safety (IPCS) and the Intergovernmental Forum on Chemical Safety (IFCS) to conduct an assessment on 12 specified POPs and to make recommendations to the UNEP Governing Council and to the World Health Assembly on the need for appropriate action. An ad hoc working group established to do the work was chaired by John Buccini from Environment Canada.

What this meant was that we were now on track for a global agreement on POPs. Nevertheless, I was quite worried about the prospects for the ad hoc working group. In May 1994, at the United Nations Commission on Sustainable Development, Canada had offered to host an international experts meeting to discuss the transboundary environmental and human health issues associated with POPs. Harvey Lerer, Hajo Versteeg (both working with Environment Canada) and I were tasked with organising the meeting. It took place in June 1995 in Vancouver as a joint initiative with the Philippines government. Siu-Ling Han from our NCP made a presentation describing the high body burdens of POPs amongst Inuit people in Arctic Canada who live far from any known sources of POPs. It resonated around the conference room. A joint statement from the meeting concluded:

> There is enough scientific information on the adverse human health and environmental impacts of POPs to warrant coherent action at the national, regional, and international level.

However, the statement then went on to record the range of opinions on what actions should be taken. These spanned production bans through the concept of virtual elimination from the environment to a range of management options still allowing for use. Governments and industry favoured the latter. This hesitance was common regardless of any country's level of economic development. Most of the POPs then known to be of concern were already either banned or strictly controlled in countries with developed economies and their patents had expired, but some had subsidiaries with manufacturing capacities elsewhere. To implement possible controls, countries with developing economies could face the need to use alternatives that would still be patented and be more expensive. In these circumstances, it was quite natural to wonder what were the true motives of those seeking aggressive controls. As I remember it at that stage, the only country with a developing economy that was unequivocal in its support for stringent POPs controls was the Philippines. In addition, at that time in North America, top-level managers in chemical regulatory agencies, such as Environment Canada, preferred self-regulation by the industry rather than legislation. The underlying concern in developed countries (including Canada) from industry and from economic ministries was not generally related to the 12 substances being focused on in the global negotiations but on the possibility of commercial implications for other newer and yet unmarketed substances.

It was with these ominous thoughts in mind that I went into the first IFCS ad hoc working group meeting held in Canberra, Australia (March 1996). It opened with much the same divergence of opinions on what should be done as we had seen in Vancouver. However, as the days went by in Canberra and at the next meeting held in Manila (June 1996), John Buccini steered the group beyond the consensus on the need for worldwide POPs controls to agreement on what the policy intent of those controls should be and therefore the nature of the actions. The final IFCS report recommended immediate international action to protect human health and the environment by actions that would "reduce and/or eliminate the emissions and discharges of the 12 POPs" (called the "dirty dozen") and "where appropriate eliminate production and subsequently the remaining use of those POPs that are intentionally produced". In January 1997, the UNEP Governing Council reviewed the IFCS report and adopted a decision to set up an intergovernmental negotiating committee to begin work in the following year.

Every country in the United Nations can take part in the negotiation of a UNEP convention. It was a rather humiliating education as to how little I knew about the many countries that exist around the world. When setting off in the conference hall to discover the views of a particular country, I passed country flags that were totally unfamiliar. I would not have been surprised to see Anthony Hope with the Prisoner of Zenda sitting behind the flag of Ruritania or Jonathan Swift at that of Brobdingnag, Lilliput or Laputa. Ignorance about a country is a major handicap in negotiations. It is so much easier to make intelligent and fair compromise if you know something of a country's background and have a feeling as to why a particular issue is held to be important. It is also helpful when you recognize that an intractable gridlock may be caused by politics that have absolutely nothing to do with the subject being negotiated.

The etiquette of international negotiations has been carefully nurtured over centuries. Many delegates at the CLRTAP and future POPs Stockholm Convention negotiations (especially the lawyers) would have felt quite at home working on the Treaty of Vienna after the Napoleonic Wars. It works like this: An intervention made by a delegate that begins with a generous panegyric praising your merits and the brilliance of your ideas forewarns you of imminent slaughter. To give you a taste, here is what happened to a suggestion made during the Johannesburg session of the Stockholm negotiations: A small group of countries put forward an idea that was simply unworkable. Fearing endless debate amongst the 150 or so countries in the plenary session, the chair of the negotiations set up an ad hoc legal group to examine the proposal and to recommend how to proceed. The legal group swiftly returned with something like this: "We carefully examined the intriguing proposal presented by countries X, Y and Z. We found it contained many fascinating, helpful and innovative features. However, it did present difficulties in implementation, and in the end, we decided to put it aside." It is like the old Germanic Mensur form of duelling, set up to conclusively decide an issue but to do so without risk of death and with the preservation of honour for all (at least on the surface). Those involved in present-day negotiations on climate tell me that such diplomacy is quickly fading and being replaced by another mode best captured by a comment made to me one day by a Swedish delegate. We were sitting in a small side session that was arguing endlessly about some little issue that had more to do with egos than substance. Overcome with frustration, she turned to me and muttered: "There is too much testosterone in this room!"

Some remarkable people were involved in the negotiations. First, there was John Buccini. His skills at negotiation are mesmerizing. I would watch the storm clouds gather over some issue and then see how he would dissect the problem into manageable segments. He then rhetorically examined the policy intent of each segment. He described his job as being the "director of a ballet of elephants"! It is a perfect metaphor because it is very difficult for all delegations to stray far from the positions agreed on in their home capitals. Just as important as John were Jim Willis and John Whitelaw, the director and deputy director, respectively, of UNEP Chemicals. They were everything you would not expect to see in an archetypical UN organisation: innovative, unconventional, very knowledgeable about POPs, inexhaustible and, most importantly, capable of attracting funding to support the negotiations. They left UNEP Chemicals soon after the completion of the agreement. Their loss greatly impoverished our lives.

Some of the delegations also included outstanding participants, but for many, the star on the floor was Sheila Watt-Cloutier (from the eastern Canadian Arctic) who at the time was president of the Inuit Circumpolar Council of Canada. At the second negotiating session, she presented an Inuit carving of a mother and child to Klaus Töpfer, the then–executive director of UNEP. The carving sat beside John Buccini for the remainder of the negotiating sessions – a powerful reminder of why we were all there. Sheila is a very remarkable person. She is a born leader who thinks carefully before she speaks and never strays from the moral high ground. She somehow leads by mediation. When others try the same approach, you are usually left with a weak hybrid that no one is happy about. This is not the case with ideas nurtured under Sheila. They are offspring that everyone can be proud of. Here is a little example: DDT remained in the "toolboxes" of public health agencies trying to control malaria, especially in Africa. A nongovernmental organisation (not representing any health agency or regional indigenous organisation) took the floor and stated that if the Stockholm Convention removed this tool, it would be responsible for the deaths of millions – many of them children. It was an old, tired and unfounded notion that resurfaces periodically in association with attacks on the work of Rachel Carson and of government regulatory agencies. It could have driven a rift through the negotiations. In their 2010 book *Merchants of Doubt*, Oreskes and Conway pointed out that the same accusation was resurrected in 2007 from several libertarian think tanks and lobby groups in the United States. In fact, no draft of the convention had ever proposed removing DDT for malaria control, and although the use of DDT had dramatically fallen from earlier years, this was (and still is) because of aggressive mosquito resistance to the pesticide. However, it was not necessary to get into a technical debate. Sheila quickly told the floor that the Inuit would not be a party to any agreement that threatens the lives of others. She could not believe that a mother in the Arctic should have to worry about contaminants in her milk as she feeds her baby – just as she could not believe that a mother in Africa should have to rely on the very same chemicals to protect her baby from disease. There must be another way. Sheila's intervention immediately brought peace to the room and we moved on.

Two people with very complementary personalities supported Sheila throughout her work on POPs. Stephanie Meakin is a very capable biologist with a charming personality whose natural inclination is to unfailingly believe in the best of another person. Terry Fenge is a political analyst specializing in Arctic affairs. He has always reminded me of Plato's famous evaluation of Socrates as being the gadfly that stings a government into action. In this role, he has often been very effective in Canadian Arctic affairs.

There were five negotiating sessions: Montreal in June 1998; Nairobi in January 1999; Geneva in September 1999; Bonn in March 2000; and Johannesburg in December 2000. All the sessions were scheduled to run over five days. As was the case with the CLRTAP protocols, information on the Arctic POPs situation packaged by AMAP was constantly being introduced – often by countries far from the Arctic and sometimes by delegations from the tropics. The final session began as usual at 10 a.m. on Friday and finished at 7.30 a.m. on Saturday. We had lost interpretation a few hours earlier, but John Buccini and Jim Willis coaxed us on, and in the end, the job was done. On May 23, 2001, the convention was signed by an initial group of 91 countries and the European Union. It entered into force on May 17, 2004, with ratification by an initial 128 parties and 151 signatories. By March 2013, 178 countries had ratified the convention and had therefore become parties (meaning their acceptance of the legal obligations of the convention). It is now known as the Stockholm Convention. At the time of writing, one Arctic Council country has not ratified either the Stockholm Convention or the CLRTAP POPs protocol.

You will remember that the emergence of POPs as an environmental and human health issue came as a surprise to agencies responsible for chemical safety. They could not believe that these substances could be evading their ways of evaluating hazards and risks. The experience warned negotiators of the danger of developing an agreement whose basic structure is frozen to reflect the state of knowledge that existed at the time of negotiation. Therefore, the POPs protocol and the Stockholm Convention were designed to be "living" agreements and both share a very similar architecture. They can respond at any time to new scientific information relevant to the objectives of the protocol or convention. The features of the agreements that provide this ability are clustered around their essentially similar three main "doorways" through which scientific information can gain access to their decision-making processes.

The doorway with the most scientifically comprehensive range of potential topics is Article 11 of the Stockholm Convention and the equivalent doorway is Article 8 of the POPs protocol. They place obligations on parties concerning research, development and monitoring. However, this doorway is not directly linked to a decision-making process in either agreement.

A second doorway is the best known. Both agreements include a process for allowing a party or parties to propose adding new substances to be controlled and for the technical review of such proposals (Article 8 in the Stockholm Convention and Article 14 in the POPs protocol). This process has been quite active. Most of the substances added since the two agreements were negotiated are there because of Arctic data assembled and brought to the CLRTAP and the Stockholm Convention by AMAP participating countries. Some were already suspected to be a problem in the 1990s, but others were hardly thought of in terms of the combination of chronic toxicity, biomagnification and long-range transport until Arctic screening studies and monitoring programmes were able to increase the number of substances studied.

When the POPs protocol and the Stockholm Convention were under negotiation, the common but not universal belief was that over time, the total number of substances controlled under the agreements might perhaps reach 20. Those who believed in a larger number were in the minority, but now both agreements are moving beyond 20 and there seems to be no end in sight. Several studies have been published that suggest a range of between about 50 to hundreds of substances presently in use that could be logical candidates for control! If this turns out to be the case, it reflects very poorly on the procedures that allow such chemicals into commerce. For a glimpse at the extent of the number of substances presently being used in commerce that to some degree possess the properties of POPs (persistence, ability to travel in the environment far from source, ability to biomagnify and chronic toxicity), take a look at the papers by Muir and Howard published in 2006 and by Martin Scheringer in 2012 – all listed in the bibliography. Table 8.2 lists the substances controlled under the CLRTAP POPs protocol and the Stockholm Convention at the time of their original negotiation and as of May 2013.

One of the concepts that appeared to gain acceptance at the 1992 Rio conference on sustainable development was that of the precautionary principle. This basically means you should not wait for absolute scientific certainty before taking action if you have reasonable evidence to believe an activity or substance is going to cause harm (my definition).

Table 8.2 *Substances controlled under the CLRTAP POPs protocol and the Stockholm Convention (as of May 2013)*

Agreement	Substances Originally Listed	Substances Added
CLRTAP Protocol	Aldrin; chlordane; chlordecone; dieldrin; endrin; hexabromobiphenyl; mirex; toxaphene; DDT; heptachlor; hexaclorobenzene; PCBs; HCH (including lindane); dioxins and furans; and PAHs	Hexachlorobutadiene; octabromodiphenyl ether; pentachlorobenzene; pentabromodiphenyl ether; perfluorooctane sulfonates; polychlorinated naphthalenes; and short-chain chlorinated paraffins
Stockholm Convention	Aldrin; chlordane; DDT; dieldrin; endrin; heptachlor; hexachlorobenzene (HCB); mirex; toxaphene; polychlorinated biphenyls (PCB); polychlorinated dibenzo-p-dioxins (PCDD); and polychlorinated dibenzofurans (PCDF)	Chlordecone; technical endosulfan and its related isomers; alpha hexachlorocyclohexane; beta hexachlorocyclohexane; lindane; hexabromobiphenyl and heptabromodiphenyl ether (commercial octobromodiphenyl ether); hexabromocyclododecane; pentachlorobenzene; perfluorooctane sulfonic acid (including its salts and perfluorooctane sulfonyl fluoride); and tetrabromodiphenyl ether and pentabromodiphenyl ether (commercial pentabromodiphenyl ether)

In other words, it is better to anticipate problems than to wait for them to occur. The principle is mentioned in a generic fashion in the CLRTAP and Stockholm agreements, but it was an Olympian struggle to get them in and it is done in a way that provides no teeth. The text does not oblige the parties to comply. However, the philosophy of the precautionary principle did make its way into the process for adding new substances. This includes the requirement in "doorway two" to demonstrate that a candidate substance meets certain physical, chemical and toxicological criteria. If any one of these criteria is not met, one can instead use monitoring data as a surrogate if it suggests that the substance is actually able to perform the activities that should be predicted by the relative criterion. For example, if a criterion for vapour pressure or fugacity that is intended to predict long-range transport is not met by a candidate substance that is nevertheless showing up in such remote places as the Arctic, clearly it has the ability to move long distances (even if we do not understand how it is happening). The parties are required to take these criteria into account during the process of approving new substances for use within their jurisdictions. If they do, then this is truly giving the precautionary principle some anticipatory teeth. However, the wording does not make it a fundamental legal obligation. The cruel fact remains that these agreements are necessary because historical arrangements to assess the safety of many chemicals have failed. Present-day arrangements for chemical safety are probably better, but they face a plethora of new substances to review. Furthermore, the fundamental paradigm of hazard and risk assessment has so far failed to meet the challenge posed by our present knowledge on the chronic health and environmental effects of many POPs, including those involving endocrine disruption. We will come back to this later.

The third doorway provided by both agreements requires parties to periodically review their effectiveness (Article 10 of the POPs protocol and Article 16 of the Stockholm Convention). This feature was not in the original draft of the Stockholm Convention. It arose from uneasiness of Canadian indigenous peoples' organisations. They were unhappy that MEA lack independent verification to prove that countries are living up to their obligations. I was very sympathetic to this notion but agreed with the lawyers that an independent verification scheme would never be accepted for this type of agreement. The more I thought about it, the more I realized what really mattered was not whether a country was "cheating" but that levels of POPs were going down. We would only know this if we had a global monitoring capability (which would probably show where any "cheating" was going on anyway). It was not initially welcomed in Canada, but Anne Daniel (Canada's most experienced lawyer at the time for environmental agreements) immediately embraced it and the indigenous peoples' organisations were supportive. I prepared a proposal that was introduced at the second negotiating session. It was built around the logic that the convention needed a way to measure whether it was being effective in driving down the levels of POPs in the environment. It was not an easy sell, mainly because of cost implications, but eventually, it became Article 16 of the convention. Bo Wahlström from the Swedish Chemicals Inspectorate was temporarily working at UNEP Chemicals and organised several workshops to see how comparable global monitoring data could be obtained and analyzed. In 2006, I was asked to prepare drafts of a global monitoring plan (a short strategic document) and of an implementation plan that was designed to evolve as monitoring activities grew. The drafts were modified between the third and fourth conferences of the parties before being adopted and the first effectiveness evaluation was completed in 2009. However, rather than being a true evaluation, it basically set out a baseline for the future detection of trends in POPs levels in people and the environment.

At the heart of the environmental parts of the effectiveness evaluation are periodic reports prepared on a regional basis by Regional Organization Groups (ROGs), which draw on results from the global monitoring programme. These are consolidated into a global synthesis that is available to the Conference of the Parties (COP) to the convention when they undertake an effectiveness evaluation. Details on how all this works is available on the website of the convention at www.pops.int.

The AMAP experience played a major role in helping design the global monitoring programme, which has resulted in the production of joint reports, such as the 2011 report *Climate Change and POPs: Predicting the Impacts*. An important element of the new global programme has been the introduction of a global network of cost-effective passive POPs air samplers. Three of the original champions of these comparatively cheap and low maintenance samplers have been Tom Harner (Environment Canada), Kevin Jones (University of Lancaster) and Ivan Holoubek (Masaryk University, Czech Republic). Tom was one of the protégés to spring from the laboratory of Don Mackay. In 2004, he organised the global atmospheric passive sampling (GAPS) network. It consists of about 60 passive monitoring sites on seven continents to measure trends of POPs in space and time in the atmosphere. This has greatly improved our ability to assess the long-range transport of these substances, although there have been some difficulties with operation in Arctic conditions. A number of other complementary passive sampler networks have since been set up. One example of the practical potential of such arrangements was given in a 2012 paper that reported on results from passive samplers deployed in five continents under the Stockholm Convention global monitoring plan. It showed remarkably high PCB concentrations in a number of African locations. This is surprising because PCBs were never heavily used in Africa. The most likely explanation is that PCBs are included in material that is sent to Africa for disposal from other parts of the world.

As a result of expanding air-monitoring activities, more and more is being learned about the propensity of cold environments to receive high deposition levels of POPs through long-range atmospheric transport. For example, Sheng and colleagues reported in 2013 on results of an air-monitoring programme that ran in the south-eastern Tibetan Plateau from 2008 to 2011. DDT levels were much higher than reported for polar regions, the North American Rocky Mountains and the European Alps. Cold condensation works wherever it is cold.

We can now take a very brief and selective look at some of the most significant recent developments in knowledge about POPs and the Arctic. For this, my main source of information has been the 2009 AMAP *POPs Assessment* and the 2009 AMAP *Assessment of Human Health in the Arctic*. We will encounter the term *legacy POPs* frequently in the following pages. It is used to group together substances no longer used or ones used only under strict controls and where present-day emissions are mainly from such secondary sources as soils and water. Such emissions are therefore a "legacy" of past uses. It is, however, a fluid definition because several of the substances referred to as "new POPs" as recently as 2008 are now moving or have moved into the category of legacy POPs following the curtailment of their use due to new international controls. At the same time, research and monitoring are resulting in new substances now being identified as "new/emerging POPs".

Most of the legacy POPs have been showing a general decline in the atmosphere at rates that would be expected given their different atmospheric half-lives and the time that has passed since their use has been severely restricted or banned. However, for some of these substances (such as PCBs and HCB), atmospheric levels increased between 2002 and 2004 at Zeppelin, the air-monitoring site on Svalbard at Ny-Ålesund. This may signal that warmer temperatures are causing these POPs to reenter the atmosphere from, for example, reservoirs stored in Arctic soils. It is similarly believed that recent ice-free winters have been allowing the ocean to "de-gas" some of its POPs inventory back into the atmosphere. This mechanism has already been observed for alpha-HCH, a component of the pesticide lindane. Episodes or pulses of increased PCBs, the pesticide chlordane and the DDT derivative DDE (dichlorodiphenyldichloroethylene) in the atmosphere have also been observed at Zeppelin and are thought to have been caused by the mobilization of these substances from vegetation during forest fires.

The frequency of such fires is increasing (fulfilling a long-held prediction of the IPCC). An emerging question for POPs and heavy metals (particularly mercury) is to understand how much a warming climate will remobilize these substances from inventories stored in the oceans, soils, sediments and vegetation. In wildlife, the legacy POPs are generally following decreasing trends throughout the Arctic, where data are available in terrestrial, freshwater and marine ecosystems. However, in some areas and for some species, it has so far been difficult to detect trends (such as HCH in marine mammals).

We are about to dive into some heavy text, so perhaps it is time to catch our breath with a little historical digression. Why is the air-monitoring site at Ny-Ålesund on Svalbard called Zeppelin? One of the foothills behind Ny-Ålesund is called Zeppelin Mountain. In 1926, a Zeppelin airship called the *Norge* left Ny-Ålesund with the famous Norwegian explorer Roald Amundsen and Umberto Nobile (an Italian explorer) on board. The *Norge* required a tall mooring mast that still stands at Ny-Ålesund. Corrosion is a slow process in the Arctic and the mast today looks as if it were constructed just a few years ago. Their aim was to land at the North Pole. However, when they arrived, the weather was so poor that they had to keep going until they landed two days later at Teller in Alaska. In 1928, Nobile again set out for the North Pole in another airship called the *Italia* but this time without Amundsen. The *Italia* crashed on the return trip, suffering loss of life, but Nobile survived and was eventually rescued. Tragically, Amundsen lost his life in a plane accident during the search and rescue.

In the next few paragraphs, we will start to encounter some chemicals with names that can be quite intimidating if you have not studied organic chemistry. An organic chemist names a substance according to its constituent parts. Numbers in the name tell you how many times a similar component may be involved in the complete molecule and where all the bits and pieces may be attached to each other. It is rather as if you were describing how to build something with a child's construction toy. I have used these names because we will be talking about some specific substances, but it is not important that you try to remember or to pronounce any of them. All you need to do is appreciate how ubiquitous POPs have become in our everyday lives, how they have been accumulating in the Arctic ecosystem, even though they may have had little use in the Arctic and how there are far more POPs in the world than listed in the new control agreements.

The Arctic atmospheric and wildlife levels of legacy POPs may be declining, but the opposite is the case with a number of substances with the characteristics of POPs that were not included in the original CLRTAP protocol and the Stockholm Convention. Brominated flame retardants are a case in point.[2] They have been widely used by society in, for example, upholstery, textiles, plastics, circuit boards and rubber. Circumpolar atmospheric monitoring results consolidated by AMAP showed that a number of brominated flame retardants are capable of reaching the Arctic from distances of several thousand kilometres, but they can also be introduced into the atmosphere through the use and disposal of products containing these substances within the Arctic, especially through uncontrolled trash burning. Once deposited on the surface, brominated flame retardants enter the terrestrial, freshwater and marine food webs and biomagnify upward to higher trophic levels in just the same way as organochlorines. Levels in top predators tend to be much lower than has been observed for PCBs, but their spatial distribution is similar to that of PCBs, with the highest Arctic concentrations being found from East Greenland to Svalbard. Levels are lower in terrestrial animals and higher in upper-trophic-level marine predators, such as some killer whale (orca) populations off Alaska and glaucous gulls from the Barents Sea. After the first generation of brominated flame retardants was banned, their replacement substances have begun to appear in biological and nonbiological media in the Arctic (such as pentabromotoluene and pentabromoethylbenzene). The Arctic Messenger must have a sense of déjà vu.

A class of substances that caught us all by surprise were polyfluoroalkyl substances (PFAS). This is a new term that replaces the former term of poly- and perfluorinated compounds (PFCs)[3]. They were heavily used as surfactants in firefighting foams and as stain repellents in textiles, carpets and food packaging. In 2002, the main manufacturer in the United States voluntarily phased out production due to PFAS being found in the blood of workers involved in the manufacturing process.

PFAS are highly persistent in the environment. Derek Muir has told me that they have been described as a molecular rebar – indestructible in water and soils under normal environmental conditions. You would think this would not lead to environmental complications because they are not volatile and thus there should be no potential for them to undergo atmospheric transport. However, it was not long before it was

---

[2] Specific examples include polybrominated diphenyl ethers (PBDEs), including penta- and octa-BDE, hexabromocyclododecane (HBCD) and polybrominated biphenyls.

[3] PFAS include perfluorinated sulphonates (PFOS) and perfluorinated carboxylic acid (PFCA), of which perfluoroctanoic acid (PFOA) is the best example.

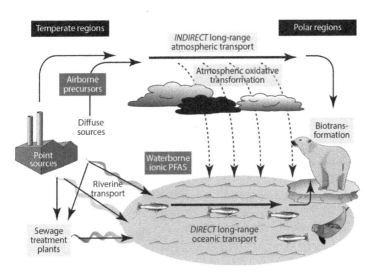

Figure 8.7 *Transport pathways for fluorinated compounds*

found that they are globally distributed and are present in Arctic air and Arctic biota. Exactly how their long-range transport is achieved is the subject of much debate and we will carefully avoid the details. It is now believed to be made up of two parts. There is the atmospheric component from volatile precursors that then degrade in the atmosphere to the acids. The PFAS found in caribou have taken this route. On the other hand, ocean biota seem to be accumulating their PFAS largely from long-range ocean transport but possibly with an atmospherically deposited contribution when the ice melts (see Figure 8.7).

In animals, PFAS do not partition into fatty tissues, as do organochlorine and other organohaline substances, but accumulate in the blood, kidneys and liver, where they remain for a very long time (especially in the liver). Little is known about the toxicity of these substances, although at high concentrations, they have been associated with a wasting syndrome that disrupts lipid metabolism and enlarges the liver. Because PFAS do not partition into fatty tissue, it was not expected that they would be capable of biomagnification. However, the AMAP 2009 assessment team reported that some studies do suggest a degree of biomagnification in the Arctic, leading up to fish-eating whales and seabirds. Müller and colleagues have provided a recent example. They measured the biomagnification of PFAS in the lichen-caribou-wolf food chain involving the Porcupine and Bathurst caribou herds of western Arctic Canada. The lowest concentrations were found in vegetation and the highest in the wolf population. Biomagnification factors were highly tissue and substance specific. Therefore, individual whole body concentrations were calculated and used for biomagnification and trophic magnification assessment. Trophic magnification factors varied but were highest for PFAS that contained 9–11 carbon atoms. Finally, we find that PFAS are also present in Inuit. A study published in 2012 by Lindh and colleagues found that blood serum levels of seven types of PFAS in Greenland Inuit were amongst the highest found in a general population anywhere in the world.

Incidentally, we should not have been surprised that brominated and fluorinated substances behave in such a similar way to organochlorines. Fluorine, chlorine and bromine sit together in the same group in the Periodic Table. You will probably have already noticed that some lessons were learned here. According to our understanding of POPs when the CLRTAP POPs protocol and the Stockholm Convention were being negotiated, the PFAS family should not have created any environmental problems. The procedure for adding new substances to the two agreements was based largely on two notions. First is the idea that to be a global problem, a substance should be able to enter the atmosphere at warm temperatures and to leave it at cold temperatures. In other words, a substance must be semivolatile. Second, for the substance to reach toxic levels in biota far from source, it would have to be capable of biomagnification, a process that depends (or so we thought) on the substance being fat soluble. Therefore, you could use predictive criteria for volatility and fat solubility to ascertain whether a substance qualified as a POP without waiting for the substance to accumulate in remote cold areas. The PFAS family failed both of the predictive criteria but can still travel long distances and accumulate in biota. In these circumstances, the monitoring data gathered and interpreted by AMAP were extremely useful because they showed that the predictive criteria were meaningless for these substances. Instead, the teams who evaluate proposals to add new substances to the agreements were able to use the monitoring data surrogates to justify taking action on these substances.

As you will have noticed from Table 8.2, members of the PFAS family are now banned under the CLRTAP POPs protocol and the Stockholm Convention agreements. Action on PFAS would not have been possible without negotiating fundamental amendments to the agreements if the flexibility to use surrogates had not been built into the agreements. The close linkages between the scientific evidence presented through AMAP with the political influence of the Arctic Council and of representatives of Arctic indigenous peoples played a very important role in the global curbing of the use of these substances.

A good deal more is now understood about the toxicity of POPs than was known back in 1994 when we compiled the CLRTAP state of knowledge report. However, as Robert Letcher and the AMAP 2009 assessment team point out, there remains a dearth of information on cause-effect relationships concerning POPs and Arctic biota and wildlife. We continue to rely heavily on extrapolations from controlled experiments with laboratory animals, on comparing body burdens in Arctic animals (mainly top predators) to effects levels established from laboratory animals and on inferring effects from various biomarkers that indicate change in immune, endocrine, reproductive and developmental systems.

The briefest of overviews of POPs toxicity would occupy the rest of this book, and even after that, I am not sure we would be much further ahead. Therefore, we will largely rely on Table 8.3 to summarize knowledge on a selected number of POPs, including the legacy organochlorines, brominated flame retardants and the PFAS family.

An important point to notice is the column under the heading "Thyroid and Vitamin A". It illustrates how many of these substances have endocrine (hormonal)-disrupting potential and it is presently believed that a number of outcomes and effects listed in the other columns may involve perturbed endocrine activity. Similarly, we could devote another book to the health implications of exposure by Arctic fish and wildlife in relation to the table of POPs toxic properties. However, we will simply note at this stage that if you look at the concentrations of POPs in the lipid-rich tissues of upper-trophic-level animals in the Arctic (particularly in marine fauna), most species cluster at concentrations where effects have been observed and/or guideline levels have been established.

Several pages ago, while we were looking at information on the toxicity of POPs to humans, I mentioned a study published in 1992 by Joseph and Sandra Jacobson and colleagues that examined various cognitive and behavioural attributes of a group of children in relation to the PCB levels of their mothers' blood. Before we take a close look, it will help if we spend a few moments thinking about how health researchers try to answer questions concerning the impacts of chemicals at very low doses. The foundation is usually animal studies, but no species is absolutely identical, and at some point, one has to ask "Am I confident that these results are applicable to humans?" This is tricky, especially when you may be concerned about potential neurological outcomes. Therefore, epidemiologists try to detect patterns in health outcomes with patterns of exposure to some potential cause. If you are finding these

Table 8.3 – A simplified overview of the toxic properties of some POPs

Substance	Reproduction Development	Neurotoxic	Liver Enzymes	Immune Suppression	Thyroid and Vitamin A	Cancer	Other
Aldrin and dieldrin	✓		✓	✓		✓ NM	
Chlordanes	✓		✓	✓		✓ NM	
DDT and metabolites (1)	✓		✓	✓	✓		✓AD
HCB (2)	✓		✓	✓	✓	✓ NM	✓ P
α HCH (3)			✓			✓ NM	
β HCH (3)	✓		✓	✓		✓ NM	
γ HCH (3) (lindane)	✓		✓		✓	✓ NM	
Mirex	✓		✓	✓		✓ NM	
Toxaphenes	✓		✓	✓	✓	✓ M	✓AD, BF
Endosulfan	✓		✓	✓		✓ NM	
Dioxins, furans and dioxin-like PCBs (4)	✓	✓ PC	✓	✓	✓	✓ NM	✓ P
Other PCBs (4)	✓	✓ PC	✓	✓	✓	✓ NM	✓AD, P
SCCPs (5)	✓	✓	✓		✓	✓ NM	
PCNs (6)	✓		✓				
PBDEs (7)	✓	✓ PC		✓	✓	✓ NM	
PFAS (8)	✓		✓			✓ NM	

Notes on substances:

(1) DDT: Dichlorodiphenyltrichloroethane and its metabolites; (2) HCB: Hexachlorobenzene; (3) HCH: Hexachlorocyclohexane; (4) PCBs: Polychlorinated biphenyls; (5) SCCP: Short-chain chlorinated paraffins; (6) PCNs: Polychlorinated naphthalenes; (7) PBDEs: Polybrominated diphenyl ethers; (8) PFAS: Polyfluoroalkyl substances; also known as perfluorinated compounds (PFCs) and including perfluorinated carboxylic acid (PFCA), perfluorooctanoic acid (PFOA) and perfluorooctane sulfonate (PFOS)

Notes on effects:

AD: Overstimulation of the adrenal gland; BF: Bone brittleness in fish; NM: Nonmutagenic tumour promoter; M: Mutagenic; PC: Permanent changes in learning behaviour and memory; P: Porphyria aspects interesting, I hope you will look at some of the source documents listed in the bibliography. If so, you will immediately need a few simple explanations on basic epidemiological methods.

A longitudinal study is one in which observations of the same variables are repeated on the same individuals (cohort) at intervals over a long period of time. A retrospective study looks backwards and a prospective study looks forwards. The Jacobson study noted earlier was a prospective cohort study. A cross-sectional or latitudinal study compares attributes of different populations at the same time, such as the prevalence of cardiovascular disease in France and Canada. In our present context, you will mainly find yourself looking at prospective longitudinal studies. We will spend more time on these related to questions and lessons raised by the epidemiological studies on the significance of very low doses of POPs and mercury later in this chapter. It may turn out in the end to have been one of the most important lessons we should learn from the Arctic Messenger. We will begin with a sample of such work drawn from epidemiological studies within and outside the Arctic.

In its 1992 study, the Jacobson team looked at PCB levels in umbilical cord blood in comparison to test results designed to evaluate cognitive processing efficiency and the ability to maintain sustained attention in a cohort of four-year-old children. The mothers came from a non-Arctic community where fish formed a major part of the diet and where aquatic PCB pollution was known to have occurred. The researchers found that "[p]renatal exposure to PCBs was associated with less efficient visual discrimination processing and more errors in short-term memory scanning but not with sustained attention." Surprisingly, although a greater exposure to PCBs must have occurred during breast-feeding, this source did not contribute to the reduced cognitive performance. Therefore, the window for susceptibility appeared to be located within the time of embryonic development. In 1996 and 2003, Joseph and Sandra Jacobson published results after looking at the same types of cognitive endpoints for the same children now aged 11 years. Prenatal exposure to PCBs via umbilical cord blood was associated with lower verbal IQ scores, memory and (this time) their ability to maintain attention. The most highly exposed children were three times as likely to have lower IQ scores and twice as likely to be at least two years behind in reading comprehension. Again, there was no relationship to PCB exposure via breast milk.

Similar cohort studies have been carried out in other parts of the world, but the results between areas and cohorts have been quite variable, perhaps reflecting different methodologies. In the Faroese cohort we will look at in relation to mercury exposure, Philippe Grand-jean and colleagues looked for neuropsychological outcomes of prenatal exposure to PCBs in seven-year-old children. PCB levels in umbilical cord blood at high concentrations were associated with similar types of cognitive deficits as reported by the Jacobsons, but there was evidence of confounding with mercury levels also present in the cord blood. In Arctic foods used by northern indigenous peoples, these two toxic substances are almost always found together.

Recently, several researchers have been using cohort studies to examine outcomes related to POPs that were not included in the original CLRTAP POPs protocol and Stockholm Convention. Most are not members of the original "legacy" organochlorines but do include several substances that are already now being addressed in the two international agreements. These studies reinforce the justification for the actions taken, but they are also exposing the need for a much more precautionary approach to chemical regulation than has been the practice in the past. For example, in a study published in 2012, Brenda Eskenazi and colleagues looked for associations between impaired neurobehavioural development in five- to seven-year-old children (whose parents were agricultural workers in southern California) and their prenatal and postnatal exposure to one of the brominated flame-retardant family of substances (PBDE). They found "associations between maternal PBDE levels during pregnancy and evidence of deficits in children's attention, fine motor coordination and cognitive functioning at both ages". In addition, the children's PBDE levels (not the mothers' blood levels in pregnancy) "were asso-ciated with lower scores for full-scale IQ, particularly processing speed, verbal comprehension and perceptual reasoning". The researchers also found that each ten-fold increase in the children's total measured PBDE levels was associated with at least 4.5 times higher odds of the child being rated by teachers as at least moderately hyperactive and impulsive.

It is very difficult to summarize exactly what the PCB cohort studies mean in the Arctic context, partly because (as we shall see) in utero exposure to mercury has been associated with similar outcomes and partly because cohort studies conducted in the Arctic have always used quite small sample sizes. Clearly, there is sufficient evidence from the studies conducted outside the Arctic to generate concern, but we will understand a little more when we complete our mercury review.

Organophosphate pesticides were considered to be an alternative to organochlorines, such as DDT. They have a much greater acute toxicity than most organochlorines, but they rapidly degrade on exposure to sunlight and in air and soil. They function by blocking the activity of an enzyme called *acetylcholinesterase* that is essential for nerve function in many species, including insects and mammals. They are not POPs in the context of the CLRTAP protocol and the Stockholm Convention because of their lack of environmental persistence and there is no evidence they are responsible for any problems in the Arctic. However, I thought they deserved a quick mention here because studies have associated them with similar cognitive outcomes following the prenatal exposure we have seen for PCBs and mercury. This clearly indicates the need for a more rigorous toxicological examination of chemicals before they reach the market and for ongoing monitoring vigilance when such products come into general use.

The epidemiological evidence for low-dose chronic exposure effects for certain POPs is not restricted to prenatal exposure or to neurological effects. For quite some time, a link has been suspected between a particular form of breast cancer and exposure to some POPs, such as DDT and its metabolite DDE (which can mimic the hormone estrogen). This is probably an example of endocrine disruption. Human fertility has also been suspected for some time to be susceptible to disruption through chronic low-dose exposure to certain POPs. For example, the prospective studies of Buck Louis and colleagues have implicated exposure to several PCBs, PFOS and DDE to reduced fecundity in couples that were followed for 12 months after they ended the use of contraception.

Another recurrent suspicion has been whether chronic exposure to POPs increases the risk of developing a particular form of diabetes, known as type 2 diabetes. Type 2 diabetes (also known as adult-onset or noninsulin-dependent diabetes) is a chronic condition. It affects the way our body metabolizes glucose by either impairing the regulatory actions of insulin that normally control the movement of sugar into our cells or by preventing the body from producing enough insulin to maintain a normal glucose level. I include this work here to give an example of a purported effect for which different epidemiological studies have been unable to reach entirely consistent conclusions. These types of confusing results probably occur when several different factors, such as genetics and environmental exposure to chemicals, are responsible for the susceptibility to certain medical disorders, such as type 2 diabetes. Nevertheless, a report published by Wu and colleagues in 2013 is a large and robust study that gives a clear result. Using a database of more than 1,000 female nurses whose blood was collected in 1989–1990 and who were diabetes free at that time, the researchers found that the incidence of type 2 diabetes by 2008 was associated with HCB and total PCB serum concentration.

This will conclude (for the moment) our short and very restricted review of the present understanding of the chronic toxicity of POPs. If you go back to Table 8.3, you will see just how selective I have been and how I have ignored many types of purported effects attributed to chronic low-level exposure to POPs, such as their possible role in the development of tumours. In general, what I have done is to try to indicate the diverse nature of health outcomes that are believed to be associated with POPs and to emphasize some outcomes that the reader may not have heard about in the past but which carry significant messages for humankind. One such message is related to these long-term cohort studies. Are you like me and wonder why we are not developing a global precautionary chemical regulatory mechanism that catches the propensity of chemical families to be associated with long-term chronic effects? We will return to the question in another page or so.

Before we leave the two POPs agreements to discuss the Minamata Convention on mercury, we should remember the part played by Arctic data to stimulate the development of these agreements. It was the presence of these substances in the High Arctic at concentrations with the potential to cause human health effects that caught the attention of the world. When the two CLRTAP protocols were being born, collaboration on circumpolar environmental monitoring through AMAP was just developing. However, by the time the Stockholm Convention reached the negotiation stage, AMAP had produced its first full assessment on these substances. Therefore, Arctic countries and AMAP itself could bring unequivocal information and arguments to the negotiating table. It is not surprising that both these agreements acknowledge the vulnerability of Arctic ecosystems and of Arctic indigenous peoples to POPs and that the need for global action on mercury was not ignored.

What did happen about mercury? In October 2000, AMAP provided an interim assessment to the Arctic Council ministerial conference in Barrow, Alaska. The key part of the report was an update on the mercury section of the 1998 AMAP *Assessment Report*. New information was presented that summarized results reported in 1997 and 1998 by Philippe Grandjean and his colleagues on the Faroese cohort of more than 1,000 children. I mentioned this study earlier. For each child, the researchers knew the mercury concentrations in umbilical cord blood, in maternal hair and in breast milk at the time of birth seven years previously. The children were divided into high and low uterine mercury exposure groups and subjected to a number of tests concerning, for example, hand-eye coordination, reaction time on a continuous performance test, intelligence scales for children, visual motor testing and verbal learning tests. The researchers found mercury-related neuropsychological decrements in language, attention and memory and in visuospatial and motor functions. The decrements were mild, but they were there and caused much concern. To put this information into a circumpolar context, AMAP was able to provide the Arctic Council ministers at the Barrow meeting with a much more detailed "map" of mercury levels in Arctic people and in Arctic marine foods than had earlier been available.

Perhaps emboldened by the way in which the CLRTAP actions on POPs had grown into a global initiative under UNEP, the Arctic Council Ministerial Conference decided to push for global action. Their Barrow Declaration records that they "[n]ote with concern that releases of mercury have harmful effects on human health and may damage ecosystems of environmental and economic importance, including in the Arctic, and call on the United Nations Environment Programme to initiate a global assessment of mercury that could form the basis for appropriate international action in which the Arctic States would participate actively." The United States brought the issue to the Governing Council of UNEP and the Inter-Organization Programme for the Sound Management of Chemicals (IOMC) was asked to provide a global mercury assessment. This document was made available for the Governing Council at its twenty-second session in February 2003. Much time was then spent on considering what type of action could be taken. However, in February 2009, at the twenty-fifth session of the UNEP Governing Council, in Nairobi, Kenya, 141 countries agreed to begin negotiations on a legally binding instrument for the global control of mercury pollution.

To make progress on developing a global convention, it is obviously essential that there is a common understanding on the pressing need for that convention. This is usually not easy, given our world mosaic of vastly different degrees of economic development, wealth and political dogma. Success is very unlikely if this aspect is ignored. One of the key people involved in this supportive work was Gunnar Futsaeter. He moved from the Norwegian State Pollution Control Agency to UNEP Chemicals in Geneva to work on putting the UNEP Governing Council decisions on mercury into action. Gunnar had for years been the Norwegian delegate to AMAP and is a passionate advocate for Arctic environmental integrity. His energy was exactly what was needed to help develop UNEP partnerships for action on major mercury sources, such as the burning of coal. This was often achieved by coordinating projects funded by the Global Environment Facility (GEF)[4] in, for example, India, Russia and China. He also played a key role in preparing the Global Mercury Assessment, much of which was delivered through a joint AMAP/UNEP collaboration.

By the time the negotiations began in 2010, we had learned a good deal more about mercury contamination in the Arctic and especially about the potential for lasting effects to children exposed in the womb. In 2011, AMAP took the opportunity to publish (in the midst of the negotiations) a major assessment report devoted to mercury in the Arctic. It was perfect timing for maximum impact. Here is a quick look at what the AMAP assessment report had to say about two of the issues related to mercury and human health in the Arctic.

Firstly, the subtle neurological effects of prenatal exposure to mercury in the Faroese cohort of children detected when the children were seven years old were still present when the children were 14. Using the same family of tests employed when the children were age seven, reports by Weihe, Grandjean and their colleagues found that prenatal methylmercury exposure was statistically correlated with deficits in domains of language, attention, memory and auditory and visual brain processing. Postnatal methylmercury exposure had no discernible effect. Olivier Boucher and colleagues who studied a cohort of Inuit children from Nunavik in Arctic Québec have reported somewhat similar behavioural results. In papers published in 2010 and 2012, they reported that mercury concentrations in umbilical cord blood even below 10 micrograms per litre could be associated with attention deficit hyperactivity disorder (ADHD)–like symptoms at about 11 years of age.

Secondly, chronic exposure to mercury through diet has also been linked by studies from northern Finland to an increased risk to the cardiovascular system, including hypertension in adults. However, Inuit (who do not live in Finland) consume very high concentrations of mercury in their diet but historically have experienced very low mortality

---

[4] The Global Environment Facility (GEF) is an independently operating financial organization that provides grants for projects related to biodiversity, climate change, international waters, land degradation, the ozone layer and persistent organic pollutants. Since 1991, the GEF has provided $11.5 billion in grants and leveraging of $57 billion in co-financing for more than 3,215 projects in more than 165 countries.

from cardiovascular disease. One explanation for this discrepancy may be that the high levels of selenium found in the Inuit is attenuating the toxic effects of mercury. High blood pressure in childhood is an important indicator of adult hypertension and methylmercury exposure is a potential risk factor. The Faroese child cohort study also found associations between prenatal mercury exposure with increased blood pressure and reduced heart rate variability at seven years of age. The effects persisted at 14 years but with less severity.

The implications of the cohort studies were a powerful incentive for the negotiators working to create the global convention on mercury under the UNEP. Not only was the Arctic ecosystem being impacted by mercury emissions from elsewhere in the globe, but health effects were also being detected in Arctic people, resulting from exposure experienced before they were born. The intergovernmental negotiating committee met five times: in Stockholm, Sweden, in June 2010; in Chiba, Japan, in January 2011; in Nairobi, Kenya, in October/November 2011; in Punta del Este, Uruguay, in June/July 2012; and in Geneva, Switzerland, in January 2013. At this last session, 137 governments – led by chair Fernando Lugris from Uruguay – agreed to the final text for a new global treaty on mercury. The resulting agreement, now known as the Minamata Convention, was formerly adopted as international law on October 10, 2013, in Kumamoto, Japan, but at the time of writing has not entered into force.

The Minamata Convention has taken a mercury life cycle approach to international controls that impacts mercury use in many diverse sectors, comprising, for example, coal-fired power generation, construction, electronics, heavy manufacturing, wastes, health and cosmetics industries. The convention also includes control measures on air emissions, a phaseout of existing mines and measures aimed at reducing emissions from small-scale gold mining and artisanal industries. There will be a phasedown of the use of dental fillings using mercury amalgam.

The preamble of the convention acknowledges the voice of the Arctic Messenger by recognition of "the particular vulnerabilities of Arctic ecosystems and indigenous communities because of the biomagnification of mercury and contamination of traditional foods, and concern about indigenous communities more generally" and "health concerns, especially in developing countries, resulting from exposure to mercury of vulnerable populations, especially women, children, and, through them, future generations". Article 3 prohibits "primary mercury mining not being conducted prior to the entry into force of the

Convention for that Party, and requires the phase-out within fifteen years of any primary mining that was being conducted within a Party's territory at the date of entry into force for it". The importance of research and environmental and health monitoring was recognized in a new article introduced late in the negotiations as Article 19. Sadly, as always with this type of topic, it is a "soft" obligation. Part of the article is devoted to human health and includes, for example, an interesting clause promoting "the development and implementation of strategies and programmes to identify and protect populations at risk, particularly vulnerable populations, including science-based health guidelines, targets for mercury exposure reduction and public education". In Article 21, parties are required "to report to the COP on measures taken to implement the convention and their effectiveness in meeting the Convention's objectives". It was pleasing to see that the negotiators considered that effectiveness evaluation is important. This is addressed in Article 22. The text is quite straightforward and should avoid many of the complications that arose over interpretation of the corresponding article in the Stockholm Convention.

Uncomfortable Realities, Including the Arctic Dilemma

As we approach the present day in the POPs and heavy metal stories, the Arctic Messenger has a few axes to grind.

The first concerns the significance of the epidemiological studies we have been looking at concerning prenatal and low chronic exposure to POPs and mercury. Much earlier, we met sixteenth-century physician Paracelsus and his assertion that as the dose of a potentially toxic substance goes up, so does the risk of effects. You could think of alcohol consumption as a classic example of this so-called monotonic relationship. The simple monotonic paradigm remains as the foundation for how the toxicity of substances is evaluated. Test animals are given different doses to ascertain the risk posed by a particular substance. The results are then plotted on a graph to arrive at values called the *lowest observed adverse effect level* (LOAEL) and the *no observed adverse effect level* (NOAEL). Usually, toxicologists design their testing in anticipation of particular types of effects, such as the appearance of tumours or birth defects. To be on the safe side before publishing guidance values for long-term chronic exposure, public health authorities will then add a safety factor to the LOAEL or NOAEL (the size of which is usually related to how confident they are about the toxicity mechanisms involved with respect to the animals tested and to humans).

This all sounds straightforward and logical. Chemists can now measure substances at ever-decreasing levels of concentration, enabling epidemiologists to examine the frequencies of occurrence of a wider variety of potential effects from low exposure levels that can be discriminated using sophisticated statistics. Furthermore, this can be carried out on human populations and the vulnerability of different parts of our life cycle can be examined. What has been found from these studies is that some substances are capable of producing effects at doses that according to the accepted NOAEL models should not pose a risk. The neurological and behavioural studies in Arctic and non-Arctic populations involving prenatal exposure to POPs and mercury are examples. Furthermore, it is becoming clear from animal studies that a number of substances (particularly those that disrupt endocrine function) may not always follow a monotonic relationship between dose and effect, particularly at low doses. For example, bisphenol A in mice can show a hump-shaped dose response curve for tumour development, where "moderate" exposure produces the highest frequency of tumour development. Other dose response curves are possible. For example, the oestrogen mimic p-nonylphenol shows a complex U-shaped curve in some species, while the pesticide atrazine can produce a greater effect at lower doses than at high doses.

Studies also show that the developing human brain is particularly vulnerable to disruption, especially at apparently short periods of exposure during prenatal development. As early as 2000, a U.S. National Academy of Sciences expert committee estimated that 3% of neurobehavioural disorders are caused directly by toxic environmental exposure and another 25% by interactions between the environment and inherited susceptibilities. In a short review paper published in 2012, Philip Landrigan and colleagues listed nine groups of substances that have been implicated in developmental neurotoxicity, leading to autism and learning disabilities. They are lead, methylmercury, PCBs, organophospate pesticides, endocrine disruptors, automotive exhaust, PAHs, brominated flame retardants and perfluorinated compounds (our PFAS). The list looks so familiar to the Arctic Messenger. All the substances – even the "generic" automotive waste – have been shown to be capable of long-range atmospheric transport. No country or region is immune from the toxic emissions of another region. Most of the substances on the list are capable of biomagnification and most have been implicated in low-dose epidemiological effects in Arctic and non-Arctic human populations. From our earlier study, we understand how physical and biological processes conspire to make this happen, but I am often asked if these neurological and behavioural effects are significant. It is a question I find downright repugnant at the personal level. Should a person be labelled a luddite for questioning the morality of allowing persistent substances to enter the environment that we anticipate will increase the proportion of children born with intellectual deficits and behavioural disorders?

There is another way the information can be presented. Think of intelligence as being an attribute that has a financial value to society. Next, consider its measurement as IQ being represented as a simple bell curve in the population of a country or region. The bulk of the population will not be affected by a five-point reduction of intelligence. However, if the entire curve is shifted a little to the left, the "tails" of the curve have also moved. The proportion of the population we once considered as gifted has decreased and the proportion we may call challenged and that will require assistance during their entire lives has increased. This can be the huge financial (not to mention moral) legacy left by our continued use of such substances. Will this financial argument influence our politicians? One might think so, but the Arctic Messenger finds it difficult to understand why the new understanding of low-dose relationships has still not entered the procedures used to evaluate the chemical safety of "new" substances. Rather than embrace the precautionary principle, our society prefers to wait for health problems to emerge.

I should say a few more words about endocrine-disrupting substances. The endocrine (hormonal) system modulates many biological processes in vertebrates, such as development, reproduction, behaviour and metabolic processes. Over the last 20 years, scientists have identified about 800 chemicals that are able (or suspected of being able) to disrupt hormonal systems and have linked them to disease outcomes in laboratory animal studies. At the same time, epidemiologists have been tracking increasing trends of endocrine-related diseases in human and wildlife populations that roughly correspond to the prevalence of these substances in the environment. Many of these substances are also POPs. This is why you have been encountering intermittent references to endocrine disruption in the preceding paragraphs and will continue to do so later in this chapter. One important characteristic of endocrine disruption is that it is often most effective during a short window of vulnerability. This is frequently located during embryonic development, but the effect may not be apparent until a later stage in the life cycle. It is probably one of the main reasons why disorders related to endocrine disruption do not follow the classic dose-response relationships. That should sound familiar to you from the work noted earlier on POPs and mercury by Eric Dewailly, Joseph and Sandra Jacobson, Olivier Boucher, Gina Muckle, Philip Landrigan and L. N. Vandenberg. If you have found the whole topic of low-dose relationships to environmental exposure to toxic substances interesting, have a look at some of the references given in the bibliography by these authors.

For nonspecialists, I recommend you start with the book by Theo Colborn and colleagues entitled *Our Stolen Future: Are We Threatening Our Fertility, Intelligence, and Survival?* Although published in 1996, it remains a compelling introduction to the topic and to endocrine disruption. For an up-to-date and more technical summary of the present understanding of endocrine disruption, there is the 2012 report edited by Åke Bergman and colleagues and organised by WHO and UNEP (WHO has also produced a summary document). It is a very comprehensive synthesis. As you make your way through the report, you will immediately notice the high degree of concordance between the Arctic disorders in people and wildlife noted earlier with laboratory and epidemiological studies conducted elsewhere. It will help you understand the Arctic Messenger's frustrations.

The Arctic Messenger's second concern relates to the so-called "Arctic dilemma". Earlier in this chapter, we saw how physical and chemical processes conspire to bring POPs and mercury to the Arctic from mid- and low-latitude man-made emissions. We then saw that biomagnification has resulted in some animals at upper-trophic levels accumulating concentrations of these substances that approach or even exceed levels at which toxic effects could be expected. Finally, we learned that these same upper-trophic-level species are frequently the foundation of the traditional diet of Arctic indigenous peoples, particularly the Inuit. From a nutritional perspective, these traditional foods are far superior to the preserved, packaged and processed foods available to Arctic communities. From a social perspective, the same traditional foods are also at the centre of indigenous culture. The persuasive power of modern advertising places a heavy impetus on Arctic peoples (especially the young) to drift towards a southern diet. This process is known as *dietary transition*. Indigenous leaders watch this trend with concern. What will the impact be on unique northern cultures that are founded on direct relationships with the life cycles of the animals that provide them with their food? Human health authorities are equally concerned. As dietary transition has progressed, epidemiologists have also noticed changing patterns of diet-related disease. For Inuit, the most dramatic relates to a postdietary transition increase in the occurrence of metabolic syndrome. This is the name given to a combination of clinical attributes, such as obesity, high blood pressure and raised blood triglycerides, that increases the risk of diabetes and cardiovascular disease. The latter in particular was historically almost unknown in traditional Inuit communities despite widespread smoking from an early age. In contrast, there is now a similar prevalence of cardiovascular disease in many Inuit communities as in mid-latitude populations. Another cost of dietary transition appears to be related to reduced efficacy of the immune system, particularly in children. Readers may remember that immune suppression has long been known as a common toxic effect of POPs.

A paper published in 2008 by Joseph and Sandra Jacobson and colleagues shows just how convoluted the Arctic dilemma has become. It looked at the levels of docosahexaenoic acid (DHA, which is an omega-3 fatty acid), other polyunsaturated fatty acids and three environmental contaminants (polychlorinated biphenyls, mercury and lead) in umbilical cord plasma and maternal milk in 109 Inuit infants in Arctic Québec. Statistical tests were then used to control for the contaminant confounders and to examine growth and development at six and 11 months in relation to DHA levels from umbilical cord blood and DHA from breastfeeding. The researchers found that higher umbilical cord DHA concentrations were associated with more optimal visual, cognitive and motor development in the infants. They believed that this demonstrated how critically important this fatty acid is during the third "trimester spurt of synaptogenesis in brain and photoreceptor development". Put simply, synaptogenesis in this context is the process when our brain neurons are connecting up with one another. Now the key point in the dilemma here is that the main dietary sources of these essential fatty acids to the Inuit are the same upper-trophic-level fish and marine mammals from which they also acquire their body burdens of POPs and mercury.

The Arctic dilemma has given leaders of indigenous communities and regional health care workers an unhappy responsibility. What dietary advice should be provided? The strategy followed has varied regionally to reflect which contaminants pose the greatest risk and to identify foods that carry the highest contaminant levels. The backbone of all strategies has been to provide Arctic residents with adequate information for them to make their own individual dietary choices. To provide this information, indigenous leaders and public health workers present the facts as comprehensively as possible and when appropriate in the regional indigenous language. Indigenous peoples' organisations and regional governments usually prepare the communication materials. There is frequently a special focus made on children and females below and during reproductive age in order to point out how individual diet choices even within an indigenous diet can reduce exposure to POPs and mercury. The bottom line was (and continues to be) that the weight of evidence points to the conclusion that the known health benefits of indigenous diets outweigh the risks posed by the POP and mercury contaminant burdens of those diets. Twenty years ago, I was never comfortable with this communications message but had to agree that it was the correct one to follow. Today, I think it is still the correct message. However, with our increased understanding of the association of prenatal exposure to POPs and mercury with the potential for lasting neurological and behavioural outcomes, I am even less comfortable. To appreciate the full measure of the Arctic dilemma, it helps to try to imagine oneself in the position of the Arctic leaders who have shown great courage in promoting the message. We must not be complacent, however, because the veracity of that message totally depends on the underlying science. Governments of Arctic Council countries must ensure the long-term well-being of the science base and never hesitate to adjust their public health message if the balance of evidence begins to shift towards different dietary advice.

Now that we are almost up to date with the stories of POPs and heavy metals from the Arctic Messenger, I have to pay homage to the hundreds of people who advanced our understanding of the behaviour of POPs. Even more praise is due to those who saw the importance of contributing to the AMAP assessments. It was this commitment, coupled with the organisational skills of Lars-Otto Reiersen and Simon Wilson, that enabled the Arctic story to gain political attention. It provided the foundation for the CLRTAP protocols and the Stockholm and Minamata conventions to control these substances. A few of these scientists have actually been the lead (or close to it) on every AMAP assessment on that topic – for example, Derek Muir and Cynthia de Wit for POPs and Rune Dietz for metals. As a Canadian, it is hard for me to imagine the Arctic Messenger without Derek. He always seems to have been at the very centre of POPs work – at least in Canada. What I appreciated most of all was that, like Robie Macdonald, any panic call for help always had a quick response that hit at exactly what I needed. I think I probably owe more to Derek for his consistent support than to anyone else. On the human health side, Eric

Dewailly was always the researcher to whom I turned for guidance. Others have appeared more recently and have played key roles in some of the latest Arctic assessments from AMAP. I am thinking, for example, of Tom Harner, Hayley Hung, Roland Kallenborn, Peter Outridge and Gary Stern. Some people just seem to have always been there to help, such as Terry Bidleman, Birgit Braune, Geir Gabrielsen, Andrew Gilman, Henry Huntington, Robert Letcher, Jay van Oostdam and Frank Riget.

Has the Arctic Messenger Been Understood?

Has the world heard, understood and acted on the insights given by the Arctic Messenger? As is so often the case with these types of issues, the answer is an ambiguous "Yes" and "No". The negotiation of the 1998 regional CLRTAP protocols on POPs and heavy metals, of the 2001 global Stockholm Convention on POPs and of the 2013 Minamata Convention on mercury suggest a "Yes" response. However, before we become too self-congratulatory, we should remember how ponderous these actions have been. I think I made the first plea to a UN organisation for action on POPs in August 1990, but it took until 1998 and 2001 for countries to agree on regional and global action (and then a number of years before they entered into force).

The objective of all three agreements is to drive down the environmental levels of these pollutants. It is too early to answer the question with respect to mercury, but what about POPs? Both of the agreements on POPs have strong legal obligations to deal with substances that have proven to be POPs and that have been shown in one way or another to be causing concern for their environmental and human health effects. These are the substances listed in the annexes to the agreements and the lists are slowly growing. Environmental levels of substances on the original annex lists are generally decreasing due to reduced or eliminated use and/or emissions and discharges to the environment. This encouraging news suggests that substances more recently added to the annexes should also be expected to eventually decline in the environment.

However, there are few (with the significant exception given shortly) strong international legal obligations to prevent new substances with the properties of POPs entering into use. This is left to the discretion of national regulatory processes, even though the case of POPs has dramatically shown how we all share the same atmospheric, marine and freshwater environments wherever we live.

Ironically, some of the substances now appearing as candidates for inclusion on the annex lists of the POPs agreements are beginning to include the "safe" replacements for "legacy" substances on the original lists. We have already learned that the potential number of substances with the properties of POPs is estimated to be somewhere between hundreds to thousands. We also learned that health scientists increasingly believe that a number of POPs, especially those with endocrine-disrupting and neurotoxic properties, are already exerting a cost to society – possibly even at the level of human embryonic development. Despite this knowledge, there are few signs that these findings and concerns are being fully addressed by national agencies responsible for regulating the commercial introduction of new chemicals. The fundamental challenge is the difficulty of implementing a precautionary approach to chemical regulation in a milieu characterized by short-term economics and the power of industrial lobby groups in decision-making processes. Until world governments get to grips with this issue, new POPs will continue to enter into commerce and into the environment.

There may be an important exception to this state of affairs. In 2007 (after seven years of negotiation), a process called Registration, Evaluation, Authorisation and Restriction of Chemicals (REACH) entered into force in the European Union (EU). No other international chemicals management legislation comes as close as REACH to embracing the lessons learned in the Arctic and the precautionary principle. It applies only in the EU, but its phased implementation should impact chemical use throughout the world. Embedded within REACH is a category called *substances of very high concern*. These are chemicals proposed for use in the EU that meet one or more of the following criteria:

- It is carcinogenic.
- It is mutagenic.
- It is toxic for reproduction.
- It is persistent, bioaccumulative and toxic.
- There is "scientific evidence of probable serious effects to human health or the environment which give rise to an equivalent level of concern". This criterion allows, for example, that neurotoxic or endocrine-disrupting substances can be regulated through REACH.

The present priorities for assessment under REACH are persistent bioaccumulative toxic substances (which include POPs), substances that widely disperse during use (which captures long-range environmental transport) and substances used in large quantities. How it is all intended to work is described in a document of more than 800 pages, but a key element is that the onus is placed on industry to show that products are safe. Therefore, the costs of REACH are largely carried by industry and ultimately by consumers. There are also a number of voluntary arrangements – the most comprehensive being the relatively new Strategic Approach to International Chemicals Management (SAICM). This is a global policy framework administered by UNEP that takes a "cradle to grave" approach to the management of chemicals and includes, for example, risk assessments of chemicals and the disposal of obsolete and stockpiled products. The SAICM is certainly a worthy activity, but it was not intended to be a legal global proxy for REACH.

The slow progress (except perhaps in the EU) of regulatory authorities to fully grasp the lessons from POPs and mercury in terms of controlling chemicals before they become an environmental and health problem is disappointing. This concern also extends to the environmental fate of high-use substances, such as pharmaceuticals, which have only recently been thought of as pollutants. Here are just two examples of why we should be worried: Where do endocrine-disrupting substances, such as ethynyl estradiol (the active ingredient in contraceptive pills), end up? The answer is in our sewers and eventually in fish, causing feminization and reduced sperm counts. One of the best-known examples of an environmental impact resulting from pharmaceuticals concerns is the familiar and popular nonsteroid anti-inflammatory drug diclofenac. A veterinary version of the drug was widely used to treat cattle in Asia and was found to be responsible for a dramatic collapse of vulture populations in the Indian subcontinent that was reversed once such use was banned. Diclofenac is now common in freshwaters around the world and has been associated with detrimental liver, kidney and gills effects in fish where the alteration of liver gene expression has also been reported. But despite this type of evidence, an EU attempt to deal with the environmental disposal of pharmaceuticals is running (at the time of writing) into very determined opposition.

Another issue that regulatory authorities find very difficult to address is that toxic substances are tested and measured in isolation from one another. However, my body undoubtedly contains a complex cocktail of substances I have acquired from my environment but have been unable to excrete. Therefore, what we may be interested in knowing is the combined risk of this cumulative body burden. The toxicity of a substance is often related to its molecular structure and shape. For example, dioxins, furans and a particular class of PCBs (known as coplanar PCBs) are sufficiently alike that toxicologists are able to estimate their toxicity in relation to one form of dioxin (2,3,7,8-TCDD). However, no similar method exists to group substances in chemical risk assessments and we still have little notion of the true impact of the complex mixture of toxic substances that are now present in our bodies even before we are born.

It is time to return to our earlier question as to whether the world has heard, understood and acted on the insights given by the Arctic Messenger. I think it has heard, understood a little, acted a little and misunderstood or ignored almost completely the crucial element. We cannot continue to complacently regard our environment as a convenient sewer into which we can dump persistent, biomagnifying toxic substances even if such substances show very little acute toxicity. The ponderous strategy of applying regulatory controls only after a huge pollutant inventory has been accumulated in the environment and is exerting its inevitable toll on wildlife, fish and human health is not only amoral but also unsustainable.

The Minamata Convention may escape some of the future problems concerning POPs because it deals with a single substance. However, we will face an unwanted legacy for many years because air-ocean modelling indicates that the oceans have not yet reached equilibrium with current atmospheric levels of mercury, which stays in the upper ocean above a depth of 200 metres for about 30 years and for centuries in intermediate and deep waters. This is much longer than the one-year residence time in the lower atmosphere. Thus, removal from the ocean takes much longer than does removal from the air, so concentrations will change more slowly.

We arrive at the same conclusion for POPs. Thirty years ago, countries began to introduce reductions in technical HCH (the mixture of HCH forms that made up the pesticide lindane). However, today, beta-HCH is still increasing in Canadian Arctic marine waters. This is an ominous indication of what the future may hold for PFOS in the Arctic. The environment has a finite capacity for benignly acting as a reservoir into which humankind can dump toxic, biomagnifying and persistent chemicals, including mercury. The limits of this capacity before biological and health effects occur has been reached in some areas, and for mercury, the physical mass balance models of the environmental inventory predict that recovery will be slow, however effective the Minamata Convention may be.

Climate Change, POPs and Mercury: A Conspiracy of Impacts

We have not quite reached the present day in our review of POPs and mercury in the Arctic. We earlier learned that when POPs are released into the environment, these persistent chemicals can pass into the atmosphere and be transported long distances very quickly. However, their volatility is temperature dependent and therefore they can condense out of the environment at colder temperatures. This process, known as *cold distillation* (or condensation), can therefore result in POPs becoming trapped in such cold ecosystems as the Arctic. One of the methods used by scientists to study the global behaviour of POPs released into the environment is known as *mass balance modelling*. The details are complicated, but the concept is simple and involves calculating how much of a given mass of a particular POP with its own unique physical properties (particularly those responsible for volatility) will move (partition) into environmental elements (reservoirs or sinks), such as air, marine waters, freshwaters, marine and freshwater sediments and soil. Once a substance is no longer being released by humankind, the inventory of substances in these compartments will move towards an equilibrium – the behaviour of which will be determined by its own physical properties and those of its "host" environment.

So far, so good, but some readers have probably already guessed where this is going. Our global climate is warming, and in the climate chapter, we will see that the Arctic is warming at twice the global rate. The key physical property controlling movement between reservoirs and to different latitudes is volatility, which is generally temperature dependent. Robie Macdonald first pointed out to AMAP that mass balance models would likely predict a reshuffling of the POPs inventories between environmental compartments and a further poleward migration of POPs. I have worked on and off with Robie since the early 1990s. Every scientific activity needs an individual like him – a person with an impeccable reputation in his own field of marine chemistry but who somehow manages to keep abreast of an eclectic range of other disciplines. Habitually looking over the horizon in unusual directions, he is always one of the first to spot new issues and alternative ways of looking at things. His mind is never trapped by accepted dogma. Robie organised an assessment team that produced an AMAP report on climate and contaminants that was published in 2002–2003. Another version was published by Robie and colleagues in 2003 in the journal *Human and Ecological Risk Assessment*. Essentially, the prediction is that as the subArctic and Arctic warm, some of the POPs sequestered in the oceans, soils, sediments and lakes will be reemitted back into the atmosphere. In the case of the oceans, this process will be further enhanced by the removal of summer seasonal ice which otherwise presents a barrier to sea-atmosphere gas exchange and heat transfer.

Eight years later, another AMAP assessment team organised by Roland Kallenborn published a comprehensive report that further clarified the predictions and reviewed the emerging evidence showing that POPs are indeed being remobilized in response to climate warming and, in particular, to warming in the Arctic. The most complete long-term Arctic air-monitoring data sets available (back to the early 1990s) are from Alert on Ellesmere Island and from Zeppelin on Svalbard. They enable trends in concentration to be detected for the legacy POPs that were the first targets of the CLRTAP protocol and the Stockholm Convention. For a few POPs (HCB and some PCBs), increasing levels have been observed in the first decade of the present century. However, most of the other legacy substances now show a general slow decrease in concentrations, presumably reflecting the decreasing use and increasing control of these substances. In 2011, Jianmin Ma and colleagues published results of a study indicating that atmospheric levels of the POP pesticides alpha-HCH, DDT and chlordane are showing an increasing trend that corresponds well with increasing mean Arctic sea level temperature and with declining Arctic sea ice cover. Although the statistical methods used were subsequently challenged, the difficult question of clarifying the impact and significance of rising temperatures on the polar distribution of POPs is now attracting much attention. The 2013 papers by Henry Wöhrnschimmel and by Todd Gouin with their respective colleagues (listed in the bibliography) will help illustrate just why it is such a challenging topic. The elements to be considered include the inventory of "old" POPs presently sequestered according to their volatility characteristics in soils, water and ice at mid- and polar latitudes that could be the source of secondary emissions. In addition, a warmer world will most probably increase the demand for pesticide use in mid-latitudes. At the same time, the trophic structure of Arctic ecosystems and their biomagnification potential can also be expected to change, but these are difficult elements to quantify. My own interpretation of the present computer modelling is that it further emphasizes the need for vigilance in the approval of chemicals with properties of POPs in a warming world. It also helps to focus attention on the key properties of substances that could be expected to undergo enhanced transport and polar deposition under given climate-warming scenarios.

The impact of climate change on mercury has been less well studied, but the potential impacts and uncertainties of these aspects were reviewed by Gary Stern and colleagues in 2012 based on the 2011 AMAP mercury assessment noted earlier. You will recall that the cycling of mercury in the environment is very much linked to the cycling of organic carbon, especially with respect to those parts of the cycle involving the biologically available and toxic methylmercury. The basic picture contains two fundamental climate-sensitive elements. First, the transformation of mercury in the cycle from one form to another and those elements concerned with the movement of mercury into and out of the atmosphere that are temperature dependent. This includes those strange atmospheric mercury depletion events (AMDEs) that are strongly linked to ice-related processes. Second, a disproportionately large slice of the environmental reservoir of mercury resulting from human activities has built up in the Arctic cryosphere (ice sheets, ice caps, glaciers and frozen sediments and tundra soils).

From a biological perspective, a key issue concerning climate warming relates to methylmercury. Warmer and longer summers will increase conditions in low-oxygen environments (including thawing tundra permafrost) that are favourable for the microbiological formation of methylmercury from inorganic mercury. Thawing permafrost alone has been estimated to be able to release 200 micrograms of mercury per square kilometre each year. To put this into perspective, it is an amount that is greater than that deposited annually from the atmosphere. At the same time, the release of nutrients from frozen soils and sediments will likely increase primary production at the base of the Arctic trophic system. We can therefore expect an enhanced movement of methylmercury to higher trophic levels through biomagnification.

In the marine environment, the reduced proportion of the Arctic Ocean covered by ice will increase the amount of particle-bound mercury deposited directly onto the sea surface and increase the exchange of elemental mercury between surface waters and the atmosphere. Warmer temperatures are thought to have the potential to alter the balance of geochemical and microbiological processes. This could increase the rate at which the large marine reservoir of inorganic mercury may be transformed into biologically available methylmercury. At the same time, the warmer waters and longer period in which light is available to phytoplankton (plant plankton) may very well increase the total annual primary production of new organic carbon and thus enhance the potential for the increased flow of methylmercury through biomagnification to upper-trophic levels. There is much here that requires further research. Even the nature of climate warming on AMDE deposition is controversial. On the one hand, warmer temperatures are expected to decrease mercury deposition from this process. However, on the other hand, the increase of exposed sea surface in spring (due to less sea ice) could increase the amount of AMDE-derived mercury that is absorbed into the Arctic Ocean. What a tangled web!

If you would like to learn more about the impact of climate warming on contaminant distribution and behaviour in the Arctic, look at some of the papers referenced in the bibliography. The reviews organised by Robie Macdonald in 2002–2003, by Roland Kallenborn in 2011, by Gary Stern (for mercury) in 2012 and by Henry Wöhrnschimmel and by Todd Gouin (both in 2013) provide a solid introduction.

The Long and the Short of It

The Arctic Messenger has shown that:

- POPs are substances rarely used in the Arctic (with the possible exception of use by the military). However, they are able to reach ecosystems and impact human populations far-distant from the areas where they have been used and released to the environment. The ability of POPs and methylmercury to aggressively biomagnify means that even in remote locations, such as the Arctic, extremely high concentrations can be built up in top predators and human populations, particularly those feeding primarily on certain species of fish and fish-eating animals.
- POPs and methylmercury are chronically toxic. Therefore, their potential for serious health outcomes is not immediately obvious and may not even appear until a generation has passed. Growing evidence supports the view that one of the most insidious impacts is mediated through prenatal exposure to the embryo that increases the probability of lasting neurobehavioural effects.
- In some Arctic communities, particularly those that heavily depend on the marine environment for their diet, human exposure had by the 1990s reached levels categorized by health authorities as indicating "concern". It was and is a devastating situation for the indigenous communities involved. Their traditional diet – the core of their social, cultural and biological well-being – has been invisibly compromised. They have played no role whatsoever in the cause and they have

no prospect for quick remedial action that does not involve other unwanted health and cultural consequences.

- With the Arctic lessons of POPs and mercury so recently given, have our national and international chemical regulatory agencies learned that prevention is better (and cheaper) than a cure? The answer is perhaps the saddest part of the story. Despite the knowledge that persistent biomagnifying substances are reaping their chronic toxicological toll inside and outside the Arctic, there is not a great deal of evidence outside the EU to show that agencies responsible for evaluating the safety and regulating the use and disposal of new substances (including pharmaceuticals) are taking into account the properties of POPs and of endocrine disruption.

- Governments can work cooperatively to address a problem once it has occurred. We now have the regional POPs and heavy metals protocols (under the CLRTAP) and the global conventions on POPs and mercury (Stockholm and Minamata, respectively). However, once again, initial progress was painfully slow and involved more activity from the "denial" lobby groups. Oreskes and Conway have shown that this opposition was still rumbling on – complete with posthumous attacks on Rachel Carson – well into the first decade of the present century. The time period that elapsed between alerting the CLRTAP to Arctic concerns on POPs and the signing of the POPs protocol was eight years (1990–1998). The UNEP Governing Council initiated the global process in 1995, but the resulting Stockholm Convention was not signed until 2001. The Barrow Declaration of the Arctic Council in 2000 asked UNEP to consider global controls on mercury, but the Minamata Convention was only agreed to in January 2013 and had only one ratification by September 2014.

9

Conducting Marine Science in the Arctic

Over the last 30–40 years, all branches of science have advanced at an amazing rate. The common theme behind this trend is the continual emergence of new technologies. They make it possible for scientists to pursue questions today that were impossible to address or to even ask only a few years ago. Arctic oceanography is no exception to this trend, but it also has another dimension. It is now much more feasible to work in the Arctic at any time of the year, although it is still not exactly easy.

It was November 1977. Fresh from graduate school and woefully inexperienced, I was in charge of an oceanographic survey in Davis Strait (between Baffin Island and Greenland). I joined the ship at the small settlement of Frobisher Bay on Baffin Island. Loading supplies took much longer than planned. After this date, ice conditions dictated that any resupply in the Canadian Arctic must wait until early July. The alternative was Greenland or Newfoundland. There was no jetty in the town and the flat-bottomed boats that usually moved cargo from ocean-going ships had been hauled onto the beach for the winter. They resembled old military landing craft. Perhaps that was their provenance. This time, everything came out by freighter canoe and the weather was grim. After a couple of days, we were almost ready to go. That evening, most of the scientific crew went ashore for one last meal of vaguely fresh vegetables. It was, of course, dark or twilight, snowing and blowing a gale and the little plates of pancake ice on the water seemed to be whispering something about hypothermia. It was no surprise that the freighter canoe took a long time to find us. I was on the bridge watching the canoe trying to come safely alongside when the skipper, always a man of few words, turned to me and pronounced: "If you don't have to go, don't go." It was the first of his many edifying maxims to come my way.

The following morning, we made our way south down the full 230-kilometre length of Frobisher Bay. The mysteriously named Meta Incognita Peninsula bounds this wide inlet on its western flank. I had noticed on the chart that halfway down the peninsula was Wynne-Edwards Bay. Vero Wynne-Edwards was my zoology professor at Aberdeen. He was a keenly intelligent but very modest man. Despite his responsibilities as the head of the department and chair of the UK's powerful Natural Environment Research Council, he was often available to us students. Periodically, we would ask him to give us extracurricular lectures and I remember him showing an old lantern slide of himself on the deck of a sailing boat somewhere in the Eastern Arctic.

In the 1960s and 1970s, he was fascinated by the widespread occurrence of behaviours in the animal kingdom that appeared to resemble altruism. He argued that animal populations with behaviours that prevented them from exhausting their resources would be more likely to succeed in raising offspring for future generations than those lacking such behaviour. Natural selection would be operating on groups rather than on individuals. However, this so-called "group selection" quickly ran into troubled waters. It was hard to show how natural selection could operate on groups to promote apparently altruistic behaviour. Any individual whose genes encouraged it to cheat and take more food would produce the most numerous offspring. There it rested until William Hamilton introduced a theory of "kin selection", in which he pointed out that close relatives (kin) share many of the same genes. Therefore, if you possess a gene that promotes an apparently altruistic behaviour, such behaviour may help transfer many of your genes into the next generation, even though this has been achieved through the agency of your close kin rather than yourself. This type of approach provided a way to explain the evolutionary advantages of altruism without taking on board the problems associated with the theory of group selection. Robert Wright gives an easily understood description of how it may work. More technical explanations for altruistic behaviour can be found in the books by Richard Dawkins. (See the bibliography.)

Before the European colonialists arrived in North America, the indigenous peoples had names for the local features of their landscape, but by the 1950s, few of them appeared on the map. More recently, there has been a revival of the original names or of names that have a meaning in an indigenous language. Baffin Island and, therefore, the inlet of Frobisher Bay are close to the eastern boundary of the huge northern territory of Nunavut, which is roughly the size of Western Europe. Most people living in Nunavut are Inuit and speak Inuktitut.

It is therefore not surprising that there has been a high priority given to the restoration of Inuit place-names. The settlement of Frobisher Bay I mentioned earlier is now called Iqaluit. In this context, it is ironic about Wynne-Edwards Bay. I feel sure that if my Aberdeen professor were still alive, he would be an enthusiastic champion of the move to restore traditional place-names.

I should say a few words about the ship. At that time (late 1977–1980), the fleet of icebreakers operated by the Canadian Coast Guard was used in winter exclusively to support shipping off Labrador and in the St. Lawrence Seaway and Great Lakes. The oceanographic fleet was operated by the Department of Fisheries and Oceans, but it included no vessel with anything other than light ice strengthening. As a result, there was a great deal of mystery about how ocean waters and biota behaved around Arctic Canada in winter. However, at this time, the OPEC-driven oil crisis spurred intense interest in exploring for hydrocarbons in the Davis Strait and Baffin Bay. It was no longer an option for Canada to ignore the physical, chemical and biological ocean-ography of these waters. There was only one type of vessel instantly available to operate in pack ice and survive if frozen in and subjected to ice pressure. These were ships built mainly in Norway in the 1930s to support the seal fishery. They were normally a little less than 60 metres (200 feet) long, had high endurance (could operate for long periods between refuelling) and were very heavily ice strengthened. They were not icebreakers but could safely nudge their way though pack ice and just "wait it out" if they became frozen in. Most importantly, they featured round-bottomed hulls, enabling them to rise above the ice when under pressure from the ice field. This was the big advance in Arctic ship design made by Fridtjof Nansen in the construction of the *Fram* at the end of the nineteenth century. Without this type of hull, ships under prolonged pressure were crushed despite how strongly they were constructed. The most famous example of the cost of ignoring the lesson of the *Fram* was the fate of Shackleton's *Endurance* in the Antarctic winter of 1915.

Ours was one of two of these tough little ships that was quickly outfitted to provide a winter science platform. Hydrographic winches to lower instruments over the side, rudimentary wet and dry laboratories and extra accommodation were installed. A wooden helicopter pad was built over the stern for emergency evacuations. Most of the cabins and work spaces had no heating, although some were fortuitously located near the engine room's cooling and exhaust pipes. Warmth at night came from using the best sleeping bags money could buy supplemented by sleeping in all your clothes. The mess could only accommodate six at a time seated at a single table. Between meals, that table was our office and lounge. If you missed your mealtime spot, you risked going hungry, as the galley had no facility to reheat or store cooked food.

In the 1930s, it had served as a research ship for the Norsk Polar Institute. Embossed in large letters below the bridge was its original name. After World War II, it worked in the Barents Sea seal fishery and around Svalbard and eventually found its way to the spring seal fishery off Newfoundland. I had one of the two most "comfortable" cabins on the ship, facing forward and directly below the bridge. Because of its rugged specifications, it was often in demand and had spent several years ferrying supplies from Cape Town to Antarctica as part of the 1957 International Geophysical Year. Pinned to the wardrobe in my cabin was a telex from Nikita Khrushchev thanking the crew for its good work.

It carried a crew familiar with winter ice conditions off Labrador and a wonderfully experienced skipper who had worked on similar vessels all his life. He could predict the nature of approaching ice (before we could see it) just by looking at its reflection in the sky or clouds. The day after leaving Iqaluit, we entered Hudson Strait and began our methodological transects back and forth across Davis Strait. On arrival off the Greenland Coast, we would move north for 100 kilometres, head east until we were back off Baffin and then repeat the pattern, moving ever northwards. This type of sampling design produces what is called a *synoptic survey* and is a necessary step if you wish to describe the characteristics of a given piece of the ocean. At regular intervals, we would stop at a "station", make physical and chemical measurements from the surface to the bottom and collect plant and animal samples from the water column to the seabed. Also at intervals, we would set up arrays of current meters. Getting the heavy railcar wheels to which the meter arrays were anchored over the side on a heaving icy deck was always a hazardous adventure. I was rapidly learning how different this would be to working on purpose-built oceanographic ships. We were equipped with a very early satellite navigation system (GPS) to record exactly where we had left these lonely arrays. Months later, we would return and try to recover them. Once we were convinced we were onsite, an acoustic signal would be sent down, which in theory would release the long array of current meters from the railcar wheel. At the upper end of the array was a brightly coloured float (with a flashing light) that would bring it to the surface. We would often spend more than a day searching for the float. The sea was usually rough and the float could be in a wave trough and invisible for much of the time. For half the year, it was at best twilight and we never encountered a float with the light functioning. Over the course of three years, I think we lost about 50% of the current meter arrays, mostly towards Baffin Island, where they were probably swept away by icebergs.

The weather was a constant problem on that first cruise. It was, of course, winter. The cold Baffin current flows south along the eastern coast of Baffin, bringing with it a growing ice pack at that time of year. Beginning at the coast, the ice consolidates into a continuous field that grows eastwards as winter deepens. It became increasingly difficult to get anywhere near the Baffin coast. The west coast of Greenland has a north-flowing and relatively warm current (an offshoot of a Gulf Stream eddy), but this gave other problems. Away from the ice, the seas were usually rough, making it quite hazardous handling heavy oceanographic equipment on the deck. The worst conditions occurred where the cold western waters rubbed shoulders with the warm waters flowing north in the east. The ship was always doing its best to imitate an Olympic swimmer doing the breaststroke. The air temperature was below freezing and the flying spray and seas breaking over the bow could very quickly encrust everything on deck with a heavy shroud of ice. You could distinctly sense the effect on the ship's stability as it sluggishly recovered from each roll. In these conditions, the skipper never countenanced a suggestion of stopping for a station. While every one of us attacked the ship's growing cloak of ice, we would skedaddle as fast as possible to the cold, calmer water of the ice pack or the warmer waters off Greenland.

The most fundamental set of information sought by a physical oceanographer is a vertical profile of the water's temperature and salinity. These two parameters control the density of the water and a unique combination of temperature and salinity is characteristic of water masses formed in specific locations and conditions. Once you have a horizontal picture of the vertical distribution of temperature and salinity (and hence of density) over a particular geographical area, you can decide how to organise your chemical, biological and current meter sampling strategy. One of the seminal technological revolutions taking place in the 1970s related to the measurement of temperature and salinity. At that time, we still largely relied on reversing thermometers to measure temperature and the postcruise determination of water conductivity to derive salinity. It worked like this: You would lower a string of water bottles open at each end until each bottle would eventually be at a preselected depth. Then, you would send a weight called a *messenger* down the wire, which would close the top bottle and release another messenger down to the next bottle. When each bottle closed, a reversing thermometer attached to the bottle would flip upside down and the mercury column would break. Once this had happened, the reading on the thermometers would not change when the bottles were brought back to the surface.

These thermometers were expensive and very temperamental. After reading the temperature, you had to carefully turn the thermometer to its original "upright" position in order to reconnect the mercury column. The thermometers objected to this being done at Arctic air temperature, they did not like to be quickly brought into a warm room, and they did not appreciate vibration. In other words, they were not happy on our ship. I found that the ideal spot for this delicate operation was in the doorway between the galley and the mess and a little forward of the engine room access. Having gone to all this trouble, you still did not know the salinity until you returned to your home laboratory. You never knew the exact depth of the sample because you had no idea how the wire angled in response to the ship's drifting and the currents below.

The path-breaking new technology on our cruise was a CTD (for conductivity, temperature and depth). These instruments started to become generally available in the early 1970s. They were lowered down to whatever depth you wished and they continually measured temperature and conductivity as they descended. Conductivity of seawater primarily depends on salinity, which makes the conversion fairly simple. The early versions of CTDs recorded internally on magnetic tape you could read later if you had the necessary equipment. We had one of these and a more advanced model that somewhat resembled those used today. The sensors on this later version sent their measurements up an electrical cable, and in theory, you could access the vertical profile of conductivity and temperature while still on station. It was a tremendous advance because the profile was continuous (not just a set of points determined by where you had decided to place the bottles). Furthermore (again in theory), there was the potential to almost instantly have the data. You could change your sampling strategy as you went along. Note that I said "in theory". These were early days and the electronics were designed for use on purpose-built oceanographic ships with carefully regulated electrical systems. The voltage on board our trusted vessel varied dramatically by the minute. It was so unpredictable that it sometimes fried the portable voltage regulators that we had brought to protect the entire CTD system! We suffered so many breakdowns that we still relied heavily on water samples analysed months later in the laboratory.

Of course, we were not just interested in salinity and temperature. The measurement of other parameters was approached by collecting samples, such as dissolved oxygen, plant nutrients, chlorophyll (as an indicator of phytoplankton biomass), benthic biota and plankton, that were then "fixed" for analysis ashore. As a result, you had very limited notions of the significance of the cruise until weeks or months after return. It was only then that you knew what you should have done and where you should have done it. Enlightenment during the cruise could have enabled changes to activities, which would have greatly increased the value of the whole exercise.

In the winter, all the northern Canadian Coast Guard operations shut down and we had to rely on Greenland stations for radio communications. The only radio channel we could get often was on the Greenland emergency frequency. Consequently, we were generally out of communication with the rest of the world for six or seven days. One characteristic of the ship was that obsolete equipment seemed to have never been removed. It was a veritable maritime museum. Behind the bridge was the radio room, which years earlier would have housed a radio operator. The equipment was still there, including a radio that was so large that a small person could have climbed through the access door. Inside were the largest valves (tubes) I have ever seen. It still worked and the skipper sometimes coaxed it into life because it was the only set that was guaranteed to find a radio station somewhere. The equipment was by this time illegal, so we could never use it to send a message.

Our refuelling and crew changes in winter were all done through Greenland because it was impossible to break through the ice to a Canadian port. It was quite an undertaking to leave or join the ship after early November. It involved a helicopter transfer from Godthab (now Nuuk) to the (then) U.S. military base at Sondrestrom (now Kangerlussuaq). The helicopters strictly operated on visual flight rules, so you faced a high probability of cancellation. The SAS (Scandinavian Airlines) flights from Copenhagen came twice a week, but Sondrestrom was itself prone to fog, leading to more cancellations. Finally, there was the trip back to Canada.

Probably the most ambitious work we attempted at that time was to measure the productivity of the epontic flora that lives on the bottom of sea ice in spring. This flora is the first major source of new food for pelagic and zooplanktonic organisms in springtime. It is therefore a very basic feature of the Arctic marine ecosystem. Measurements had been conducted by others (such as Rita Horner and Vera Alexander) in shore-fast ice off Alaska. They used divers who isolated an area of the flora in situ and injected carbon-14 into the water contained in the isolated sample. They then recovered the entire sample and measured how much of the isotope had been incorporated into the epontic flora over the sample period. The rate of epontic primary production was calculated from this rate of isotope uptake. For safety reasons, we were prohibited from following the same procedure in our far-offshore location of mobile pack ice. Tom Windeyer (a colleague) designed and built a flexible cantilevered arm that we lowered through a hole drilled in the ice. It could isolate and pluck epontic samples from the ice and then move them to the surface. In order to see what we were doing, we used the longest veterinary fibre-optic scope we could find. Tom's instrument was also used to deploy a light meter to measure the under-ice light regime. On the foredeck cargo doors, another colleague, Jacquie Booth, set up an array of cells in which the samples were incubated for a period of time after the introduction of carbon-14. Light filters were used to mimic the under-ice light environment. It was not the best way to do things, but it was remarkable that the results were in line with those from Alaska and supported the growing appreciation of the important role of epontic production in the Arctic marine food chain.

On one occasion in late April, we were using this gear close to the ice edge when the skipper ordered us all back on the ship. Visibility was failing. A rising wind was catching the ship and driving it and our attached ice floe towards open water. A radar deflector was quickly set up to help us find and return to the same floe if the weather improved. We had noticed a polar bear several times through the haze that afternoon, so the last items of gear to be hauled on board were the gun and ammunition. David Gillis (the team member on whom I relied most on that cruise) and I then released the anchor lines that held the ship to the floe. However, the wind was so strong that we had no time to climb on board before the ship was blown away from the floe. Several attempts were then made to reach us, but each time, the strengthening wind kept us apart. Finally, it came in with such force that cracks appeared all over the place and the floe split in half. For a few seconds, David and I were separated until he jumped across the widening gulf. After a few minutes, we noticed that the ship had fallen silent and was slowly drifting off. Someone came to the stern and yelled "Engines quit!" It was not good news. We knew that the ship itself was their only option to reach us and our floe was now sailing off into the increasingly rough open water. The ship soon vanished into the chill mist. We began to wonder about that bear, the new cracks beneath our feet and the waves that were starting to upset the equilibrium of our diminishing floe.

Fortunately, David had been wise to join my part of the floe because it had the radar reflector on it. Although we could not see the ship, they knew exactly where we were. After some hours of worsening weather and further loss of ice, that wonderfully handsome profile of our ship emerged and picked us up at the first attempt. The next morning was calm and had better visibility. The ship had drifted far back into the pack ice. Less than 500 metres away was our radar reflector. It was stoically sitting on a much smaller floe than the one abandoned the previous evening.

Ships were not the only floating platforms for conducting Arctic oceanography. Ice sheets mainly on the north coast of Ellesmere Island periodically shed fragments. Their size was measured in miles or kilometres. Some drifted into the circumpolar circulation. The most famous was designated as T3 (Fletcher's Island) and intermittently carried a research station until 1983, when it exited via Fram Strait into the Atlantic. They were extremely stable platforms, but, of course, there was no control over where they went. A year before the loss of T3, a large part of the Ward Hunt Ice Shelf disintegrated. On one of the resulting fragments, the Canadian Ice Island Research Station, or Hobson's Choice, was set up. George Hobson was a neighbour in Manotick, south of Ottawa, Ontario, and as director of Canada's Polar Continental Shelf Project (PCSP), he was a very important person to know. He controlled the logistics used by most scientists to physically travel, work and survive in the Canadian Arctic. The research station was operational from 1985 until the early 1990s. Marty Bergmann was a later director of the PCSP.

Unfortunately, the 24-square-kilometre Hobson's Island never entered the circumpolar circulation, and after drifting very slowly westward, it eventually exited the Arctic Ocean through the Arctic Archipelago. I visited the island in February 1990, arriving (as did everyone else) by a Twin Otter on skis. At the time, it was home to several projects in our quest to understand the cold condensation processes by which POPs reach the Arctic and then biomagnify (described in the previous chapter). To me, the island resembled a fortress with a high defensive wall. The surrounding multiyear sea ice was reticulated by pressure ridges caused by ice floes pressing against each other in a fashion resembling the creation of the Himalayas through plate tectonics. It was 50 metres thick. Therefore, ice floes under pressure at the perimeter of the island were simply thrown up haphazardly against this solid mass to build the "fortress wall". It is unlikely that we will see large floating ice island-based research stations again. The ice shelves from which they were born are rapidly disappearing in response to global warming.

In 1995, Canada still had no winter-capable Arctic research vessel. Canadian scientists looked with envy at the polar-capable ships of other countries, such as the amazingly equipped and resilient *Polarstern* operated by the Alfred Wegener Institute in Germany. However, in that year, the Coast Guard and oceanographic fleets came under joint management within the Department of Fisheries and Oceans. Gradually, Coast Guard ships began to be used as scientific platforms – some of which ventured to the Arctic in winter. Then, in June 2002, a consortium of Canadian universities and federal agencies produced a plan to convert the decommissioned icebreaker *Sir John Franklin* into a truly dedicated Arctic-capable research vessel. The proposal was accepted by the International Joint Ventures Fund of the Canada Foundation for Innovation (CFI). With the help of similar contributions from other agencies (particularly the Department of Fisheries and Oceans), the ship's transformation took place. In 2004, it reentered service as the research icebreaker *Amundsen* in honour of the famous Norwegian, Roald Amundsen. As a general rule, it is available for scientific research from May to December and spends the rest of the year on icebreaking duty in the St. Lawrence.

The people who recognized the quiescent opportunity of the decommissioned *Franklin*, connected this vision to potential funding mechanisms and then pursued their dream should be justly proud of their efforts. They were responsible for one of the two initiatives that have led to a renaissance in Canadian Arctic environmental science at a time when government investment in such activities has been shrinking. The other initiative has been ArcticNet, launched in 2003. ArcticNet is a large research network funded by the Canadian government through the Networks of Centres of Excellence (NCE) programme. ArcticNet's executive director, Martin Fortier, has developed a very effective management strategy that identifies, evaluates and addresses information needs for northern decision making. The science is organised in a way that depends on the benefits of interdisciplinary studies being able to cross-fertilize each other. The key has been the concept of the Integrated Regional Impact Study (IRIS). There are four IRISs – each dedicated to a different region of Arctic Canada (Western and Central Arctic; Eastern Arctic; Eastern Subarctic; and Hudson Bay) and each focused on the consequences of environmental change to society and ecosystems.

What is it like to work on an icebreaking Arctic research vessel, such as the *Amundsen*, in the twenty-first century? The last time I sailed on an icebreaker in ice was on a sister ship to the *Sir John Franklin* in the mid-1980s in Lancaster Sound. I have only been on the *Amundsen* when it was in port. It is capable of maintaining a speed of 3 knots in continuous ice that is a metre thick. If the ice is thicker, it can back up and ram its way through. Therefore, under most conditions, it is able to go wherever scientists want to be taken. It carries a "resident" helicopter that further extends the working range of the ship and which can carry equipment to temporary scientific camps. Larger helicopters can land on the helicopter pad if needed and assist, for example, with crew changes. There are comfortable accommodations for more than 40 scientists. In addition to the plethora of onboard laboratories, there is a cafeteria, dining room, lounges and plenty of hot showers! The ship is more or less in constant communication with the outside world and you can even send e-mails. GPS enables pinpoint navigation. On deck, there is a bewildering array of winches, cranes and A-frames for handling heavy gear. There is a moon pool that enables sampling gear to be lowered through a hole inside the ship, removing the need to be constantly pushing ice aside when trying to lower or retrieve water samplers. There are launches, barges and vehicles (to help move around on continuous ice).

With all this sophistication and apparent onboard comfort, it is easy to forget that any oceanographic work in polar regions is hazardous. The *Amundsen* and the *Polarstern* have endured fatal helicopter accidents in recent years. Excluding these two incidents, I have personally known five scientists who were killed while working in the Canadian sector of the Arctic.

Let us imagine repeating our 1977 winter cruises in Davis Strait using the *Amundsen*. The standard physical and chemical oceanographic survey uses a CTD mounted inside a rosette of sampling bottles. The lowering of the whole rosette assembly is managed in a CTD control room. The operator has the option of real-time monitoring for temperature and salinity as the rosette descends and can remotely control the depth at which each water bottle takes its sample. Other sensors may be included on the rosette, enabling, for example, the measurement of dissolved oxygen and chlorophyll fluorescence. There is an onboard salinometer that can be used for instant calibration of the CTD. While someone is taking samples, there are no worries as to whether the ship had drifted off station because it is fitted with a dynamic positioning system. The phytoplankton nutrient analyses can be completed with an autoanalyser in one of the many specialist labs on board. In the absence of ice, you could even consider using a moving vessel profiler (MVP). This device is towed behind the ship and can carry a variety of sensors, including a CTD and a fluorometer (for chlorophyll measurements). While being towed, it can maintain a sequence of falling to a determined depth (or until it senses the bottom) and returning to the surface.

Plankton sampling may be undertaken using a multinet system, enabling the samples to be collected from preset depth ranges. Taxonomic analyses can be conducted on board with the aid of computerised reference material. It is possible for primary production studies to be completed on the ship. There are special facilities available for trace metal analyses and a portable laboratory for mercury studies. Current meters resembling those used in the 1970s are still a modern-day workhorse, but other options are available, such as those that use acoustic Doppler profiling.

However, there is a growing problem with modern deep-sea oceanographic ships wherever they may be operating. They are very costly to run. A 70–75-metre-long non-Arctic ship costs about US$36,000 per day. It is more for an icebreaker. Scientists are often reluctant to prepare proposals that involve such ships, and consequently, many of the present vessels are underutilized. To further aggravate the situation, most of the world's fleet is now elderly and in need of replacement. At the same time, there is the impact of the technological revolution that has transformed oceanography. We glimpsed this on our tour of the *Amundsen*, which was equipped with an MVP. The pivotal year is said to have been 1978, which saw the launch of *Seasat*, the first satellite to be devoted to remote sensing of the ocean. It unfortunately had a premature end, but in its three months of operation, it is said to have collected as much data on sea surface temperature, wind speed and ice conditions as had been collected by all ships in the preceding 100 years.

After the advent of *Seasat*, more and more of the data used by oceanographers today do not depend on the use of a deep-sea oceanographic vessel. The capacity of modern remote sensing from satellites is quite astonishing. For example, the Moderate Resolution Imaging Spectroradiometer (MODIS) on NASA's *Aqua* satellite has provided monthly maps of global surface chlorophyll concentrations, which can be used to infer abundance of phytoplankton. The same satellite provides global maps of sea surface temperature, while NASA's *Aquarius* mission has been doing the same for surface salinity. *Aquarius* also carries the Advanced Microwave Scanning Radiometer (AMSR), which can measure differences in microwaves emitted from sea ice and water to provide maps of ice concentration.

The technological revolution is not just confined to space. It is also in and on the water. Probably the best-known example of an ocean-going sampling system is the *Argo* programme. Since 2004, 3,500 free-drifting Argo floats have been measuring temperature, pressure and salinity. They do so as they slowly rise and sink over a 10-day cycle. When at the surface, they send a package of data to an orbiting satellite. The Argo "fleet" collects about 120,000 vertical profiles a year! The next generation of Argo floats will even have the capacity to measure phytoplankton primary production.

A somewhat related development has been the use of "gliders". These resemble a fat surfboard with several antennae and sensors on top. They can move following directions given by an operator ashore and many have no fuel requirements. Power comes from a set of solar cells attached to the upper surface and from wave energy. A payload of sensors is carried mainly on the lower surface. They can measure, for example, sea surface temperature and salinity at subsurface depths just beneath the level that is "sampled" by orbiting satellites. Data are again downloaded to a passing satellite. Today, these platforms frequently work in association with satellites equipped to measure sea temperature and salinity in studies designed to investigate the global water cycle and its linkages to climate change.

Of course, ship-based observations remain essential, particularly in the Arctic, but the previous examples will have given a taste of the exciting opportunities that now exist. In practice, the new autonomous systems (such as gliders) often work together with a ship-based programme.

To sum it all up, if Fridtjof Nansen (from the third *Fram* expedition to the Arctic between 1893 and 1896) could have been transported through time to join us in Davis Strait in 1977, he would have been thoroughly familiar with a good deal of our equipment. This included the reversing thermometers. After all, he designed the water-sampling bottle we were using (appropriately called *Nansen bottles*). On my bookshelves, I have an account of the 1925–1927 Antarctic Ocean expedition of the research ship *Discovery*, written by Sir Alister Hardy. The crew of the *Discovery* would have recognized perhaps as much as 50% of our gear in 1977. The pace of oceanographic technology development had been so slow that Nansen and Hardy would have quickly become perfectly competent members of an oceanographic team in the 1960s and 1970s. However, if oceanographers could magically be moved from that time to a modern oceanographic vessel, they would be utterly bemused. The microchip revolution has changed everything. They may indeed find there is no need to go to sea to do their work!

10

Climate Change in the Arctic

"There in the north where ice, water and air mingle is, without doubt, the end of the earth. There I have seen the lung of the sea."

Pytheas, fourth century BCE

Until the mid-nineteenth century, the Great Powers were obsessed with discovering whether ships could pass from the Atlantic to the Pacific and vice versa through or north of North America. In pursuit of this obsession, many ships and lives were lost in the eighteenth and nineteenth centuries. Included were the entire crew of the famous Franklin expedition that sailed into oblivion in 1845. It was the unsuccessful search for Franklin, launched from the Atlantic and the Pacific, that essentially led to the discovery and mapping of the Northwest Passage, but it was not until 1906 that a ship managed a complete transit. That honour went to the Norwegian Roald Amundsen in the little sloop *Gjøa*. His three-year voyage was completed in 1906, but the lesson learned was that even at the most favourable time of year, the passage had no commercial potential. As late as the 1970s, a transit could only be attempted by a heavy icebreaker in late summer or with the constant assistance of one or more icebreakers (as was the case with the experimental transit made by the tanker *SS Manhattan* in 1969). The single exception came when nuclear submarines began to patrol the Arctic Ocean.

You can imagine my excitement when, in September 1979, I sailed to the Northwest Passage on the maiden voyage of the icebreaker *Kigoriak*. Its design owed much to features developed in Finland that were quite revolutionary for Canada. It was built on the Atlantic coast in New Brunswick but was to operate in the Beaufort Sea. At exactly the same time but coming in the other direction was the more traditionally designed *Sir John Franklin* – also on its maiden voyage. Impassable multiyear ice blocked the M'Clure Strait and choked the upper part of the Prince of Wales Strait between the Banks and Victoria islands. Travelling on an icebreaker in heavy ice is quite an experience. Depending on where your cabin is, the noise can make you feel you are in a steel barrel being rolled down an uneven flight of stairs. It is impossible to predict the motion of the ship. Icebreakers do not cut their way through ice but ride up on top and break it with their own weight. The *Kigoriak* made it through the Prince of Wales Strait to the Beaufort Sea unscathed, but a propeller of the *Sir John Franklin* was knocked out of action and it had to be escorted to safety.

The crews on the *Kigoriak* and the *Sir John Franklin* could never have imagined that only 30 years later, summer cruise ships would sail unescorted through the Prince of Wales Strait. Historically, the geographical extent of Arctic sea ice expanded in the winter and retreated in summer, leaving a core of multiyear ice roughly at the highest latitudes. Since 1979, which coincidentally was when satellite observations became available, the size of the seasonal sea ice minimum has been declining at a rate of about 13% per decade and much less ice now survives for more than one or two years. To be technically correct and using the terminology of the National Snow and Ice Data Centre (NSIDC), I should say that the linear trend of decline since 1979 relative to the 1981–2010 average was –13.7% in 2013. Scientists refer to the difference between a past mean and a present mean for a particular parameter (such as mean global temperature) as an anomaly. The anomaly relative to the 1981–2010 average was –44% in 2012. Current projections tell us we can expect to see an Arctic Ocean without summer ice somewhere between 2030 and 2050. (See Figure 10.1.)

Why is the extent of summer sea ice receding? The answer is quite simple. Our world is getting warmer. In the words of the fifth assessment (AR5) of the Intergovernmental Panel on Climate Change (IPCC): "Warming of the climate system is unequivocal, and since the 1950s, many of the observed changes are unprecedented over decades to millennia. The atmosphere and ocean have warmed, the amounts of snow and ice have diminished, sea level has risen and the concentrations of greenhouse gases (GHGs) have increased."[1] The same

[1] IPCC, 2013. "Summary for Policymakers." In Stocker et al. (eds.), *Climate Change 2013: The Physical Science Basis*, p.4. Cambridge, UK, and New York: Cambridge University Press.

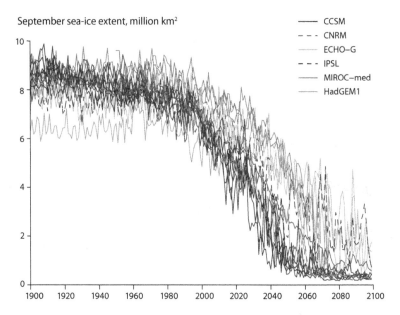

September sea-ice extent, million km²

— CCSM
− − − CNRM
⋯⋯ ECHO–G
− − − IPSL
— MIROC–med
— HadGEM1

Figure 10.1 *Northern Hemisphere September sea ice extent as simulated by six models using the IPCC A1B emission scenario. The results show the chaotic nature of natural variability and the long-term trend (decline) due to anthropogenic forcing.*

assessment reported: "Each of the last three decades has been successively warmer at the earth's surface than at any preceding decade since 1850. In the Northern Hemisphere, 1983–2012 was likely the warmest 30-year period of the last 1400 years (medium confidence)." Furthermore, "the globally average combined land and ocean surface temperature data as calculated by a linear trend, show a warming of 0.85°C over the period 1880–2012 when multiple independently produced datasets exist".[2]

However, the Arctic is warming much faster than almost anywhere else on the planet – at about twice the global rate. It is a key point we will return to again and again in this chapter. The difference between a temperature range that supports the survival of summer ice (as the *Kigoriak* experienced in 1979) and one that does not (as seen today) is small. (See Figure 10.2.)

To understand why the Arctic is warming, we must first take a look at global climate change. One of the big steps forward in the IPCC

---

[2] IPCC, 2013. "Summary for Policymakers." In Stocker et al. (eds.), *Climate Change 2013: The Physical Science Basis*, p. 5. Cambridge, UK, and New York: Cambridge University Press.

Surface temperature anomaly 2005–2007 vs 1951–1980

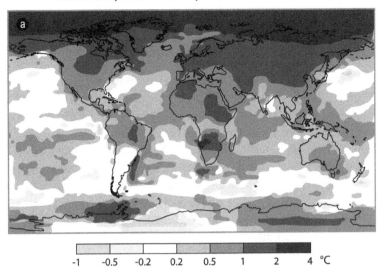

Zonal mean temperature anomaly, °C

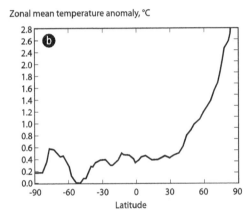

Figure 10.2 *Global warming: The map (a) shows the surface temperature anomaly 2005–2007 relative to 1951–1980. The graph (b) presents the same data as zonal means. Note that Northern Hemisphere high latitudes and particularly the Arctic have experienced the greatest degree of warming.*

assessment process has been the increasing confidence with which it is judged that the observed increase in global surface temperature is due to anthropogenic GHG emissions. Here the IPCC is addressing what is known as the "attribution problem": "It is *extremely likely* that more than half of the observed increase in global average surface temperature from 1951 to 2010 was caused by the anthropogenic increase in GHG concentrations and other anthropogenic forcings together. The best estimate of the human-induced contribution to warming is similar to the observed warming over this period."[3]

To understand the scientific "mechanics" that underline this statement, we need to take a look at GHGs and be familiar with their role in maintaining or changing the world's climate.

The sun is very hot and consequently emits energy (short wavelength radiation – visible light). This radiation is absorbed at Earth's surface, which is consequently warmed. The warmed surface then emits long wavelength infrared radiation back into space in a way that is described by the Stefan-Boltzmann law. That is a relationship that most readers have probably not thought about since their last school examination in physics. However, it is easy to follow – even for a biologist such as me. In an imaginary case of a naked Earth without an atmosphere, the Stefan-Boltzmann relation simply means that outgoing infrared energy loss to space takes place at a rate that depends on temperature. This is very important because it means that the warmer a naked Earth is, the faster it loses heat until a balance is achieved between the incoming and outgoing movement of energy. If Earth did not have reflection (albedo) or an atmosphere, it would have a surface temperature of –18°C.

Of course, the average temperature of Earth's surface is much warmer than –18°C because Earth has an atmosphere. Some of these atmospheric gases capture much of the outgoing infrared radiation and only about 38% is lost to space. The remaining energy is recycled back to warm Earth and its atmosphere, giving us our more familiar average global surface temperature of about +14°C. This is known as the *greenhouse effect* and the responsible atmospheric gases are known as *greenhouse gases* (GHGs). The most important are:

- Carbon dioxide (CO_2).
- Methane (CH_4).
- Nitrous oxide (N_2O).
- Sulphur hexafluoride(SF_6).
- Hydrofluorocarbons (HFCs).
- Perfluorocarbons.[4]

In addition to concentrations of a GHG in the atmosphere, two other factors are also important: their relative atmospheric lifetimes

---

[3] IPCC, 2013. "Summary for Policymakers." In Stocker et al. (eds.), *Climate Change 2013: The Physical Science Basis*, p. 17. Cambridge, UK, and New York: Cambridge University Press.
[4] The last three of these substances do not occur naturally.

and their efficiency as GHGs. Together, these define their global warming potential (GWP) relative to CO_2. The most common GHG in the atmosphere is water vapour. However, it is not thought of as a forcing gas in the context of global warming (and therefore not included in the list) because it has an atmospheric lifetime of only about 10 days and, furthermore, its atmospheric concentration is a direct function of temperature. Nevertheless, it is a very important player in the warming mechanism.

By far, the most important GHG in terms of the combination of atmospheric concentration and GWP is carbon dioxide (CO_2). Of course, our atmosphere has contained CO_2 for billions of years without any anthropogenic input. The question, therefore, is what is unusual about events at the present time? The answer is that humankind is upsetting the way in which our planet maintains carbon in a number of linked "reservoirs". Very briefly, these reservoirs include the atmosphere, the oceans, the biosphere and the lithosphere (or outer and solid layer of Earth's crust). Carbon moves between them in a cyclic fashion, such that CO_2 can dissolve into the oceans and be incorporated into organic matter, sediments and, ultimately, rocks. In the course of geological time, these rocks will be thrust above the sea. Subsequent erosion leads to the carbon finding its way back into the atmosphere by, for example, biological uptake followed by respiration. Under certain conditions of heat and pressure, sediments containing large accumulations of organic matter give rise to carbon-rich products that are collectively called *fossil fuels*, such as oil, gas and coal. When we burn them, we are short-circuiting the carbon cycle.

The reservoirs vary in their capacity to hold CO_2. Most CO_2 is contained in the lithosphere. The transfer rate (flux) between reservoirs also differs. For example, fluxes into and out of the lithosphere normally run at a snail's pace. Prior to the onset of the Industrial Revolution (about 1750), the dynamics of this system had settled for a very long time into a status quo, where the atmospheric concentration of CO_2 was held at about 280 parts per million (ppm). It was this concentration that maintained our historical average global surface temperature of about +14°C.

The situation that has evolved since the 1750s is that the engine of the Industrial Revolution and of our present-day world economy runs on fossil fuels. Inadvertently, we have found a way to rapidly move massive quantities of carbon stored in the lithospheric reservoir into the atmosphere. We do it every time we burn fossil fuels. Our ability to release CO_2 into the atmosphere is much greater than the ability of natural

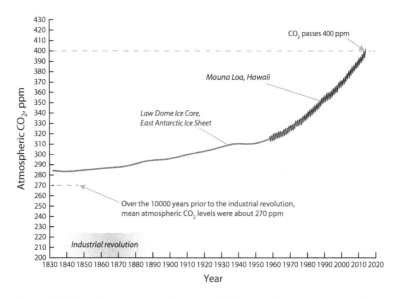

Figure 10.3 *Trends in atmospheric CO_2 since 1830 based on Law Dome ice core and Mauna Loa observational data*

mechanisms to remove CO_2 into another reservoir, mainly the oceans and, ultimately, sediments. By analogy, before 1750, the tap and the drain of my bathtub were set at the same flow rate, but now the tap is flowing faster than the drain. In fact, between 1750 (the start of the industrial era) and 2011, humankind has injected about 375 billion tonnes of carbon as CO_2 emissions into the atmosphere. When the Kyoto Protocol to the United Nations Framework Convention on Climate Change (UNFCCC) was adopted in 1997, the annual average CO_2 concentration measured at the observatory on Mauna Loa in Hawaii stood at 363.76 ppm. By 2012, it stood at 393.84, and in 2013, it had reached 396.48 while intermittently overshooting 400 ppm. The consequence of this increasing accumulation of anthropogenically derived CO_2 in the atmosphere is that the average global temperature is moving upwards as dictated by the Stefan-Boltzmann relationship (which we investigated earlier). Figure 10.3 shows the atmospheric concentration of CO_2 since 1830.

There is a very important Arctic connection to the carbon cycle. The only part of the world's oceans that can take up atmospheric CO_2 is the wind-mixed surface layer. The ocean is highly stratified, meaning that changes in seawater density with depth caused by differences in water salinity and temperature act as barriers to vertical water movement. The way to force vertical movement is to create conditions that lead to large increases in water density by elevating salinity and/or

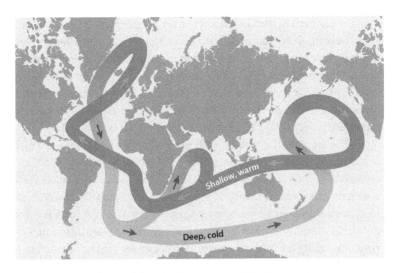

Figure 10.4 *Global thermohaline circulation and meridional overturning*

decreased temperature. Almost all the deep water in the world's oceans is produced in only two regions of the world where these conditions exist. They are the North Atlantic/Arctic Ocean (Greenland-Norwegian Sea, Labrador Sea) and around Antarctica (Weddell Sea, Ross Sea). A review by Stefan Rahmstorf in 2006 has calculated that the North Atlantic's contribution to deep water is about 17 sverdrups (Sv).[5] For the Antarctic, the contribution to bottom water is given as 15 Sv, although studies using chlorofluorocarbon (CFC) as a tracer gave an estimate as low as 5 Sv. These are the only areas in the world where the surface waters are subject to sufficiently intense cold temperatures, coupled with salt rejection from the creation of sea ice, to manufacture huge quantities of heavy (high density) cold and saline water. The deep waters formed in the Arctic and Antarctic flow towards the equator. The Atlantic deep water penetrates far into the Southern Hemisphere. The movement of surface waters to depth in the Arctic and Antarctic is called *meridional overturning* and is an example of thermohaline circulation. Without meridional overturning, the ocean's capacity to efficiently transport CO_2 and heat from surface waters to the deep would be dramatically reduced. (See Figure 10.4.)

It is time for another little digression. Sometime in the fourth century BCE, a geographer called Pytheas from the Greek colony of

[5] To describe the volume of water transported by an ocean current, oceanographers use a unit called a *sverdrup*, or Sv. One Sv is equivalent to a volume transport of 1 million cubic metres per second.

Massilia (modern Marseilles) sailed through the Straits of Gibraltar and turned north. Only fragments remain of his records, but they contain the first surviving written description of a frozen sea. It included the little poetic gem that opens this chapter. Its visual imagery perfectly captures the nexus between North Atlantic and Arctic Ocean waters. It is tempting to dream of "the lung of the sea" as a prescient allusion to meridional overturning in the Nordic seas and Arctic shelves!

The fact that the carbon cycle has limitations to the rate at which it can remove carbon from the atmosphere to the oceans and from the oceans to consolidated sediment leads to a concept that climate scientists call "commitment". The oceans have already removed about half the anthropogenic carbon released to the atmosphere since the Industrial Revolution. However, as the oceanic concentration of CO_2 rises, its ability to take up more CO_2 decreases partly because it is approaching saturation, the saturation concentration for CO_2 is lower in a warm ocean than in a cold ocean, and it is related to certain characteristics of carbonate chemistry. Climate models tell us that even in the impossible emission scenario of global economies moving tonight to carbon neutrality, Earth's mean global temperature will continue to rise well into the future because our past emissions have committed the globe to this outcome. As every year goes by, the size of the commitment increases. It is because of this concept of commitment that many climate scientists believe our politicians should be aiming for achieving carbon-neutral economies within the next 50 years if we are to maintain temperatures at manageable levels. IPCC AR5 summed it up with these words: "Cumulative emissions of CO_2 largely determine global mean surface warming by the late 21st century and beyond. Most aspects of climate change will persist for many centuries even if emissions of CO_2 are stopped. This represents a substantial multi-century climate change commitment created by past, present and future emissions of CO_2."[6] In other words, warming and rising sea levels would continue for centuries even if GHG emissions were reduced and concentrations stabilized.[7]

Why is the IPCC so confident in attributing the warming global temperature trend of the last 50 years to the increase in anthropogenic GHG emissions? Of course, there is the very obvious coincidence

---

[6] IPCC, 2013. "Summary for Policymakers." In Stocker et al. (eds.), *Climate Change 2013: The Physical Science Basis*, p. 27. Cambridge, UK, and New York: Cambridge University Press.

[7] A more detailed review of commitment can be found in Appendix V.

that GHG levels have been on the rise ever since the advent of industrialization. However, IPCC's confidence is founded on basic physical principles and the fact that climate scientists have a wide range of quite independent tools with which to investigate attribution and they all lead to the same conclusion. Our global temperature is going up because of the increasing accumulation of GHGs in the atmosphere.

Some of the main tools used to investigate attribution are climate computer models that are developed from models used for weather forecasting. They are built by dividing the world's atmosphere into a three-dimensional grid, within which data on such parameters as radiation, relative humidity and heat transfer can be added. By using basic laws of physics (such as thermodynamics and hydrodynamics), the interactions with adjacent cells can be investigated mathematically. The most sophisticated such models are coupled atmosphere-ocean global climate models (AOGCMs) that combine this approach with the atmosphere and the ocean, allowing the two to interact with each other.

When the models are run for time periods before the industrial era with only the natural external forcings (changes in the solar luminosity and the effect of explosive volcanic eruptions), they are perfectly able to replicate the global temperature record derived from proxy data. If the models are run for the last 100 years, taking into account all natural and anthropogenic external forcing, they replicate the observed warming shown in the global climate record. However, this ability to replicate the 100-year temperature record vanishes if the anthropogenic forcing is removed. The key element of this anthropogenic forcing over the last 100 years has, of course, been the dramatic rise in radiative forcing concentrations in the atmosphere. When these modelling experiments are run, they show that the internal climate-forcing factors, such as the El Niño Southern Oscillation (ENSO), are important only for short periods of time. It is the external influences that matter in terms of climate trends. You will have probably already spotted one of the most valuable contributions of this type of study. They are not structured as such to explain any particular global climate change. They simply use the laws of physics and of atmospheric and marine geochemistry to see what will happen according to a particular scenario of increasing concentrations of GHG emissions.

The ability of these techniques to replicate what our climate was like before and during the industrial period means we have very good reasons to be confident in attributing the abnormal and rapid global warming of the last 50 years to anthropogenic GHG emissions. That message increases in strength the more we look at it. The same basic results tumble out regardless of which model is used. Furthermore, the models are able to reproduce the differences in the observed amplitude of temperature change seen between the continents of North America, South America, Europe, Africa, Asia and Australia. They even replicate the instrumental record to show that the warming trend is most intense in the mid- to high latitudes of the Northern Hemisphere.

Here are the results from the use of some of the other attribution tools. We have already noted that the most important GHG is carbon dioxide. Completely independent methods show there is no doubt where this extra atmospheric carbon dioxide is coming from. Carbon has two stable natural isotopes – ^{12}C and ^{13}C – that occur naturally at a ratio of roughly 99:1. Carbon-13 is less common in vegetation and in fossil fuels because plants prefer to take up the lighter isotope (^{12}C) and it is more abundant in the oceans and in volcanic or geothermal sources. Therefore, we would predict that if the "extra" carbon dioxide in the atmosphere is derived from the burning of fossil fuels and vegetation, the relative amount of atmospheric ^{12}C to ^{13}C should be increasing. This is indeed what is found and this trend in carbon isotope ratios is also found in the coral core record over the last two centuries. Another independent verification is that of the ratio of oxygen to nitrogen in the atmosphere. This ratio is going down as carbon dioxide has increased – just as would be expected, as oxygen is used up when fossil fuels are burned.

Another family of techniques used to investigate attribution (which we will not investigate here) is to compare the geophysical fingerprint of the expected response to a certain GHG forcing with the fingerprint that would be expected from an alternative explanation (such as changes in solar luminosity).

What natural processes could "force" the climate into a warmer or colder mode? Studies of climate in the geological past (paleoclimate) have identified a number of phenomena that are associated with changes in the global climate.

Some of the candidates relate to how Earth orbits the sun. This is known as *orbital forcing*. The three best-known types of orbital forcing are collectively called the *Milankovitch cycles*. They have played a key role in the starting and ending periods of glaciation. Although the seasonal and higher-latitude climate forcing by orbital cycles is large, the net forcing is close to zero in terms of the global mean. They operate over thousands of years. Together with our present situation of being in relatively "benign" parts of the cycles, this makes them very unlikely candidates for being responsible for the present global warming reported by the IPCC. If you are interested, you will find a fuller description of orbital forcing in Appendix IV.

Variations in solar intensity (brightness) are natural processes that can make our world warmer and colder. Could they be responsible for any climate changes being observed today? Solar intensity is roughly linked to the sunspot cycle that has a periodicity of about 11 years. A spotty sun is a warmer sun, and overall, the cycle alters the intensity by about 0.08%. The Little Ice Age, which ran from as early as 1350 and that some experts consider did not end until about 1850, is believed to have been a time when the sun carried fewer sunspots. Within this period is the Maunder Minimum, from 1645 to 1750, when no sunspots were recorded. When Sir John Franklin was sent off by the British admiralty to find and transit the Northwest Passage in 1845, it did not pick the most auspicious time in terms of varying solar intensity. Franklin overwintered in 1845–1846 beside Beechey Island at the south-west corner of Devon Island. Three young crewmen died there and were buried with marked graves. Their bodies were given autopsies in the 1980s, when they were found to be carrying unusually high levels of lead, probably coming from the lead solder used to seal their canned foods and from the water distillation equipment carried on the ships. The men were later returned to their graves. I have only been ashore there once, landing in a Twin Otter equipped with skis. The gravestones were just visible above the soft snow. It is an evocative spot and a poignant reminder of the follies of history. In the summer of 1846, the Franklin expedition left Beechey Island on the *Erebus* and the *Terror* and sailed into oblivion. Although Franklin's fate is now somewhat understood, it was not until 2014 that one of his ships (*HMS Erebus*) was found.

But getting back to the topic, can the sunspot cycle explain our warming climate? Variation in solar intensity over the past 30 years can be measured and therefore evaluated in terms of its potential to explain current changes in global temperature. When this is done, it is found that the change in energy received by Earth over a complete sunspot cycle is roughly equivalent to Earth's temperature response to only 15 years' worth of present-day anthropogenic emissions of carbon dioxide. Clearly, our present climate trends are not being primarily driven by changes in luminosity.

To conclude this section on attribution, we will listen to the words of IPCC AR5: "GHGs contributed a global mean surface warming likely to be in the range of 0.5°C to 1.3°C over the period 1951–2010 with the contributions from other anthropogenic forcings, including the cooling effects of aerosols, likely to be in the range of –0.6°C to 0.1°C. The contribution from natural forcings is likely to be in the range of –0.1°C to 0.1°C, and from natural internal variability is likely to be in the range of –0.1°C to 0.1°C. Together these assessed contributions are consistent with the observed warming of approximately 0.6°C to 0.7°C." They sum it all up with this: "It is *extremely likely* that human influence has been the dominant cause of the observed warming since the mid-twentieth century."[8]

The reason I have gone into some depth on global climate attribution and global climate projections for the future is because I want it to be quite clear that amongst the scientific community, there is absolutely no doubt that our globe is warming and that this is a response to increasing atmospheric concentrations of anthropogenic GHGs. Furthermore, it will help clarify what is going on in the Arctic and to explain how a changing Arctic is very likely to lead to changes elsewhere on the globe. Now that we understand how and why our global climate is warming, we can move back to the situation in the Arctic.

The Arctic Messenger and the Arctic Council

How did the Arctic Messenger gain a voice on climate warming? It is a long and complicated story. The IPCC was established in 1988 and its first assessment report was published in 1990, the year before Arctic Monitoring and Assessment Programme (AMAP) was born. Despite being undeniably responsible for bringing a consensus view on climate change to the minds of the scientific community, it was not until its fourth assessment in 2007 that the IPCC began to take a hard look at the Arctic. In 1994, Lars-Otto Reiersen, Lars-Erik Liljelund and I met in Stockholm to decide what to propose as the subject matter for the first AMAP assessment. By this time, we already knew that the second IPCC assessment would again be largely silent about the Arctic. We found this troubling, and a short time later, the AMAP working group agreed that the first full AMAP assessment (published in 1998) should include a section on climate and stratospheric ozone depletion. The climate chapter was organised by Elizabeth (Betsy) Weatherhead at the University of Colorado, who, with Michele Morseth, was also responsible for editing

---

[8] IPCC, 2013. "Summary for Policymakers." In Stocker et al. (eds.), *Climate Change 2013: The Physical Science Basis*, p. 17. Cambridge, UK, and New York: Cambridge University Press.

the final chapter. Elizabeth had a very challenging task persuading the circumpolar climate science community to jump on board. However, she persevered with the patience of the biblical Job. We all owe her a great debt because after the first assessment was published, the Arctic Council and the circumpolar science community began to take notice.

Based on the findings of the 1998 AMAP assessment, the Arctic Council asked AMAP, the Conservation of Arctic Flora and Fauna working group (CAFF) and the International Arctic Science Committee (IASC) to "provide a comprehensive evaluation of Arctic climate change and the impacts for the region and for the whole world". The United States funded an Arctic Climate Impact Assessment (ACIA) secretariat under Gunter Weller at the University of Alaska. A team of more than 300 specialists and representatives of circumpolar indigenous peoples prepared the assessment under the integration management of Bob Corell from the American Meteorological Society (United States) and Pål Prestrud from the Centre for Climate Research (Norway). With others on the AMAP working group, I watched Lars-Otto Reiersen and Simon Wilson quietly conduct the ACIA orchestra. Their personal contribution and commitment were critical, but the ACIA is a testament to everyone who took part.[9] The result was the 1,042-page *Arctic Climate Impact Assessment [ACIA] Report*. This foundational report was published in 2005. It provided the basis for a "plain language" synthesis report released a year earlier under the title of *Impacts of a Warming Arctic*. We will take a closer look at what the ACIA had to say later, but at this stage, we will continue with an unbroken narrative of how the Arctic Council has addressed climate change.

In Reykjavik, in 2004, the Arctic Council ministers reviewed the ACIA synthesis and made a number of decisions that included the following. They recognized that the Arctic climate is a critical component of the global climate system, with worldwide implications. In response, they endorsed a set of policy recommendations concerning such things as mitigation, adaptation and research. Although useful, these actions carefully avoided implying commitment to any particular policy towards, for example, GHG emission reduction. They

---

[9] Having said this, I would be wise not to mention any particular individuals, but some names keep cropping up, as the ACIA and the subsequent SWIPA continue to stimulate new understanding. They include David Barber, Jim Berner, Terry Callaghan, Dorthe Dahl-Jensen, Henry Huntington, Margareta Johansson, Harold Loeng, Gordon McBean, Walter Meier, Mark Nuttall, Morten Skovgaard Olsen, James Overland, Terry Prowse, James Reist, Hjálmar Vilhjálmsson, John Walsh and Betsy (Elizabeth) Weatherhead.

did acknowledge the need to consider the findings of the ACIA and other relevant studies in implementing their commitments under the UNFCCC and other agreements. They also supported Arctic climate research "so that the exchange of expertise at the global level through the IPCC can better reflect unique Arctic conditions and that global decision-making can take Arctic needs into account".

At the time, many of us were disappointed with the response from the Arctic Council. However, it is easy to forget that the ACIA was quite a political hot potato, coming as it did within the first half of the George W. Bush administration in the United States. In hindsight, I think we could not realistically have expected more. It was not a bad start. The eight Arctic governments had formally recognized that here was information of great regional and global significance and that such information should influence their national energy policies and their actions under the UNFCCC. Thanks to the ACIA, which by 2011 boasted more than 3,700 citations in the scientific and technical literature, Arctic climate warming was now under the political spotlight.

An effort was begun by AMAP (and still continues) to promote the results of the ACIA and of subsequent Arctic climate studies to relevant meetings and organisations all over the world. I think the leaders in this "missionary" work were and remain Lars-Otto Reiersen, Bob Corell and Sheila Watt-Cloutier. We previously met Sheila in the chapter on persistent organic pollutants and we will meet her again before our tale is fully told. These three targeted every conceivable element, including the general public, educational institutions, the scientific community and international legal frameworks, such as the UNFCCC (including the Kyoto Protocol), the Convention on Long-Range Transboundary Air Pollution (CLRTAP) (under the United Nations Economic Commission for Europe) and the European Union.

At the same time and working with other organisations, including the World Climate Research Programme (WCRP) and the International Polar Year (IPY), AMAP embarked on coordinating a new series of studies. This time, the aim was to measure and understand how the Arctic cryosphere (those parts of Earth where water is frozen) is responding to climate change. A project that approached ACIA dimensions was launched under the title of *Climate Change and the Cryosphere: Snow, Water, Ice and Permafrost in the Arctic* (SWIPA). Preliminary results of the Greenland ice sheet (and sea level rise) components of the SWIPA assessment were presented at the 15th UNFCCC meeting of the Conference of the Parties (COP 15) in Copenhagen in December 2009. Final SWIPA results were presented at the 3–6 May 2011 Arctic

Messenger conference in Copenhagen and a week later were put before Arctic Council ministers in Nuuk, Greenland. It was at this meeting that the idea of the Arctic Messenger as presented in this book was born.

We will take a deeper look later, but in brief, SWIPA contained enough to hold the attention of council ministers. It confirmed that Arctic snow and ice are melting much faster than expected. The consequential impacts on Arctic ecosystems, Arctic indigenous peoples and Arctic infrastructure are painful to contemplate. In response, the ministers decided to strengthen the council to address the challenges that lie ahead. Furthermore, they recognized that substantial cuts in emissions of carbon dioxide and other GHGs "are the backbone of meaningful global climate change mitigation efforts". They noted "with concern the accelerated change in major components of the cryosphere and the profound local, regional and global effects of observed and expected changes" as described in the SWIPA report. They saw the need for forward-looking Arctic cooperation to increase Arctic resilience and to enhance Arctic Council leadership to minimize the human and environmental impacts of climate change. Their senior Arctic officials were instructed "to consider how best to follow up on the SWIPA recommendations in the future work of the Arctic Council".

At the same time, the ministers in Nuuk urged all parties to the UNFCCC to take urgent action to meet the long-term goal of holding the increase in global average temperature below 2°C of preindustrial levels. So far, so good. But these proved to be brave and overambitious sentiments that came back to haunt at least two of the Arctic Council governments when they sat down at the conference of the parties to the UNFCCC in Durban, South Africa (2011), and Doha, Qatar (2012). It was clear that despite their words to the Arctic Council, some Arctic countries were far from being prepared to adjust how their economies obtain or use or sell hydrocarbon-based (fossil fuel) energy.

Very substantial carbon dioxide emission reduction is essential for the long-term stabilization of the global climate. However, a number of short-lived climate forcers (SLCFs), such as black carbon, tropospheric ozone, hydrofluorocarbons (HFCs) and methane, collectively have a similar temperature impact at the present time as carbon dioxide. Therefore, it is possible that emission reductions aimed at these substances could "buy time" for slowing Arctic warming until carbon dioxide reductions can be agreed on and take effect. AMAP produced several reports before the 2011 council meeting that explored the feasibility of these policy options in technical and non-technical formats. We will examine the physics and chemistry of

SLCFs in a later section entitled "Arctic Implications of Short-lived Climate Forcers, Including Black Carbon".

In earlier chapters, we saw how certain contaminants (notably persistent organic pollutants and the heavy metal mercury) are brought to the Arctic by atmospheric and marine circulatory patterns, mainly from far-distant mid-latitude sources. We also saw that temperature plays a very important role in these pathways, particularly with respect to impacts on the direction and speed of flux between environmental compartments or reservoirs (for example, between air and soil). The next question was rather obvious. What would the impact of global warming be on exchanges of contaminants between such compartments? AMAP has published two reports (in 2002 and 2011), plus another jointly with the Stockholm Convention on persistent organic pollutants (also in 2011), that have examined this question and we have already reviewed present knowledge in the persistent organic pollutants (POPs) and heavy metals chapters.

Climate warming and associated changes in the global carbon cycle are causing the CO_2 content of the world's oceans to increase. This has resulted in a decline in seawater pH, particularly in near-surface waters. In 2013, the AMAP produced an *Arctic Ocean Acidification Assessment*. An accompanying summary for policymakers was presented to the Arctic Council at its eighth meeting in Kiruna, Sweden, in May 2013. This reinforced the SWIPA conclusion urging member states "to reduce the emission of carbon dioxide as a matter of urgency". In the subsequent Kiruna Ministerial Declaration, the ministers stated that they "recognize that Arctic States, along with other major emitters, substantially contribute to global GHG emissions and confirm the commitment of all Arctic States to work together and with other countries under the UNFCCC to conclude a protocol, another legal instrument or an agreed outcome with legal force no later than 2015, and urged all Parties to the Convention to take urgent action to meet the long-term goal aimed at limiting the increase in global average temperature to below 2 degrees Celsius above pre-industrial levels". Once again, the big question is whether these words will lead to the actions that are needed. For other issues faced by the Arctic Council, such as dealing with POPs and mercury, Arctic Council countries did call for, supported and then quite quickly implemented international mitigative actions. Thus far, however, this step has proven to be elusive as far as climate is concerned. This political paralysis is a vital issue we will come back to in the last two chapters. However, before we reach that point in our story, we must hear in greater detail what the Arctic Messenger has to say about Arctic warming.

Arctic Temperature Change and Arctic Amplification

What temperature changes have been seen in the Arctic? Proxy temperature data from lake sediments, tree rings and ice cores show that for
2,000 years prior to about 1800, the Arctic was slowly cooling. This is a
trend that is consistent with knowledge on how Earth's climate has
responded in the past to the state of the sun's energy output (luminosity)
and to Earth's orbital characteristics. However, since 1800, the temperature record reveals a warming trend, with Arctic summer surface air
temperatures now higher than at any time in at least 2,000 years. These
recent trends since the Industrial Revolution cannot be explained by
orbital variations or by solar luminosity. In fact, these two natural external forcing agents should be gently nudging us into a cooling phase.

As was foreseen by climate models as early as 1980, the Arctic is
warming faster than any other region on the planet (with the possible
exception of some parts of the Antarctic Peninsula). Since the late 1960s,
the surface temperature has been increasing at a rate of approximately
0.4°C per decade. In Alaska and western Canada, the ACIA noted that
average winter temperatures have risen by between 3 and 4°C.

The air surface temperature anomalies in the Arctic are more
than twice those recorded at most lower latitudes and are found over
land and ocean and in summer and winter. Looking further back,
Martin Tingley and Peter Huybers have recently published a study using
paleoclimate proxies (such as tree ring, ice core isotopes and lake
sediment records) to reconstruct previous Arctic temperatures. They
concluded that the summers of 2005, 2007, 2010 and 2011 were warmer
than those at any other time since at least the year 1400. Regional
patterns of Arctic surface air temperature anomalies according to the
seasons are also emerging, with some areas showing more intense
warming than others. This seasonality is illustrated in Figure 10.5.

Despite this trend, there is always variability and October 2011 to
August 2012 saw smaller warm temperature anomalies over the central
Arctic and therefore did not conform to the strong positive surface
temperature anomalies experienced between 2003 and 2010. However,
as we will see shortly, Arctic warming has achieved such momentum
that the 2011–2012 surface air temperatures appeared to have little
impact on record sea and glacial ice loss occurring over the summer
of 2012.

What temperature changes can we expect in the Arctic in the
future? The crude answer is that we can expect warming of the Arctic
to continue throughout the twenty-first century. The extent of this

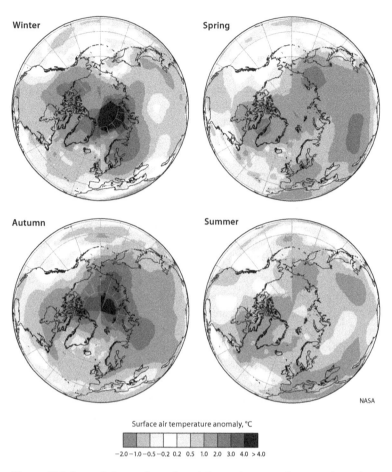

Surface air temperature anomaly, °C

−2.0 −1.0 −0.5 −0.2 0.2 0.5 1.0 2.0 3.0 4.0 > 4.0

Figure 10.5 *Seasonal circumpolar surface air temperature anomaly averaged over the period 2005–2009 (relative to the mean for 1951–2000)*

warming depends on what actions (if any) governments may take in the near future to reduce GHG emissions. In order to study these different possibilities, the IPCC published a *Special Report on Emissions Scenarios* (SRES) in 2000. Each global scenario describes a different combination of possible demographic, economic, social and energy-use trajectories that could characterize humankind in the twenty-first century.[10] Six different computer models were used to increase

---

[10] If you are interested, there is an expanded discussion on the use of scenarios and models in Appendix II entitled: What will happen in the future if we do nothing or if we try very hard to aggressively reduce GHG emissions?

confidence in the resulting SRES scenarios that are grouped according to a storyline and associated families:

- The A1 storyline and family explore a future of very rapid economic growth, a global population peaking in mid-century, followed by a decline and a rapid introduction of new and more efficient technologies. There is a general global convergence in levels of economic development and of per capita income. There are three families within the A1 scenario. Each family describes different possible paths of energy production and use: fossil intensive (A1FI), nonfossil energy sources (A1T) or a balance across all sources (A1B).
- The A2 storyline and family explore a future with pronounced differences in regional economic development, lower per capita economic growth and a continuously growing global population.
- The B1 storyline and family portray a similar future as the A1 storyline but with rapid changes towards less energy-dependent economies and the intensification of clean and resource-efficient technologies and environmental sustainability.
- The B2 storyline and family portray a future characterized by local solutions to economic, social and environmental sustainability. Economic development and technological change is less rapid and more diverse than in the B1 and A1 storylines, while the global population continues to increase but at a rate lower than in A2.

Altogether, 40 scenarios were developed under the SRES framework. The GHG emissions anticipated from these scenarios have been fed into AOGCMs. The model runs usually begin in preindustrial times using observed natural forcing agents (such as changes in solar luminosity) and observed values for GHG and aerosol concentrations. They then proceed into the future, "feeding" on the GHG and aerosol atmospheric concentrations computed for whichever of the SRES scenarios is under study (such as A1FI). At least 20 different climate models have been used to examine these scenarios.

When AOGCMs are run using the SRES scenarios, we can see what the future may hold depending on how aggressively we continue to pump GHGs into the atmosphere. When the models are focused on the Arctic north of 60°, we find that Arctic air mean temperatures are projected to be 7°C and 5°C above present-day climate for the A2 and B2 scenarios, respectively, by the year 2100. It is sobering to remember that A2 and B2 are not the most fossil fuel–intensive economic scenarios. In fact, now that we are in the second decade of the twenty-first century, A2 and B2 are beginning to look sadly optimistic. The mean temperature increases for 2100 show considerable regional variation. Under the B2 scenario, the increase for Scandinavia and East Greenland is projected as about 3°C, for Iceland as 2°C and for the Canadian Arctic archipelago and the Russian Arctic as up to 5°C. The mean summer temperature increase in the central Arctic Ocean by the end of this century is projected at below 1°C but to be up to 9°C between autumn and winter due to interaction between the air and warmer surface waters that have lost their continuous ice cover. In contrast, warming over northern Eurasia and Arctic North America for the same period is projected to be more pronounced in summer than in winter.

The SRES scenarios are now (post 2013) being superseded by four representative concentration pathways (RCPs) that will enable the consideration of global mitigation actions. However, I opted to devote several paragraphs to the SRES-based approach because it was used in most of the available relevant scientific literature up to 2013. The first IPCC assessment to use the RCPs was the AR5 that appeared in 2013. Each of the four pathways has a different signature of possible levels of radiative forcing in the year 2100. Therefore, we have RCP2.6, RCP4.5, RCP6.0 and RCP8.5. The numbers refer to the different possible levels of radiative forcing in watts per square metre (W/m^2) in the year 2100 relative to preindustrial values.

Here are the primary characteristics of the RCPs relative to the present century:

- In RCP8.5, the GHG emission increase over time leads to increasing radiative forcing without stabilization and a global temperature anomaly (relative to preindustrial times) that passes through 4.9°C (again without stabilization) by 2100.
- In RCP6.0, the GHG emission reduction leads to a stabilization of global radiative forcing soon after 2100, giving a global temperature anomaly that stabilizes at 3.0°C.
- In RCP4.5, a more aggressive GHG emission reduction leads to a stabilization of global radiative forcing soon after 2100, giving a global temperature anomaly that stabilizes at 2.4°C.
- In RCP2.6, radiative forcing rises to about 3.1 W/m^2 by the middle of the century and then falls to 2.6 W/m^2 by 2100. This scenario requires a very aggressive GHG emission reduction to begin within the next few years. It would lead to a global temperature anomaly that stabilizes at 1.5°C.

This approach indicates that hopes of the world achieving a total temperature rise by 2100 of less than 2.0°C are fading and we may be tracking a path that exceeds RCP6.0 or perhaps much more.

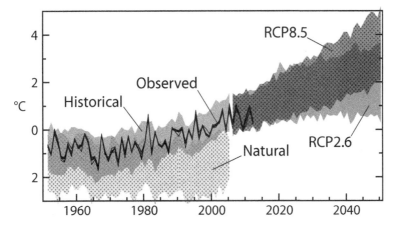

Figure 10.6 *Observed and simulated variations in past and projected future annual average temperature over land areas north of 60°N*

An example of the use of the RCPs in an Arctic context is provided in Figure 10.6, which shows historical, observed and projected temperature anomalies specifically for the Arctic. Using the HadCRUT4.2 dataset,[11] it can be shown that Arctic land has warmed by 1.5°C during the 1973–2012 period, while the globe (land and ocean) has warmed by 0.69°C. That is a factor of 2.2, which coincides well with observational data. This increases our confidence in the projections for the future.

I have mentioned several times already that the Arctic is warming faster than elsewhere. Several factors are at play here that collectively contribute to this phenomenon, known as *Arctic amplification*. The most dominant are associated with positive feedback resulting from the progressive loss of ice and snow. A positive feedback is a process in which an initial change will bring about an additional change in the same direction (leading to amplification).

The property of albedo is the ability of a surface to reflect the sun's energy back into space without changing its wavelength. Snow and ice have a high albedo, resulting in a high proportion of the incoming solar radiation being reflected back into space without being absorbed by (and warming) the surface. Water and terrain have a low albedo. Historically, much of the incoming Arctic solar radiation was reflected back into space and it was not until late summer that the sun

[11] The HadCRUT data sets are standardized collections of monthly temperature records over a global grid. There have been several iterations of these data sets. The number 4.2 specifies which iteration is being referenced. Their use enables climate scientists from all over the world to use the same data.

could significantly melt sufficient ice for areas of open water and terrain to appear. As the climate has warmed, there has been a trend in which open water and terrain have appeared earlier in the year. More solar radiation is now able to reach and heat the low-albedo surface waters, which, together with warmer Arctic winter temperatures, delays the production of new winter sea ice. This results in less ice being formed before the next summer. Consequently, less and less first-year ice survives into its second year and recruitment into the inventory of multiyear ice is diminished.

Once larger and larger portions of the Arctic Ocean lose the insulating protection of summer sea ice, the warming seawater can evaporate more and more of that often overlooked GHG water vapour into the Arctic atmosphere. Much of the presently observed Arctic warming is caused by the positive feedback of this greenhouse effect due to increased evaporation producing more moisture in the air. In an Arctic Ocean insulated by ice, there is no such feedback. The bottom line is that much more water can evaporate from near freezing water than can sublimate (go directly from ice to atmosphere) from ice.

Similar terrestrial feedback mechanisms are operating over longer periods (such as decadal time scales) associated with plant growth. As permafrost layers are lost, the tundra vegetation changes to larger shrubs, resulting in dark foliage that decreases Arctic albedo even with moderate snowfall. This supports further positive climate warming.

Readers will have noticed that at least two of the mechanisms responsible for Arctic climate change amplification are often linked and feed on one another in a cyclical manner. A climate scientist would say they are coupled. This is an extremely important concept that under-pins a key finding of the SWIPA report: "There is evidence that two components of the Arctic cryosphere – snow and ice – are interacting with the climate system to accelerate warming." One could argue for a third component to this coupled system if you consider melting permafrost releasing the GHG methane.

Sea Ice Loss in the Arctic

It is astonishing how rapidly the Arctic is changing due to loss of sea ice. One of the elements of that seminal speech given by Mikhail Gorbachev in Murmansk in 1987 was his offer to open up the north-east marine passage from Novaya Zemlya to the Bering Strait to international ship-ping. The big attraction was (and still is) the potential fuel savings of up to 40% in comparison to the conventional southern routes from Western Europe to the Far East. Between 1991 and 1999, I was involved in the work of the International Northern Sea Route Programme (INSROP), a collaborative research effort led by the Central Marine Research and Design Institute (in Russia), the Fridtjof Nansen Institute (in Norway) and the Ship and Ocean Foundation (in Japan). The INSROP studied technical aspects concerning natural conditions and ice navigation; the environment; trade and commercial shipping; and political, legal and strategic matters related to this proposal. All the issues were related directly or indirectly to the problems presented by just one formidable hazard: sea ice. The overall conclusion of the INSROP was that although it would be feasible to build the ships and infrastructure needed to safely operate the northern sea route, it would not be economically viable under the environmental and commercial conditions that prevailed at that time. I remember one report that estimated the monumental costs of marine insurance for an Arctic sea route.

In 2007 (a then-record year for ice loss), the northern sea route was still blocked by a tongue of ice that extended to the coast throughout the summer, but by mid-August in 2012, the route was open. Even when the first AMAP climate assessment was completed in 1998, it never occurred to me that only 14 years later (in 2012), the sea ice barrier would be sufficiently weakened for a fully loaded liquid natural gas tanker to sail from Norway to Japan following the Northeast Passage. In late summer 2013 (which was not a record year for summer sea ice loss), the ice-strengthened Danish bulk carrier *Nordic Orion* carried 73,000 tonnes of coal from Vancouver to Finland through the Northwest Passage. The transit took four days less than the usual route through the Panama Canal and carried 25% more coal because of depth restrictions in the canal. The voyage was reported to save more than $200,000 in comparison to the canal route. Also in 2013, a small container ship travelled the Northeast Passage from China to Rotterdam.

International interest in the shipping and resource development opportunities offered by summer Arctic ice loss is gathering momentum and is not restricted to countries with claims to Arctic waters. The recent series of conferences and meetings led by the Korea Maritime Institute (KMI) in cooperation with the East-West Center in Hawaii gives a good flavour of the issues being considered. The perspectives discussed are being made available in a series of publications, such as that edited by Oran Young, Jong Deog Kim and Yoon Hyung Kim in 2012. At the same time, Russia hosted three International Arctic Forum events, entitled "The Arctic – Territory of Dialogue", that heavily focused on the opening of the seaways and of associated resource development. The first meeting was in Moscow, the second in Arkhangelsk and the most recent in September 2013 was in Salekhard in the heart of the Yamal region, which is the location of 90% of Russia's natural gas production.

Of course, ice hazards will be around for a long time because the Arctic will keep producing winter seasonal ice under foreseeable climate warming scenarios, and for as long as there are tidewater glaciers, there will be icebergs. The last days in the life of a thawing iceberg are spent as growlers. They are usually sea-green or almost black in colour and measure less than 15 metres in diameter. They degrade from slightly larger iceberg fragments called *bergy bits*. Ships' crew fear growlers much more than they do icebergs. With modern equipment, only incompetence or instrument failure will result in a collision with an iceberg, but growlers are close to being invisible to the eye and to radar because they project less than 1 metre above the sea surface. That is often well below the prevailing wave height in most northern offshore seas and oceans, which is why they can be so difficult to see or detect on radar. What makes them even more of a hazard is that (being the last dying remnant of a berg), you can come across them when no visible icebergs are in sight. I could never see them until they were almost alongside the ship – even when crew had pointed them out beforehand. Over the last 40 years, I can think of at least three ships with fully ice-strengthened hulls that have been lost after striking a growler: two in sub-Arctic Atlantic waters and one in open water off West Antarctica. All three carried highly experienced polar crews and two were passenger vessels.

Before the 1970s, a large proportion of sea ice centred very roughly over the central Arctic survived the summer and continued to exist through several summers as multiyear ice until it was advected (carried by currents) to lower and warmer latitudes within the Arctic. However, since at least 1979, satellite microwave data and sonar observations from nuclear submarines show that sea ice thickness has been declining as less and less ice survives from one year to the next and multiyear ice becomes increasingly less common. In fact, the amount of circumpolar Arctic multiyear sea ice decreased by 42% between just 2004 and 2008. For the month of March, multiyear ice has decreased from 26% of the ice cover in 1988 to 19% in 2005 and reached only 7% in 2012. (See Figure 10.7.)

Until 2012, the sea ice minimum record that was set on 18 September 2007 remained unchallenged. However, this record was broken on 26 August 2012 during a summer with comparatively weak warming over the high Arctic Ocean. Nevertheless, the ice continued to

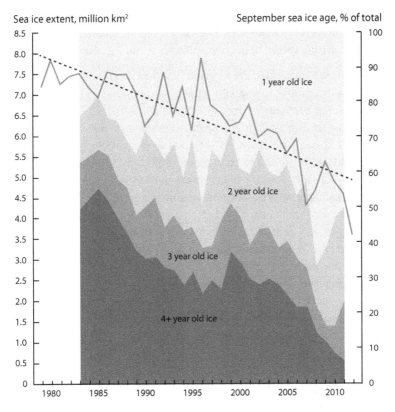

Figure 10.7 *The changing proportions of Arctic September sea ice according to age since 1985*

recede after this date until it reached a new record minimum on the 16 September 2012 of 3.41 million square kilometres (1.32 million square miles). The 2012 record is 18% below the 2007 record and is 49% below the mean for the 1979–2000 minimum. As already noted, we appear to be on course to see an ice-free Arctic in late summer by as early as 2030–2050. For more details on the history of Arctic sea ice decline and a prognosis for the future, try the papers by Wang and Overland (2009) and by Overland and Wang (2013).

Positive Feedback From Sea Ice Loss

We have already touched on this topic when we looked at the amplification of climate change in the Arctic, but now we will take a more detailed look. The Arctic is reaching a situation where the melting trend is enabling more positive feedback mechanisms to interact and support each other to accelerate ice loss even when atmospheric temperatures are not setting record highs. We will look at three examples:

1. The 2007 ice loss record occurred in a year when winds, clouds and air temperatures were all favourable to promoting ice loss. However, although 2012 was warmer than the climatic mean, it was markedly cooler than in 2007. It is probable that the primary reason that the 2007 record was broken in 2012 was a consequence of the ongoing trend for the spring to be dominated by thin, comparatively fragile first-year ice. This ice melts much more quickly in summer. The increasing proportion of open water allows more solar energy to heat Arctic surface ocean waters for a longer period of time. Increased evaporation again leads to more water vapour being able to warm the Arctic atmosphere, resulting in even less ice being able to survive the season.

2. Large areas of open water increase the fetch of wind-generated waves that can then attack the now thinner and more fragile remaining seasonal ice. Therefore, strong low-pressure systems, such as the ones that occurred in the central Arctic in August 2007 and 2012, can accelerate the rate of sea ice loss. More moisture in the atmosphere also has an impact on regional weather, but we will get to that later.

3. Another positive feedback related to the expansion of open water is increasing cloudiness resulting from evaporation. Clouds in winter mean more insulation and less heat loss to the atmosphere. Nothing acts better than clouds to give us a warm night and winter in the High Arctic is a very long night. Even in summer, most of the energy input to the near-surface atmosphere is from advection of air masses from lower latitudes and from the ocean. If you wish to warm the Arctic, the best thing to do is to melt off the sea ice and let the exposed ocean warm the atmosphere and evaporate moisture to form clouds. Once there is a cloud layer, it will trap the energy below.

For those who would like to learn more about the mechanisms of Arctic amplification (particularly with respect to the loss of sea ice and ocean atmosphere interaction), a good place to start would be the papers by Deser and colleagues (2010), Ghatak and Miller (2013), Overland, Wood and Wang (2011), Screen and Simmonds (2010) and Serreze and Barry (2011).

There is one more consequence of melting sea ice that we should consider. The surface of the Arctic Ocean is becoming much less saline. This is a result of a number of responses to climate warming, including the melting of sea ice (remember that when water freezes, it expels its salt content) and the delivery to the ocean of more meltwater from snow, glaciers, ice sheets, ice caps and permafrost. The low-salinity water also has a low density. Consequently, it floats at the top of the Arctic marine water column and has become known as *Arctic Ocean freshwater*. The subsurface density difference that results (called the *halocline*) inhibits vertical mixing. Surface waters therefore spend more time exposed to the sun. Because warm water has a lower density than cold water, we now have another mechanism that creates a density difference (called the *thermocline*), which also restricts vertical mixing. The depth at which the density differences caused by temperature and salinity combine is called the *pycnocline*.

The Arctic Ocean freshwater tends to be concentrated by anticyclonic wind forcing (due to the Coriolis effect) in the Beaufort Gyre of the Canada Basin, which has accumulated more than 5,000 cubic kilometres of freshwater between 2003 and 2012.[12] The full implications of this increasing accumulation of Arctic Ocean freshwater are poorly understood, but we will consider some later in relation to ocean acidification. At this point, we will just note that the resultant strong density stratification of the water column restricts summer replenishment of nutrients from deeper depths and therefore restricts the magnitude of summer phytoplankton growth, which is the basis of the Arctic offshore food chain.

Clearly, the Arctic Messenger is telling us that Arctic sea ice is vanishing in summer and is doing so with increasing speed. We are also being told that the loss of ice is fundamentally changing how the Arctic Ocean interacts with the atmosphere. The immediate casualty is the Arctic marine ecosystem that has evolved to depend on the existence of a summer sea ice "landscape" that will cease to exist within the lifetime of anyone born after about 1970. This is not an aspect we have so far addressed nor have we considered the cultural and economic implications for those peoples who for hundreds or thousands of years have been an integral part of that ecosystem. Like all cultures in the world, they have evolved historically to be in harmony with their regional climate. Now they will have to adjust to a new climate that is alien to their culture.

---

[12] Moving objects are deflected to the right of their path in the Northern Hemisphere (to the left in the Southern Hemisphere). It is known as the Coriolis effect and plays an important role in determining the movements of air and water. A more detailed explanation is given in Appendix III.

Melting Ice Sheets, Mountain Ice Caps and Mountain Glaciers

Ice sheets: There are two major ice sheet systems in the world. The largest covers most of Antarctica (about 14 million square kilometres) and contains 30 million cubic kilometres of freshwater ice. This amounts to 61% of all the freshwater on Earth. The second is the Greenland ice sheet, which contains about 2.93 million cubic kilometres of freshwater ice. According to the NSIDC, a complete meltdown from the Greenland and the Antarctic ice sheets would produce a total global sea level rise of about 6 metres (20 feet) and 60 metres (200 feet), respectively. Given our knowledge of GHGs and global warming, should we be worried about the stability of these ice sheets?

Yes, we definitely should be worried. As this section will show, the Arctic reservoirs of ice are melting. The 2005 ACIA and the 2011 SWIPA reports include two integrative assessments of the state of knowledge of the Greenland ice sheet and of the 402,000 square kilometres of mountain glaciers and mountain ice caps that exist in the Arctic. They are well worth a careful study because like all AMAP products, they provide a peer-reviewed synthesis of an enormous amount of the published scientific literature.

The Greenland ice sheet is a massive accumulation of ice mainly resting on bedrock. Snow accumulates at higher altitudes, where it is gradually compressed into ice by the weight of subsequent snowfalls. On the flanks, an altitude is reached – known as the *equilibrium line altitude* (ELA) – at which the rates of ice accumulation and loss are roughly equal. Below the altitude of the equilibrium line is the ablation zone, which experiences a net loss of ice. The puzzle in answering our question as to the fate of the Greenland ice sheet is related to the balance between the rate at which new ice is being added (by the compression of new snow) relative to the rate at which ice is lost. This is known as the *total mass balance*. It tells us whether the ice sheet is in equilibrium, is growing or is receding. A simplified summary of the ice dynamics of the Greenland and Antarctic ice sheets is shown in Figure 10.8.

The total mass balance is made up of two elements: the surface mass balance (SMB) and the solid mass (ice) discharge from marine terminating glaciers (producing icebergs). The SMB is the net mass of ice added or removed (flux) from the surface of the ice by such processes as snowfall, sublimation and melt runoff but excluding solid removal via glaciers (solid mass discharge). Solid mass discharge begins within the ice sheet. Ice is driven downwards by gravity into a number of drainage basins that focus the flow either into narrow mountain-

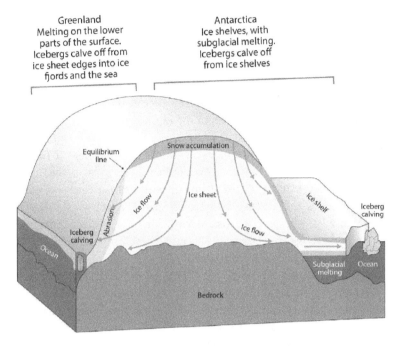

Figure 10.8 *Comparison of the dynamics of the Greenland and Antarctic ice sheets*

constrained outlet glaciers or into fast-moving ice streams separated from each other by relatively immobile ice. The ultimate fate of much of the ice of the Greenland ice sheet is to be lost through solid mass discharge as icebergs into marine fjords.

At the time of the SWIPA report, total mass balance estimates indicated that the Greenland ice sheet was losing ice at an increasing rate: from 50 Gt/year[13] between 1990 and 2000 to 205 Gt/year between 2005 and 2006. The latter seemed to be made up of roughly equal amounts of surface mass balance loss and solid mass discharge.

A more recent study used a method to detect changes in ice mass by looking at how the loss (or gain) in ice is reflected in changes in the local gravitational field. This is made possible by the NASA gravity field satellite mission *GRACE* (*Gravity Recovery and Climate Experiment*). A recent analysis of *GRACE* satellite gravity data is roughly consistent with the SWIPA results and estimated that the Greenland ice sheet lost a total mass at a rate of 222 (+/–9) Gt/year for the period 2002–2010. This is equivalent to a global sea level rise of 0.62 (+/–0.03) mm/year. At the end of 2012, a "reconciled" estimate of ice sheet mass balance was published

---

[13] A gigatonne (Gt) is 1 billion, or 10^9 tonnes.

by Andrew Shepherd and colleagues. The word *reconciled* means that the study utilized a variety of quite different methodologies. It estimated the following changes in mass between 1992 and 2011 in Gt/per year for the world's major ice sheets: Greenland: –152 (+/–49), West Antarctica: –65 (+/–26) and the Antarctic Peninsula: –20 (+/–14). It is generally believed that East Antarctica is not losing ice at the present time. The combined impact of mass loss from the ice sheets was a contribution of 11.1 mm (+/–3.8) to the total global mean sea level rise since 1992. Greenland maintained a rate of ice loss that is five times that of 1992.

Undoubtedly, 2012 was an extraordinary year to be on the Greenland ice sheet. Melting started about two weeks earlier than average at low elevations and occurred over a longer period at any given elevation than in 2010 (the previous record year) for most of June until mid-August. Once melting began, the surface albedo dropped even further (an example of positive feedback due to melting ice having a reduced albedo). Some areas in north-west Greenland between 1,400 and 2,000 metres above sea level had nearly two months more melt than during the 1981–2010 reference period. At low elevations in some parts of south-west Greenland, melting lasted for 20–40 days longer than the mean. On one day (12 July), satellite observations showed that 97% of the ice sheet surface was in a melting condition. For the technically inclined, the ELA was 3.7 times the standard deviation above the 21-year mean ELA value. Ice cores from the ice sheet suggest that similar melting events have occurred in the past, with a frequency of about once in 150 years. The last was in 1889, so the big question is: "Will this frequency change?"

I suspect it will. The entire oceanographic and atmospheric environment around Greenland seems to have been changing in a way that brings warmer air over the ice sheet and warmer water under the floating outfall glaciers (that is, the floating portions of glaciers that reach and extend into the sea). The atmospheric conditions that helped promote the record 2012 surface melt of the Greenland ice sheet was a prolonged negative North Atlantic Oscillation (NAO) index. The North Atlantic Oscillation is the name given to variations in the relative strengths and locations of the Azores High and the Icelandic Low areas of atmospheric pressure. A large difference in pressure between the two systems (known as an index of NAO+) leads to increased moist westerly winds, giving cool summers and mild wet winters in Europe. In the later section on "Climate Change Impacts on Arctic Ecosystems", we will see how a NAO+ condition can lead to feeding difficulties for reindeer in winter in Fennoscandia.

When the pressure difference is low (NAO⁻), the moist westerly winds weaken, resulting in cold winters in Europe. In spring 2012, a negative NAO index resulted in lower snow cover over much of the Arctic, including Greenland. This reduced the surface albedo of the surface of glaciers and the Greenland ice sheet. In summer, the area was again under the influence of a negative NAO index for the entire season – a situation perhaps encouraged by the general lack of surrounding sea ice. Consequently, sea level atmospheric pressure was unusually high over the ice sheet, causing warm air to be swept from the South over western Greenland, as has happened every year since 2007. This helped create the perfect conditions for the record-breaking summer temperatures over the ice sheet.

The fate and activity of glacial meltwater within the ice sheet and its exit glaciers is an interesting question. A portion of the melt is frozen in pore spaces within the percolation zone (immediately below the snow) where firn (compacted snow that has been through some freeze-thaw cycles) transforms into ice. There is evidence (see Harper and colleagues) that the percolation zone's ability to "absorb" water may act as a sort of buffer by delaying the loss of meltwater from the glacier. However, if this is the case, its capacity to soak up melt will vanish once all the pore space is occupied. It is also thought that meltwater falling through moulins (downwardly directed meltwater channels) in glaciers may reduce the drag of the glacier on its bedrock (basal drag), but this proposed effect has so far been less conclusive than it is for mountain glaciers.

We earlier noticed that in addition to surface mass balance, there is a second component to the total mass balance of the Greenland ice sheet. This is solid mass discharge from marine terminating glaciers. Anyone who has seen a glacier calving icebergs into the ocean will not forget the sight, particularly if large bergs are being born. It is spellbinding. However, what you see is only a part of the story because the terminus of the glacier is floating in the sea and is therefore melting from the bottom throughout the year as well as from the top in summer. Calving of icebergs and sub-ice melting are the major mass loss mechanisms for these glaciers. It has been estimated that the floating portion of the Peterman glacier in north-west Greenland loses 80% of its ice mass from undersea melting before it calves. This reinforces the importance of knowing the temperature of the seawater that is sitting below tidewater glaciers.

The big problem with quantifying solid mass discharge is that the basic mechanics (known as *dynamical forces*) for ice behaviour under the enormous pressures of the ice sheet and in the constricting conditions of an outlet glacier are very poorly understood. This is compounded by large variations in flow that can be seen in individual glaciers. It is an enormous headache for glaciologists trying to relate glacial observations to climate warming and even more so when thinking about projections into the future. A little earlier, I mentioned the Peterman glacier. The floating part of this glacier was about 70 kilometres long and 15 kilometres wide. In August 2010, an iceberg measuring 260 square kilometres broke free, reducing its area and volume by about 25% and 10%, respectively. This was followed in July 2012 by the calving of a 130-square-kilometre berg. How can trends be quantified in these types of discharge? You can see that small errors when dealing with such things as the Jakobshavn Isbra, or Sermeq Kujalleq (which drains 6.1% of the ice field), can have overwhelming ramifications. The former is one of the most studied glaciers in Greenland and each year sends about 35 Gt of iceberg down the fjord towards the little town of Ilulissat and the open sea. Like most fjords, this one is very deep until it reaches a shallower sill (located close to the village). Here, the larger bergs from the Jakobshavn Isbra ground on the bottom, where they may remain for months or even years in a colossal traffic jam before losing mass by melting and escaping into the bay. In the bay itself, you can usually see a fleet of up to 100 bergs that have finally escaped to embark on their long voyage – first north and then south as they follow the ocean currents of Baffin Bay and Davis Strait. The *Titanic* berg is thought to have been a child of the Jakobshavn Isbra.

When an ice sheet is in contact with the ocean – either directly or via outlet glaciers – it is vulnerable to oceanic forcing (heating from the seawater below). Ice breaking off from a floating ice shelf or floating glacier is carried away by ocean currents, where it is subject to the enormous heat capacity of the ocean and may finally melt hundreds of kilometres from its source. The resulting loss of lateral resistance associated with the now broken-off ice is thought to increase the flow of ice from the feeding ice sheet as it reoccupies the void. In the face of these technical difficulties, climate scientists have historically tended to not include solid mass discharge in their estimates of future ice loss and sea level rise. It was, for example, noted as a major information gap but not included in IPCC AR4. However, as concluded in the SWIPA report, solid mass discharge from the Greenland ice sheet is at least as important as SMB in contributing to ice loss from Greenland.

Here are some general conclusions concerning the importance of solid mass discharge reached by the 2011 SWIPA team and reinforced by some more recent publications:

1. Solid mass discharge could well be the more significant driver of total mass balance loss. Glaciological studies, such as a 2012 aerial photographic survey in north-west Greenland (by Kjaer and colleagues), show not only a substantial mass loss but also attribute this loss primarily to short-lived dynamic events rather than changes in the surface mass balance. As a result, it is being suggested that given further increases in GHG forcing, we may see nonlinear trends of ice sheet disintegration (relative to temperature), which would mean higher estimates of ice mass loss and of sea level rise (and drastic increases in stress levels for the modellers).

2. Solid mass discharge is widespread. It is also accelerating with increasing rates, gradually moving northwards along the west coast of Greenland in parallel with the gradient of regional warming.

3. The acceleration of fast-flowing marine terminating glaciers coincides with retreat of the terminus.

4. The increase in speed of these glaciers is believed to be related to the warming of ocean water underlying the floating tongue (ocean forcing). When the weakening effect of the latter has resulted in calving, the flow resistance is reduced for the ice upstream.

Therefore, previous estimates of ice sheet loss that did not take account of solid mass discharge (such as from iceberg formation) must now be considered as significant underestimates.

The previous paragraphs may be a little too technical for some readers. Therefore, here is the "take home message": The Greenland ice sheet is presently losing mass at a rate of at least about 200 Gt/year. If this loss increases to follow present model predictions, we will have a rise in eustatic sea level by 2100 of between 10 and 19 cm (3.9 and 7.5 inches, respectively). I will define "eustatic" a little later, but for now, just think of it as a theoretical sea level that irons out the effects of local and regional factors that can influence sea level, such as variations in gravity and water density. If we try to account for our lack of understanding of dynamical processes and solid mass discharge, the maximum goes up to about 40 cm (15.7 inches) of eustatic sea level alone. Remember that this is just for the amount of melting we can expect to have occurred from the Greenland ice sheet by 2100.

One final nugget to make the reader sit up: At what global mean temperature would we expect to see a complete loss of the Greenland ice sheet (resulting in an increase in sea level of 7.0 m)? The best estimate is 3.1°C above preindustrial times. In 2012, Alexander Robinson and colleagues reported that the ice sheet is more sensitive than previously thought and suggested that an ice-free state could occur in a temperature range of 0.8 to 3.2°C, with a best estimate of 1.6°C. The global mean temperature increase over the period 1880–2012 reported in IPCC AR5 was 0.85°C. It is generally accepted that the Greenland ice sheet may not vanish according to present predicted warming for perhaps up to 1,000 years, but remember that the Arctic has already seen a warming that is a good deal higher than 0.85°C. At the back of my mind, I am also recalling that RCP6.0 is projected to give us a global temperature anomaly of 3.0°C by 2100.

Mountain glaciers and mountain ice caps: The global storage of frozen ice is not restricted to the Greenland and Antarctic ice sheets. What about the state of the Arctic mountain glaciers and mountain ice caps?

Well, they contain sufficient ice to raise the sea level by 0.41 metres (16.1 inches) if they were all to melt entirely. Of 20 Arctic glaciers and ice caps (reported by Sharp and colleagues in 2011 and 2012) for Alaska, Arctic Canada, Iceland, Svalbard, Norway and Sweden for 2009–2011, 19 glaciers and ice caps had a negative mass balance. The satellite *GRACE* measurement of mass loss from all the glaciers and ice caps in the Canadian Arctic Islands was 96 (+/–49) Gt in 2010–2011. This is the largest loss calculated for this region since *GRACE* observations began in 2002. It is interesting that the region has been sitting in the same North Atlantic Oscillation and Arctic Oscillation atmospheric circulation pattern of warm air that has caused significant melting on the Greenland ice sheet. Total mass loss from Arctic mountain glaciers is thought to have probably been more than 150 Gt/year over the last 10 years. It is therefore similar in magnitude to the total mass loss so far from the Greenland ice sheet. Interestingly, in a recent global study by Gardner and colleagues, it was estimated that the 19 glaciated regions of the globe (excluding Greenland and Antarctic ice sheets) have lost ice at a rate of 259 (+/–28) Gt/year between 2003 and 2009. This is roughly equivalent to the combined ice sheet loss from Greenland and Antarctica and could account for 29% (+/–13) of the observed sea level rise over the same period. Of course, although the contributions to sea level rise from these two sources (Antarctic and Greenland ice sheets versus glaciers and mountain ice caps) are currently of comparable magnitude, mountain glaciers contain much less ice than ice sheets, so the mountain glacier contribution will ultimately diminish and cease long before the ice sheet contribution. A summary of this section is provided in Figure 10.9.

Mass balance, Gt/y

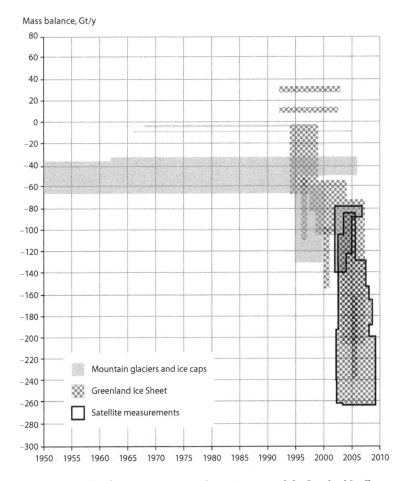

Figure 10.9 *Ice loss from Arctic mountain glaciers, ice caps and the Greenland Ice Sheet since 1950*

Sea Level Rise in the Arctic and Around the Planet

We have seen in the previous section that many of the large reservoirs of water held on land in the form of glaciers and ice sheets are melting. The ultimate fate of this water must be to enter the sea. If we think of the sea as a simple bathtub, then unless the ocean basins can expand in some way, we can predict that the mean global sea level must be going up. However, measuring sea level is much more complicated than meets the eye. For example, Earth's crust is not a stable and inflexible entity. In some places, it is being thrust up or pushed down by plate tectonics. In others, it is being pushed down by the weight of an overlying ice sheet, and in other areas, it is bouncing up because of the removal of such ice. Therefore, measurements taken at a coast may be subject to geological subsidence (such as the east coast of the United States around New York) or postglacial uplift[14] (as in coastal western Canada and in much of western Scandinavia). In some areas, sea level will be elevated because of the gravitational pull of a mountain range and/or an ice sheet. In addition, atmospheric and ocean current interaction can result in regional elevation and depression of the sea surface (such as is associated with the ENSO), while the Coriolis effect can result (in the Northern Hemisphere) in the right-hand side of currents being higher than the left (for example, the Gulf Stream off the east coast of North America). The latter is a tricky problem because ocean currents are not straight highways. They meander around – often with back eddies – to further confuse us. These difficulties can be overcome, and since 1993, satellite data has become available to measure the term *global mean sea level*. This is the area-weighted mean of all the sea surface height anomalies measured by the satellite's altimeter on a single 10-day flight track repeat cycle. If you follow up this book with some further reading, you will also probably come across the term *eustatic sea level*. This is a theoretical sea level that overcomes local effects by representing the level of all the water in the oceans as if they were contained in a single basin. Eustatic change is an alteration to the global sea levels due to changes in either the volume of water in the world oceans or net changes in the volume of the ocean basins. It is not concerned with density changes of water, such as thermal expansion. Warm water occupies more space than cold water. The importance of this often does not reach the public eye.

Given all these regional factors that can influence sea level, it is not surprising to learn that there is a great deal of global regional variation in measurements of sea level rise. For example, sea level rise in north-western Europe is lower than in many other areas. We will come back to this later, but for the next few paragraphs, we will just consider estimates of global mean sea level. Remember, though, that the global mean sea level is a hypothetical entity even if you are measuring it today because in reality, wherever you are, the real sea level is likely to be above or below the mean.

What do we know about recent changes in global mean sea level? Prior to 1880, the global mean sea level had been more or less stable for several thousands of years. This is no longer the case. According to the

---

[14] This is also known as *postglacial isostatic adjustment*.

U.S. Environmental Protection Agency's (EPA's) 2012 *Report on Climate Change Indicators*, the global mean sea level rose by about 22.9 cm (9 inches) between 1880 and 2011. The rate of rise is increasing. IPCC AR5 estimated that "[i]t is *very likely* that the global mean sea level rise was 1.7 (range 1.5 to 1.9) mm per year between 1901 and 2010, 2.0 (1.7 to 2.3) mm per year between 1971 and 2010, and 3.2 (2.8 to 3.6) mm per year between 1993 and 2010."[15]

The present estimate is that about 57% of the observed mean global sea level rise since 1993 is due to the thermal expansion of water as the sea warms up, 28% from melting mountain glaciers and the remainder from melting loss from the large polar ice sheets (West Antarctica and Greenland). Notice the size of the contribution from the thermal expansion of water. Even if the polar ice sheets could miraculously be prevented from melting, the thermal expansion will still lead to a rising global mean sea level.

What does all this mean in relation to global mean sea level for the future? At the time of writing, IPCC AR5 estimates that "sea level rise for 2081–2100 relative to 1986–2005 will likely be in the ranges of 0.26 to 0.55 m for RCP2.6; 0.32 to 0.63 m for RCP4.5; 0.33 to 0.63 m for RCP6.0; and 0.45 to 0.82 m for RCP8.5 (medium confidence)"[16]. This is still less than that suggested by many climate scientists. Take a look at the papers referenced by Bamber et al. (2013), Rahmstorf (2007), Rahmstorf et al. (2012) and Schaeffer et al. (2012) that indicate values of 70–120 cm (27–47 inches). Remember that when we looked at projected temperature rise in the light of the current trends in global emissions using RCPs, Earth appears to be on a path that is above RCP4.5 or even RCP6.0. Indeed, some projections suggest that our present emissions may be tracking RCP8.5. This suggests sea level rise by 2100 that is probably at or above the 1.0 metre projections. Robert Corell has shared with me the results of some recent work he has undertaken with a number of colleagues. Instead of using IPCC scenarios or RCPs, they have been using the targets of the 193 nations in the world that have registered in the Copenhagen Accords under the UNFCCC. In this modelling approach, the climate system tracks above RCP8.5 and indicates a global mean total surface temperature rise of

[15] IPCC, 2013. "Summary for Policymakers." In Stocker et al. (eds.), *Climate Change 2013: The Physical Science Basis*, p. 11. Cambridge, UK, and New York: Cambridge University Press.

[16] IPCC, 2013. "Summary for Policymakers." In Stocker et al. (eds.), *Climate Change 2013: The Physical Science Basis*, p. 25. Cambridge, UK, and New York: Cambridge University Press.

4.5°C by 2100. If you find this disturbing, consider how few countries are actually on course to reach their emission targets. Finally, the modelling of Arctic polar amplification suggests that the Arctic could see warming in the range of 9°C or more by the end of the century. What will this do to the Greenland ice sheet?

All these ranges imply formidable socioeconomic ramifications for human populations in low-lying areas around the world. This is especially so when you remember that in some areas, the regional rise could be much higher, such as in the New York region, where the land surface is slowly sinking. According to the Surging Seas foundation, in the United States alone, nearly 5 million people live in homes that are less than 1.22 m (4 feet) above high tide level. The socioeconomic costs of sea level rise will be enormous. For many small island states and developing countries whose population is concentrated in river delta locations, such as Bangladesh, the implications of emerging scenarios must be hard to comprehend.

In mid-2013, two papers – one led by Anders Levermann and the other by Benjamin Strauss – pointed out that in addition to the sea level rise that is occurring today and is projected to occur over the next decades, there is a long-term commitment to sea level rise on much longer time scales (at least 2,000 years). This second element is driven by the inertia of the global climate and carbon systems. As discussed earlier, they commit us to future warming over long time scales because of the stockpile of excess CO_2 we have added to the atmosphere. According to Strauss, our long-term sea level rise commitment is growing 10 times faster than the sea level rise we actually observe today. At present, the long-term commitment is growing at about 30.5 cm (12 inches) per decade. According to Levermann, even the much touted but elusive target of restricting warming to 2°C still implies a long-term commitment for 4.8 m of global mean sea level rise.

If the concept of commitment to future sea level rise is making you feel uneasy about the future, there is more unsettling news to come. I want to return to my earlier promise concerning global variability in sea level. Imagine you are a city engineer or government policymaker entrusted with ensuring the long-term safety of people, infrastructure and property in low-lying coastal areas. You will need to know more than just projections for global mean sea level. You will be asking if your geographical region is one that is expected to conform with the global mean or is a region that can expect greater sea levels than the mean. In the case of the latter, you will need to know the size of the expected deviation over a certain time frame associated with a given ice sheet melting scenario.

It is a tall order, but estimations of geographic variability in global sea level rise (and of its causes) have been carried out and they are quite the eye-opener.

The following will be a very brief summary of one of the causes of variability. For further reading, I recommend the two references by Tamisiea and colleagues (2011) and Mitrovica and colleagues (2009). To make these estimates, scientists cannot just quantify the familiar elements responsible for the global mean (such as the thermal expansion of water and the output from identified glacial melt sources). Now they must also estimate the causes of geographic variability. The easiest source of variability to understand and to measure concerns vertical movement of Earth's crust along the coasts caused by plate tectonics or isostatic adjustment following glaciation or de-glaciation. The big surprise (for me at least) was to learn about the possible influence of the gravitational pull of the Greenland and West Antarctic ice sheets on regional sea level. They cause a gravitational tide that pulls water towards them, thus creating a regional elevation of sea level. I say "regional" because the hinge line, which separates sea level elevation to sea level depression is believed to be about 2,000 kilometres from Greenland. Therefore, Greenland's gravitational influence is estimated by Mitrovica and colleagues to reach as far as Scotland. In a lecture given in Washington in March 2011, Mitrovica presented a study (see also the Tamisiea reference) that concludes that if the Greenland ice sheet catastrophically collapses, the loss of ice mass would relax the gravitational pull of the ice sheet on the surrounding sea. The result would be an astonishing 100-metre drop in sea level around Greenland.

Not many scientists presently expect the Greenland ice sheet to catastrophically collapse. Therefore, a more realistic scenario presented by the Mitrovica team is for the Greenland ice sheet to melt to the extent that it would lead to the equivalent of a 1-metre elevation of mean global sea level. In this case, they estimate that depression in the sea level around Greenland could be 20–25 metres and that the hinge line would again extend from Scotland northwards. Of course, if sea level is depressed at high northern latitudes, then elevations above the mean would have to happen far from Greenland.

The same methods can be used to study the impact of other sources of melting ice. The studies reported by Tamisiea and by Mitrovica on melting of the West Antarctic ice sheet showed that the Pacific and Atlantic coasts of North America would experience a substantial enhancement of sea level rise, with only the extreme tip of South America enjoying the benefit of sea level depression. Each such study shows a unique geographical pattern of sea level elevation and depression, known as a *sea level fingerprint.*

To sum up this entire section, the Arctic Messenger is telling the world that the Greenland ice sheet, along with the Arctic mountain ice caps and glaciers, is losing significant proportions of its mass to the ocean. The Arctic landscape is changing dramatically before the eyes of those who live in these regions. But the Arctic Messenger is also telling us that if we live close to sea level (and many of us do), we need to pay very close attention to the rate of melting of the Greenland and West Antarctic ice sheets and of mountain glaciers wherever they may be. It is a paradox that studies on gravitational tides around the two large ice sheets suggest that the greatest elevations in sea level will occur far from the ice sheets, while Greenland and West Antarctica will experience a depression of sea level!

For readers who would like to learn more about climate change and rising sea levels, I can thoroughly recommend the book *High Tide on Main Street: Rising Sea Level and the Coming Coastal Crisis* (second edition) by John Englander, published in 2013. It is a masterly summary and is easy to follow. For a more technical summary, one can turn to the IPCC *Working Group 1 AR5 Report* (chapter 13).

Snow and Permafrost

A great deal of freshwater in the Arctic is locked up in soils as permafrost and in snow. How are these two reservoirs faring with a general circumpolar trend towards more precipitation and shorter and warmer winters? The quick and general answer is that substantial decreases in snow cover are taking place.

The largest and most rapid decreases in snow water equivalent (the amount of water contained within the snowpack) and in snow cover duration has occurred in Arctic maritime regions, with the highest precipitation being recorded in Alaska, northern Scandinavia and the north-east Pacific coast of Russia. However, although this trend has been obvious in the North American Arctic record since the 1950s, it has only been evident in some Eurasian records since about 1980. In some Eurasian regions, snow depths are actually increasing.

Most of the decrease in snow cover duration occurs in the spring, which is the response expected with regional climate warming, snow albedo effects (albedo decreases as soon as melting begins) and the albedo-reducing impact of accumulations of black carbon. Once again,

we are looking at positive feedback mechanisms that promote accelerated warming. As soon as the snow has gone, the darker surface reflects
little of the sun's energy back into space and the darker terrain absorbs
more energy.

These reductions in Arctic spring snow cover also have wide
implications for the Arctic ecosystem, including growing season length,
the timing and nature of spring river runoff, wildlife ecology and the
ground thermal regime. The upper layer of soil and rock in the High
Arctic that can freeze and thaw seasonally is called the *active zone*. It can
be up to 10 metres thick. Below this is the permafrost. This is a layer of
frozen ice, rock and soil/sediment that remains below 0°C for two or
more consecutive years. You will find it underlying most of the terrestrial Arctic and even find it below some Arctic shelf seas. Permafrost
varies in thickness from a few centimetres at its southern limits to as
much as 1,500 metres in Siberia. As you travel south, it does not
abruptly end but continues as diminishing islands of ice known as
discontinuous permafrost. Saline water in soil freezes below 0°C. This
creates taliks. They are unfrozen zones in permafrost below parts of
the Arctic shoreline or below some lakes or they exist as soil brine
pockets.

Climate, especially air temperature, is one of the main factors
controlling the spatial extent and depth of the permafrost layer and
we have already seen that the annual mean surface air temperature over
Arctic land areas has risen by about 2°C since the mid-1960s. It is
therefore not surprising to see that the 2011 SWIPA team found that
permafrost temperatures since the late 1970s have increased at monitoring sites throughout the Arctic, including North America, Russia and
Scandinavia, by 0.5 to 2°C. Similarly, active layer thickness over the last
two decades has increased at sites in Scandinavia, Russia and (in the
last few years) in Alaska but not in Canada.

From the earlier paragraph, you will have noticed that not all
permafrost is at the same temperature. As would be expected, warmer
permafrost with temperatures close to 0°C occurs at southern permafrost margins and in the discontinuous zone. More northerly cold
permafrost can have temperatures as low as –12°C to –15°C. Interestingly, warm ice-rich permafrost shows lower warming rates than areas
with cold permafrost. This is because of latent heat effects associated
with the phase changes of thawing that dominate an ice-rich soil environment. Nevertheless, because of the huge extent of deep, cold permafrost, it is the warm permafrost regions that are more prone to loss
under present conditions of Arctic warming.

The vertical profile of a column of permafrost reflects its history. Therefore, the upper and youngest part of the column dates from the Little Ice Age, below which we find ice from the late to mid-Holocene, while at the bottom rests the oldest permafrost, dating from the late Pleistocene.

What can we expect in the future? Forward-looking projections of ground temperature in the circumpolar Arctic suggest that late Holocene age permafrost will be thawing at the southern limit of the permafrost zone by the end of the twenty-first century. At that time, even late Pleistocene permafrost could be thawing at some locations. Regional models add a further perspective. In Russia, which is home to most of the Northern Hemisphere's permafrost, increases in ground temperature of 0.6 to 1°C have been projected by 2020. In Canada, it is projected that the top 2–3 metres will thaw more than 16–20% of the permafrost zone by 2100, at which time widespread permafrost degradation is also anticipated throughout Alaska. What I find really sobering is to remember that (as we saw a little earlier in the paragraphs on sea level change) world GHG emissions may be following a path that is close to the highest RCP given in IPCC AR5 (RCP8.5). Add this to polar amplification and it means there is cause to be concerned that we may well see higher Arctic temperatures than anticipated in AR5 in this century. This would lead to a faster thaw of Arctic permafrost.

There is understandably a great deal of concern about the practical implications of a thawing tundra. Some of the potential impacts to infrastructure are easy to understand. For example, the easiest way to transport heavy loads on Arctic land is to do so in winter, when ice roads can support the weight. The season when this is possible is declining significantly. Existing infrastructure, such as buildings and airports – even if designed for late twentieth-century Arctic conditions – is in danger of being destabilized. Degraded permafrost leaves landforms much more susceptible to marine and freshwater erosion. Other worries are much more difficult to assess. Some of the latter are related to new observations and estimates on the amount of carbon frozen in certain Arctic soils, which suggest that the total Arctic carbon pool could be 1,400 to 1,850 Gt. This is more than is presently in the global atmosphere. About 1,024 Gt may be stored in the top 0–3 metres. The obvious question is whether thawing of these layers will move the Arctic tundra from being a carbon sink to a carbon source for the global carbon cycle. At this stage, there is no clear answer.

There are two other observations related to the same possible positive feedback issue. High nitrous oxide (N_2O) emissions have been detected from permafrost regions in Russia and in north-eastern Greenland. It is not known how ubiquitous such emissions may be in the Arctic. N_2O is a powerful GHG with a GWP 300 times that of CO_2 and it has a residence time in the atmosphere of 114 years. It occurs naturally as part of the nitrogen cycle but is also produced by human activities, such as agricultural fertilizer manufacture and use and automobile exhaust.

Terrestrial wetlands are the largest single source of another GHG: methane (CH_4). It has a lifetime in the atmosphere of 12 years and a GWP 21 times that of CO_2. In most countries, it is the second-most prevalent GHG in terms of emissions, but for quite some time, there has been concern that it could be released in very large quantities from thawing permafrost. Huge reservoirs of methane are also believed to be trapped beneath subsea permafrost, particularly off eastern Russia. More recently, high concentrations of methane have been measured through the water column and in the atmosphere over the Laptev Sea. This has added questions to an already complex puzzle that many climate scientists still consider to be unresolved.

During the IPY, a team that routinely measured summer methane release from a wetland in southern Greenland had the opportunity to extend its observations by one month. It was astonished to find that a large increase in methane emissions occurred as the wetland began its winter freeze-up. Even when scaled up for the circumpolar Arctic, the release probably did not alter global emission estimates, but it did strike me as an excellent example of the challenges of monitoring the Arctic environment. Without automated monitoring, there is little understanding of variability and the all-important seasonal processes can be missed. Perhaps the best example was the discovery of the impact of polar sunrise on mercury that we looked at earlier.

Arctic Climate Interconnections and Stratospheric Ozone Depletion

Readers who are still with me will have noticed the degree of interconnectedness that pervades our global climate system as it moves "excess" heat from equatorial regions towards the poles, which then sends cool air and water back to lower latitudes. We can revisit some of the tricks that a melting Arctic Ocean and the Arctic Oscillation (AO) and its sibling the North Atlantic Oscillation (NAO) may be playing. (See Figure 10.10.)

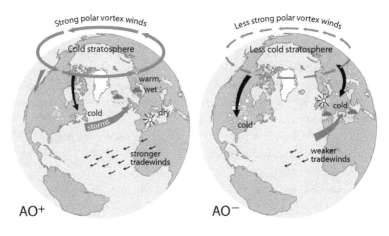

AO⁺

AO⁻

Higher-pressure air mass over North America, Europe and Asia confines extremely cold air to Arctic.

Air pressure systems weaken allowing colder air to move south and warmer air to move north.

Figure 10.10 *The Arctic Oscillation and the polar vortex*

These two recurring atmospheric patterns have long been implicated as drivers of variability in the Northern Hemisphere, including the Arctic. During the AO⁺ phase, lower than usual atmospheric pressure dominates the Arctic. This helps the subpolar jet stream blow strongly from west to east with low amplitude meanders and to maintain a strong polar vortex that locks cold air in the polar region.[17] Is it possible that climate warming may influence whether we live in an AO⁺ or an AO⁻ world? The answer is "possibly".

The argument goes like this: The progressive loss of more and more sea ice results in the excessive heating of the upper ocean relative to historical times. Two positive warming feedback mechanisms now become active. Some of the excess heat goes into speeding up warming through our now familiar ice-albedo feedback. What is new for us is the realization that much of the remaining excess heat is slowly transferred to the atmosphere through evaporation and radiation. This increases Arctic atmospheric moisture content. At the same time, the temperature gradient between the Arctic and middle latitudes is decreased. We now have just the right conditions to encourage the development of AO⁻ conditions in the winter. With a weakened jet stream and a

[17] A description of the Arctic Oscillation (AO) and the North Atlantic Oscillation (NAO) is given in Appendix III, together with a caution on how these two terms or indices are used.

weakened polar vortex, cold Arctic air, with its now increased moisture content, can more easily escape into mid-latitudes, bringing intense cold and snow to the eastern United States and to Western Europe. This is indeed what appears to have occurred in the winters of 2009–2010, 2011–2012 and 2013–2014. The AO index in February 2010 of –4.26 was the largest negative anomaly since records began in 1950.

We have here a plausible mechanism that is consistent with present observations and that links a warming Arctic and diminished ice cover with (as the proponents say) "stacking the deck in favour of severe weather outbreaks" in some mid-latitude regions. Only time will tell whether this or other plausible explanations best describe the meteorological consequences that follow from the increasing direct exposure of the Arctic Ocean waters to the atmosphere.

A weakened jet stream is also a "loopy" jet stream (prominent Rossby waves). This means it follows a path with greater amplitude meanders, reaching deeply southward with short distances between peaks that progress much more slowly from west to east and may even stall. This situation increases the persistence of mid-latitude weather patterns, locking them in place and also setting the stage for so-called meteorological blocking. It was a high-pressure system stuck in a blocking pattern over Greenland in October 2012 that sent Hurricane Sandy inland. It could well be that the situation that prevailed from the mid-1960s to the mid-1990s, which saw strong AO$^+$ conditions maintaining a strong polar vortex, will become increasingly rare. Once again, only time will tell if this explanation is the one that will prevail. Of course, because of the interconnectedness of our global climate system, the weather we actually experience in mid-northern latitudes is influenced by more than one long-distance connection or teleconnection.

Near the end of the ozone chapter, we noted that early in 2011, a very severe ozone-depletion event occurred in the Arctic. The impact of this spread beyond the Arctic. For example, in the period including January to March 2011, the UV index over southern Sweden was 20% above the mean. I have delayed taking a closer look at Arctic ozone depletion until we had reviewed Arctic meteorological and climate change processes. In the ozone chapter, we also saw that a trinity of circumstances is needed to occur simultaneously for ozone depletion to occur in the stratosphere. These are (1) a supply of ozone-depleting substances, (2) extremely cold temperatures that allow for the formation of polar stratospheric clouds and (3) solar ultraviolet light (UV). Unlike the Antarctic, the Arctic stratosphere is usually not sufficiently cold enough at the time of polar sunrise to support the formation of polar stratospheric clouds and the cascade of depleting chemical reactions will not occur or will occur only slowly. For this reason, the Antarctic has experienced severe stratospheric ozone depletion, but the situation in the Arctic, with a generally warmer stratosphere, has been less severe. Consequently, Arctic UV irradiance has not reached the very high levels seen in the Antarctic ozone hole. We also saw how implementation of the Montreal Protocol is making progress in reducing the concentration of ozone-depleting substances in the stratosphere. But what caused the 2011 event?

Despite the Montreal Protocol, there is still a large reservoir of ozone-depleting substances aloft. In spring 2011, an AO$^+$ condition that generated a robust polar vortex was set up in March and April. This constrained Arctic air in the region. At polar sunrise, the trinity of favourable conditions was established and a new record set for Arctic stratospheric ozone loss. A closer look using modelling by Strahan and colleagues published in 2013 estimated that the extremely cold conditions maintained by the vortex caused two-thirds of the ozone loss and that the remainder resulted from the vortex blocking the transport of less ozone-depleted stratospheric air from lower latitudes.

It is still not understood exactly what caused these atmospheric conditions to be established. However, could we see periodic erosion of the slow ozone recovery achieved through implementation of the Montreal Protocol? The massive Arctic depletion event was not repeated in the winters of 2011–2012 or 2012–2013, but the Arctic Messenger will be watching carefully in the coming years.

It turns out that there may also be other links between climate change and stratospheric ozone depletion. Before the age of CFCs, stratospheric ozone absorbed so much of the energy from incoming solar ultraviolet light that it warmed the stratosphere. Less of this very energetic UV radiation managed to reach into the troposphere below. Ozone depletion has changed this picture. Less ultraviolet radiation is now absorbed in the stratosphere. Consequently, the stratosphere is cooling. This leaves more UV radiation to heat the troposphere below.

Arctic Implications of Short-Lived Climate Forcers, Including Black Carbon

Agents capable of warming the atmosphere can be divided into two categories: those that remain in the atmosphere for hundreds or thousands of years, such as carbon dioxide (CO_2), and those with a much shorter half-life (days to a few years), known as *short-lived climate forcers*

(SLCFs). Just before I retired in 2004, I went to a CLRTAP meeting in Berlin, where I picked up a paper then just published in the *Proceedings of the National Academy of Sciences of the United States of America*. It examined soot as a climate-forcing agent via its impact on snow and ice albedo. It was followed by another paper by the same authors the following year. The potential significance for the Arctic was clear, but I gave it little attention until three years later, when Lars-Otto Reiersen told me about a small assessment that was being carried out as an initiative of AMAP to scope out the possible adaptation and mitigation opportunities relating to SLCFs and the Arctic climate.

The results were published by AMAP in 2008 as two small reports. For their size, they must be two of the most effective of AMAP's products. Another more comprehensive assessment initiative of the secretariat dealing exclusively with black carbon (BC) impacts was published in 2011 and was complemented at the same time by an Arctic Council report on BC emissions and mitigation options.

There is no dispute that the growing concentration stockpile of CO_2 will stay in the atmosphere and oceans for at least hundreds of years. Therefore, we cannot hope for a long-term solution without real and meaningful reductions in total CO_2 emissions that are explicitly designed to reduce atmospheric levels by an identified date. Simply reducing, for example, the emissions for a given distance travelled by a motor vehicle will not reduce CO_2 levels in the atmosphere if we simultaneously build more cars. Sadly, building and selling more "things" that require energy seems to be implicit in that holy political imperative of maintaining a growing economy. But while we wait for our global leaders to muster enough courage to act on CO_2, should we also expect them to act on SLCFs?

The answer is an emphatic "Yes", providing they remember that no effort should be spared to address the CO_2 issue before we completely run out of time. The warming effect of SLCFs is short lived because they only survive in the atmosphere for a few days or weeks or, at most, a few decades. Therefore, we should see beneficial effects much faster if we can reduce their emissions. To give an idea of the opportunity posed by the prospect of actions on SLCFs, it has been estimated that they may account for about 30–40% of the anthropogenic warming so far experienced. According to UNEP, firm action on SLCFs has the potential to reduce global warming between 2012 and 2040 by as much as 0.5°C. Substances falling under the SLCF umbrella include HFCs, tropospheric ozone, BC and methane. What can we say about them?

Ozone-depleting substances and some of their replacements are also GHGs. Today, we are particularly concerned about one family of these substances: HFCs. They first entered our everyday lives as replacements for CFCs, the substances implicated in the mid-1970s in stratospheric ozone depletion. Some HFCs are capable of very aggressive radiative forcing. As an example, HFC-134a – the most widely used of these compounds – has a GWP of 1,300 and has an atmospheric lifetime of 13.8 years. Remember that on the same scale, CO_2 has a GWP of 1. HFC emissions in many countries are the fastest-growing contributors to global warming. They are growing globally at 10–15% per year. Alone, they could sabotage global efforts to halt acute climate change by overwhelming the effect of other reductions of GHG emissions now anticipated under the UNFCCC process until 2050. Fortuitously, the Montreal Protocol has been tasked not only with controlling ozone-depleting substances but also with addressing any "adverse effects" arising from their elimination. The negative contribution to global warming arising from using HFCs as alternatives to other ozone-depleting substances can obviously be considered an "adverse effect". By summer 2012, 108 countries had come out in favour of action against HFCs under the Montreal Protocol. The Rio+20 Conference in 2012 gave cause for great optimism (but as we will see, the prize seems as elusive as ever). One of the final declaration documents states: "We recognize that the phase-out of ozone depleting substances is resulting in a rapid increase in the use and release of high global warming potential hydrofluorocarbons to the environment. We support a gradual phase-down in the consumption and production of HFCs."

It really did look as if world governments at the 25th anniversary meeting of the Montreal Protocol in November 2012 were poised to seize this opportunity to slow the rate of global warming by setting up negotiations for the phasing out of HFCs. This would normally begin with the formation of a "contact group" to prepare the ground before any negotiations of actions could begin. Momentum for the initiative vanished overnight when it was firmly opposed by three of the largest countries with developing economies. Instead, a discussion group has been established and the protocol's Technology and Economic Assessment Panel (TEAP) was asked to prepare a report on currently available and developing alternatives.

It is easy to be cynical and agree with the saying that committees and discussion groups are wonderful places to send good ideas, where they can quietly die. However, if there is a persistent champion, discussion groups sometimes act as a nurturing cocoon from which the strengthened idea can eventually see the light of day. The Arctic Messenger watches rapidly retreating sea ice, eroding permafrost and the shrinking Greenland ice sheet and hopes this embryo is allowed to quickly grow before the polar positive-feedback mechanisms gain complete control.

We have already discussed how ozone warms the stratosphere by absorbing the sun's highly energetic ultraviolet radiation and by absorbing upwelling infrared radiation from the troposphere. Ozone is also present in the troposphere, where it is formed by photochemical reactions involving nitrogen oxides (NOx) and volatile organic compounds (VOCs). VOCs and NOx can be produced naturally but are also produced by, for example, internal combustion engines. The warmer it is, the more ozone is formed. Within the troposphere, ozone performs a GHG forcing role by absorbing infrared radiation and, to a lesser degree, by absorbing the small amount of incoming ultraviolet radiation that has survived its passage through the stratosphere and through tropospheric water vapour. Ozone is thought to have an atmospheric tropospheric lifetime at high northern latitudes of one to two weeks in the summer and about two months in winter.

BC consists of small dark particles that result from the incomplete combustion of biomass and fossil fuels. The most common sources are diesel cars and trucks, wood-burning stoves, forest fires and agricultural open burning. It can influence the climate in at least two ways. Firstly, BC absorbs incoming and reflected solar radiation and therefore warms its immediate environment. This direct radiative forcing by BC is important globally and regionally. (It ranks third globally behind carbon dioxide and methane.) Secondly, if deposited on snow and ice, BC directly decreases the surface albedo and increases the amount of heat absorbed. We are already familiar with the consequences. Melting is accelerated and the darker tundra and sea surfaces exposed result in a continuing positive snow/ice albedo feedback. Once deposited on a glacier or ice sheet, BC can remain for years before it is removed by surface runoff or incorporated into glacial ice. Therefore, it can exert its impact on albedo reductions over more than one seasonal melt cycle. However, nothing is ever simple. Some important aerosols, such as sulphates, are co-emitted with BC and can be strong negative climate forcers. Without them, we would currently be experiencing greater warming!

Climate models show that BC contributes to climate warming globally. However, the intensity varies regionally because of its relatively short residence time in the atmosphere and because of regional patterns of atmospheric circulation. Therefore, BC originating from Arctic and non-Arctic sources contributes to Arctic warming. The AMAP team concluded that the BC snow/ice radiative forcing per unit of BC emitted increases with latitude and is larger for the Arctic Council countries than for the rest of the world. As a result, the Nordic countries are associated with the largest forcings per unit of BC emission due to those emissions occurring at higher latitudes and therefore closer to the areas of Arctic deposition. There is some question as to the overall net effect of a given BC source to global warming, but as far as warming in the Arctic is concerned, there seems to be little doubt. BC can be 680 times more effective as a climate forcer than CO_2 locally in the Arctic. A study published in 2007 attributed about a 0.24°C temperature increase to BC-derived radiative forcing within the Arctic atmosphere. For the deposition of BC on highly reflective snow and ice surfaces, the attributed surface temperature increase was about 0.56°C.

At the time of writing in 2013, a very large synthesis study on BC was published in the *Journal of Geophysical Research: Atmosphere*. Led by Tami Bond, the researchers estimated that BC is presently responsible for a total climate forcing of +1.1 W per square metre for the first year after emission. This is about twice that estimated in the fourth IPCC assessment and puts BC second only to carbon dioxide. However, BC sources also simultaneously release other short-lived substances, such as sulphur dioxide, that can cool or warm the climate. When the team took account of these other agents, they arrived at an industrial age net climate forcing of +0.22 (–0.50 to +1.08) W per square metre during the first year after emission. They also estimated that present-day climate forcing from BC resting on snow and sea ice is about +0.13 W per square metre. The open burning of forests and savannah vegetation is the largest global source of BC to the atmosphere, but the ranking of other sources varies regionally. In Europe, North America and Latin America, diesel engines used for transportation contribute about 70% of emissions. Not all emissions are equal in terms of their contribution to climate warming, partly because of the co-emitted substances that exert negative forcing from some sources. The study highlighted diesel engines used for transportation and possibly residential biofuels as being the two BC source categories that would produce the greatest degree of climate cooling if their emissions could be eliminated.

The magnitude of the impact of forcing by all SLCFs depends on seasonal patterns of atmospheric transport and deposition, solar radiation and snow/ice melt. Generally, the atmospheric transport of pollutants to the Arctic from mid-latitudes is most efficient in winter and early spring. Without sunlight, there is a buildup of tropospheric ozone (another SLCF) and of ozone precursors. Arctic BC concentrations also go up in winter because their transport is associated with Arctic haze. When solar radiation increases in spring, the season of forest fires begins in sub-Arctic regions and photochemical reactions involving methane and other substances (including nitrogen oxides) result in more ozone production. The warming effects of BC are at a peak in spring and summer, when atmospheric radiative warming and the snow/ice albedo feedback (both enhanced by BC) are maximized. The ozone is attacked by solar radiation, so at this time, its concentrations in the troposphere decline.

It follows that the heat-trapping effect of the ozone is greatest in the spring and that most of the ozone's contributions to the greenhouse effect occur within 45° latitude from the equator. Nevertheless, it is estimated that the seasonally averaged temperature response attributable to the ozone for 60–90° N is 0.43°C.

Methane is a GHG with an atmospheric lifetime of about 12 years and can therefore also be thought of as a SLCF. It is actually a much more potent GHG with a GWP relative to CO_2 of 21 but is present in the atmosphere at much lower concentrations. The other SLCFs we have studied so far have such short residence times in the atmosphere that differences in source strength have major implications on impacts. However, methane remains in the atmosphere for about 12 years, which enables it to be globally well mixed throughout the atmosphere. Therefore, its impact on global warming has no strong regional signals (as distinct, for example, to BC). Since the Industrial Revolution, atmospheric concentrations have roughly doubled and have now reached about 1,770 ppb (parts per billion). Ice core measurements indicate that this is higher than at any time in at least the last 650,000 years. It is estimated that the seasonally averaged temperature response attributable to methane today for 60–90° N is about 0.34°C.

Anthropogenic activities, such as coal and gas production, animal and other waste disposal, biomass burning, rice production and enteric fermentation from animal husbandry, account for about two-thirds of methane emissions. They are expected to rise in the future in consort with the increasing intensity of industrial and economic activity. However, the single largest source is believed to be natural wetlands. The wetlands north of 60° N alone may contribute about 13% (although there is much uncertainty with these estimates). Measurements over the last decade have suggested that northern wetland emissions are increasing as temperatures warm, soils become wetter and permafrost thaws – often being replaced by thermokast lakes, through which bubbles of methane can be seen coming to the surface. The ultimate source of this Arctic methane is biological degradation of organic matter that was trapped below permafrost – in some cases taking up a hydrate form within sediments.

The future behaviour of methane sequestered beneath permafrost is a very controversial topic. First, the extent and size of methane reservoirs in the Arctic is at present poorly known. However, the estimate by Shakhova and colleagues in 2010 of 3.75×102 Gt C (carbon) in methane hydrates just on the East Siberian Arctic shelf gives a good idea of the scale being discussed. What makes these Arctic methane reservoirs such a formidable wild card in climate science is the question of whether climate warming will lead to a slow and more or less linear release. At present, this scenario is probably supported by most scientists, but there remains a low probability of the extraordinary risk associated with a possible sudden release. IPCC AR4 estimated that if only 0.1% (1.8 Gt C) is instantaneously released to the atmosphere, CH_4 concentrations would immediately increase to about 2,900 ppb from the 2005 value of ~1,774 ppb. Imagine the impact this would suddenly have on radiative forcing! The paper by Carolyn Ruppel explains in greater depth the physics and uncertainties involved, while the short review by Gail Whiteman and colleagues gives a good summary of the socioeconomic consequences. They estimate that the release of methane from the East Siberian Sea alone would cost global society about $60 trillion.

These thoughts bring our quick survey of SLCFs to an end. These are the five points the Arctic Messenger hopes you will remember:

1. SLCFs comprise a very significant proportion of the total radiative forcing that is ultimately driving warming of the Arctic.
2. By conspiracies between Arctic climate, seasonal solar cycle, atmospheric circulation and the chemistry of SLCFs, Arctic spring is the time when the impact of these substances is at its greatest strength. As a result, by accelerating spring snow and ice melt, they decrease the snow/ice albedo effect and thus accelerate Arctic warming.

3. Even if we are callously ambivalent to the Arctic impacts of SLCFs, we do not escape impacts wherever we live. Remember that the Arctic is an essential element of the global climate system.

4. Almost by definition, the contribution of SLCFs to climate warming would dissipate quickly if their emissions were significantly reduced. This is an opportunity we do not have with CO_2 because of its very long residence time in the ocean. We should seize it with both hands. The CLRTAP Gothenburg Protocol (by addressing small particulate air pollution – for example, BC) is a good start, but it will be toothless until countries implement the new obligations and, of course, it is restricted to the UNECE region. Why not also find a way to increase its geographic scope?

5. Taking action on SLCFs should not distract from the necessary efforts to reduce anthropogenic CO_2 emissions; the two are complementary and not alternatives.

Arctic Ocean Acidification

When carbon dioxide dissolves in seawater, it forms carbonic acid (H_2CO_3), which dissociates, releasing hydrogen ions and bicarbonate:

$$CO_2 + H_2O \leftrightarrow H_2CO_3$$

$$H_2CO_3 \leftrightarrow H^+ + HCO_3^-$$

Although this process enables more carbon dioxide to dissolve into the sea from the atmosphere, it also sets in motion very important changes in seawater acid-base chemistry that are collectively known as *ocean acidification*. This does not mean the ocean is becoming acidic; it is becoming less alkaline.

The concentration of hydrogen ions in a solution is measured according to the pH scale, where

$$pH = -\log [H^+]$$

Therefore, the more hydrogen ions, the lower the pH. Notice that it is a log scale to the base 10. Every unit decrease on the pH scale means that the hydrogen ion concentration has increased tenfold. Solutions with a pH of less than seven are acidic, while those with a pH greater than seven are alkaline.

Until recent times, global mean seawater had a pH of 8.16. However, we now have excess carbon dioxide dissolving into the ocean surface waters, which increases the concentration of hydrogen ions. This process is estimated to have led to an increase of global ocean acidity of about pH 0.1 since the Industrial Revolution. That is the equivalent of an increase in hydrogen ion concentration of 25%. It is actually the rate of change that is of prime concern, given our knowledge of the amount of carbon dioxide known to be entering into the atmosphere and "waiting" to enter the oceans. It is thought that the oceans may have never undergone such a rapid acidification in the past. The IPCC estimated in the AR5: "Earth System Models project a global increase in ocean acidification for all RCP scenarios. The corresponding decrease in surface ocean pH by the end of the 21st century is in the range of 0.06 to 0.07 for RCP2.6, 0.14 to 0.15 for RCP4.5, 0.20 to 0.21 for RCP6.0, and 0.30 to 0.32 for RCP8.5."[18]

The conspiracy worsens because the free hydrogen ions can also combine with carbonate (CO_3^{2-}) ions in the water to form bicarbonate:

$$H^+ + CO_3^{2-} \leftrightarrow HCO_3^-$$

This has major ecological implications because it removes carbonate ions from the water, making it more difficult for some marine organisms (especially molluscs and corals) to form the $CaCO_3$ they need for their shells and skeletal structures. At present, the surface ocean is generally saturated with respect to calcium carbonate (including aragonite and calcite). This means that under present surface conditions, these minerals have no tendency to dissolve. Therefore, marine organisms do not have to work hard to take up and retain the calcium they need. However, the solubility of calcium carbonate increases with pressure and decreases with temperature. The pressure dependence dominates over the temperature dependence. As a result, the deep, cold waters of the oceans are calcium carbonate undersaturated. The depth at which the upper-saturated waters meet the deeper undersaturated waters is called the *saturation horizon*, below which it becomes difficult for organisms to find calcium carbonate. If you look into the literature, you will also find references to something called the $CaCO_3$ saturation state (Ω). In regions where Ω_{arag} or Ω_{cal} is > 1.0, the formation of shells and skeletons is favoured. Below a value of 1.0, the dissolution of pure

[18] IPCC, 2013. "Summary for Policymakers." In Stocker et al. (eds.), *Climate Change 2013: The Physical Science Basis*, p. 27. Cambridge, UK, and New York: Cambridge University Press.

aragonite and unprotected calcite shells will begin to occur. Studies have shown that because of the increased movement of anthropogenic carbon dioxide into the surface waters, the zone occupied by the calcium carbonate saturated surface waters is growing smaller. In some regions, the aragonite saturation horizons have moved upwards to shallower depths by 50–200 metres compared to their positions before the Industrial Revolution (particularly in upwelling zones). By 2050, this saturated surface zone may begin to completely disappear in some areas.

This overall story has some finer detail. It has been estimated that by the end of the twenty-first century, a 60% decrease in the concentration of surface ocean calcium carbonate and an increase in acidity of 0.4 pH may occur. Will organisms be able to adapt to these changes in such a short time frame and what will be the impact on the marine ecosystem if such fundamental food web building blocks as planktonic coccolithophores are significantly less productive? Some marine animals with shells have always occupied waters where Ω_{arag} or Ω_{cal} is <1.0. They can do this because building and maintaining their skeletal structures is a continuous and dynamic process. However, it requires the expense of devoting more energy to this task as against, for example, reproduction. Therefore, adaptation may be feasible, but whether it can happen given the rate of change is unknown.

AMAP completed an *Arctic Ocean Acidification Assessment* in May 2013. It was a difficult task because there is so little Arctic information available. Nevertheless, the assessment showed that the Arctic region is subject to the same overall trend towards acidification that is occurring elsewhere around the world, which leads to a lowering of pH from the former global mean of about 8.16. The assessment reported that in the Barents Sea and around Iceland, seawater pH has been declining at a rate of about 0.02 per decade since the late 1960s. However, once again, the Arctic and Antarctic ecosystems have been singled out by nature for special treatment of the kind we would rather not see. These two ecosystems are preconditioned to be especially vulnerable to acidification. This is taking place because a number of natural conditions and processes are amplifying the ocean acidification in the Arctic. First, pH decreases with temperature and Arctic waters are cold. Second, the Arctic Ocean receives a great deal of freshwater that has a highly variable pH, but it is always less than that of seawater and generally has a lower alkalinity (ability to resist pH change upon addition of free hydrogen ions). Also, remember the freshwater that results from melting sea ice and that a huge accumulation of freshwater is presently sitting in the Beaufort Sea. Other processes, such as the nature of coastal erosion, the location of rivers carrying organic matter and sediment and seasonality, all result in a complex pattern where acidification is not uniform across the circumpolar Arctic seas.

How these all conspire together has been studied in some detail on the coastal shelves around Alaska, especially in the Bering and Chukchi seas.

In the spring, the combination of sea ice melt and the arrival of freshwater river runoff results in these lower-density waters lying above the more salty waters below. The water column is said to be highly stratified, which results in some of the most intense phytoplankton primary production known anywhere. On the eastern Bering Sea shelf, this intense photosynthetic activity removes most of the CO_2 from the surface waters, which allows for yet more CO_2 to enter from the atmosphere. A great deal of the organic matter produced by the phytoplankton during the spring phytoplankton bloom falls to deeper depths, where it eventually decays – either directly or indirectly. This leads to an injection of CO_2 into these waters. This further lowers the pH and significantly depresses the Ω by about 0.2. According to a study reported in 2010 and 2011, these subsurface waters became undersaturated with respect to aragonite (but not calcite).

Undersaturation has also been observed in the Chukchi Sea and in the surface waters of the Canada Basin. Here, it was attributed to the new ice-free state, which allowed more atmospheric CO_2 to dissolve into the ocean. This is thought to have lowered alkalinity and calcium ion concentrations of the low salinity freshwater that has been accumulating in the basin.

The AMAP assessment judged that the short and simple food web structure of Arctic marine ecosystems leaves them potentially vulnerable to acidification. They considered that shellfish, fish eggs and early fish larval stages are likely to be adversely impacted. However, how organisms will cope with ocean acidification in the Arctic (and elsewhere) remains a topic in need of answers. As a general rule, organisms that require calcium grow more slowly under acidifying conditions in laboratory experiments. This is thought to indicate that they are diverting more energy resources to the task of acquiring and maintaining calcium.

The eastern Bering Sea supports one of the most intensive fisheries in the world. It supplies about 47% of the U.S. fish catch. However, it is looking as if it may hold the unfortunate distinction of being the first ocean area adjacent to North America to be adversely impacted by climate-induced ocean acidification. How much or when the fishery will be impacted is unclear, but it is perhaps safe to assume that shellfish and organisms that harvest them (including the walrus) are facing troubling times. In a paper published in *Nature* in 2005, Orr and colleagues predicted that "high-latitude surface waters, already naturally low in calcium and carbonate ion concentration, will be the first to have undersaturated surface waters with respect to aragonite. Undersaturations for the calcite phase of calcium carbonate are expected to follow 50–100 years later". Unfortunately, it seems we are right on course.

Climate Change Impacts on Arctic Ecosystems

In "Personal Beginnings," I explained why I have concentrated on the physical, chemical and toxicological actors in the story of the Arctic Messenger and have said comparatively little about biological and human aspects. This focus is partly an artefact of my lack of expertise in these topics, partly because popular treatments usually emphasize wildlife impacts and partly because I have wanted to emphasize the fundamental drama that the physical, chemical and toxicological actors are playing, which causes stress on the Arctic ecosystem. They are ultimately responsible for the biological changes presently under way in the Arctic. These in turn have led to the consequential cultural and socioeconomic impacts to Arctic indigenous peoples and nonindigenous residents. My neglect of the biological and cultural issues is a shortcoming I hope others will remedy. However, this delinquency is so serious in relation to climate change that you will find in subsequent paragraphs a threadbare attempt to sketch out the breadth of these impacts. I hope others will soon do a much better job. In the meantime, I can thoroughly recommend the ACIA summary written by Susan Hassol; the ACIA itself (especially chapters 10 to 17); chapters 10, 11 and 12 of the 2011 *SWIPA Report*; the *Arctic Report Cards* for 2011 and 2012; W. G. Cramer et al. (2014) on the detection and attribution of observed impacts from the IPCC AR5 working group report; and the 2002 compendium of indigenous observations edited by Igor Krupnik and Dyanna Jolly.

This last-mentioned work provides quite an innovative perspective. It describes the result of a project where Alaskan and Canadian Inuit from the Bering Sea to Baffin Bay and Davis Strait collected their own observations as Inuit on changing environmental conditions. For as long as indigenous peoples have lived in the Arctic, they have depended for their survival on a vast corpus of knowledge about their environment. It has been handed down orally from generation to generation and fine-tuned by personal experience. Krupnik and Jolly's book is a very compelling read.

Why is it compelling? First, you will notice the scope and detail of the observations, which range from the general to the particular. Second, circumpolar indigenous peoples' organisations brought their own environmental observations related to change into the ACIA assessment. This not only increased the geographic scope of indigenous observations of environmental change reported through Krupnik and Jolly but also added new Arctic ecosystems (such as the lands of the Sami). It is remarkable how much convergence there is in the observations of indigenous peoples separated by thousands of kilometres and occupying very different habitats. Third, you will quickly notice that indigenous observations of Arctic change and those of the formal scientific studies rather resemble two people taking two different roads but who nevertheless arrive at the same place, particularly with respect to the observation of ecosystem change. Finally, you will find yourself developing a sense of pathos that is gripping and depressing. As the title from Krupnik and Jolly implies, modern indigenous people find themselves living in a physical and biological environment that differs from that described by the cumulative knowledge passed on by their ancestors. Their world has changed so much in their lifetime that they have, for example, even lost confidence in their ability to predict the weather. This is a fundamental failing for someone many kilometres from home in the Arctic.

What it really means is that the basic parameters that define what is understood as the Arctic ecosystem are becoming invalid for a region that is warming at twice the rate of the global average. As we look to the North, are we watching the initial stages of the collapse of the Arctic ecosystem that has existed for the past several thousand years – something that a climate scientist would call a regime change? We will return to this question later.

The following section is a summary of some observations drawn from indigenous knowledge and formal science on how the Arctic ecosystem is reacting to Arctic warming. Before going further in this chapter, please take a good look at Figure 10.11, provided by Dáithí Stone. Any ecosystem is defined by the connections between its various components. Ecosystems that have developed under

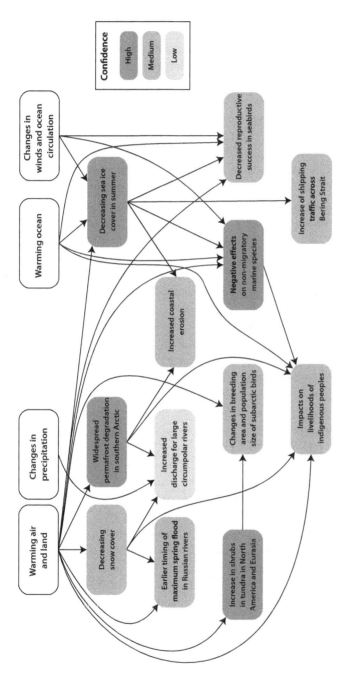

Figure 10.11 *Selected cascades of observed impacts in the Arctic attributed to observed climate change*

physical conditions that are challenging to biological processes (such as the Arctic) tend to be simple and much less resilient to changes in any of their components than ecosystems that enjoy more friendly conditions (such as temperate woodland). The figure provides a simplified view of the cascade of changes that are being observed today in the Arctic ecosystem and that are attributed to climate change. The shading indicates the degree of confidence that observed climate change has played a major role in the observed impacts. Note that we humans are a part of the ecosystem and are ourselves impacted.

It all begins with the physical reaction of the cryosphere and the resulting domino effect through ecosystems to indigenous and non-indigenous Arctic communities. According to Derksen and Brown, by 2012, the extent of summer snow had declined 17.6% per decade relative to the 1979–2000 mean. Snow cover now vanishes earlier and is reestablished later in each year. The next domino to fall is permafrost, which, lacking protection from the summer warmth, is thawing in many areas, while its southern limit is migrating northward (for example, 30–80 km in Russia between 1970 and 2005 and 130 km in northern Québec over the last 50 years). At the same time, the extent of summer sea ice is declining (by 2012) at a rate of 13.0% per decade relative to the 1979–2000 average and the magnitude of the trend has increased every year since 2001. An exposed Arctic ocean develops weather systems that lead to greater wave activity and to an increase in precipitation of about 8% in some regions. A variable proportion of this precipitation is now appearing in some widely separated regions as winter rain.

When we combine summer marine storms, rising sea levels and melting permafrost, we have coastal erosion and shoreline retreat occurring with unprecedented severity along vulnerable parts of the Beaufort, Laptev and East Siberian coastlines (at rates of more than 2 metres per year). In the words of a leader of the Inuit Circumpolar Council, "[s]ome of our communities are eroding into the ocean in front of our eyes." For example, about 400 people live in the Alaskan village of Kivalina located on a spit of land jutting out into the Bering Sea. It is no longer protected by land-fast ice in winter and storms are progressively eroding the spit. It is believed that in less than 10 years, the spit and Kivalina will have disappeared.

Cryospheric change has also resulted in terrain instability away from the coasts, leading to widespread impacts on Arctic infrastructure. These range from damage to buildings, permanent roads, sewage systems, annual freshwater supply and oil and gas transportation systems. Many Arctic communities and industries rely on winter roads for much of their annual supply of fuel, heavy materials and even certain foods. These winter roads follow lakes, rivers and wetlands. The length of the Arctic ice road season is shortening (particularly in Canada and Russia), becoming unpredictable and, in more southerly locations, is simply no longer an option in some years.

The same alliance of conditions – further complicated by increased freshwater flow from melting snowfields, glaciers and ice sheets – is resulting in the flooding of many existing coastal wetlands and consequential loss of habitat. However, the same conditions are creating new wetlands in some areas, while the longer warmer summers are drying out peatland ecosystems in other locations.

How are Arctic terrestrial plants responding? There are three major Arctic vegetation zones. In the far north lies the polar desert. Here, mosses, lichens and ground-hugging vascular species, such as purple saxifrage, mountain avens, Arctic poppies and moss campion, dominate the sparse flora. There are very few shrubs. Often, the most abundant vegetation is found beneath seabird cliffs. Moving south is the tundra biome. It is treeless in the North and characterized by the appearance in the North of dwarf shrubs, such as *Salix polaris* and *Salix arctica*, while still further south, larger species, such as *Salix reptans*, and *Betula nana* and *Betula exilis* appear. The ground storey is carpeted by an increasing diversity of mosses, sedges and grasses. Continuing south, the northern parts of the taiga biome (boreal forest) are considered to be within the Arctic. At its northern extent in North America, it is characterized by trees, such as white spruce (*Picea glauca*) and black spruce (*Picea mariana*), together with shrubs that can grow relatively tall, such as green alder (*Alnus viridis*). With the increasing floral diversity from north to south comes a corresponding richness in the diversity of insects, birds and mammals.

In the past, the temperature and precipitation environment of the coastal tundra in summer has been dominated by cold air related to the oceanic pack ice. This is changing rapidly as the summer ice recedes and the ocean warms, thus enhancing the warming of the neighbouring tundra. A warming climate increases the time in which low vegetation will be uncovered by snow and exposed to the light needed for photosynthesis. It also extends the time when liquid water will be available.

Therefore, botanical ecologists have anticipated two fundamental changes. First, they are expecting an overall increase in the growing season and a corresponding increase in plant growth. Second, they expect that the vegetation zones will shuffle northwards. The northern-most species of the boreal forest will creep into the tundra, which itself will eat into the southern parts of the polar desert.

There is clear evidence that the first of the biologists' expectations is well under way. Satellite data extending from 2000 using an instrument called a Moderate Resolution Imaging Spectroradiometer (MODIS) has indeed shown that the tundra growing season is beginning earlier and ending later. Another satellite-borne instrument called the Advanced Very High Resolution Radiometer (AVHRR) has run transects above the North American and Eurasian Arctic, from which an index of "greenness" has been derived. This index is closely related to aboveground plant biomass. Therefore, it is possible to track changes in the overall quantity of vegetation (biomass) all over the Arctic without being bitten by a single mosquito. To cut a long story short, between 1982 and 2011, the index used for greening shows an increase of 15.5% and 20–26% in different parts of the North American Arctic and 8.2% in the Eurasian Arctic. This generally reflects shrub growth.

The second of the botanists' expectations is a more complicated story. Although there are reports of a northward expansion of some boreal zone species into the tundra, it seems more common at present for tundra species that in the past have kept close to the ground to now grow much taller, creating a new form of shrubby deciduous woodland. In North America, one of the last species of conifer to disappear as one passes from boreal forest to tundra is the black spruce, so you would expect it to be one of the first species to march into a warmer tundra. However, studies in Alaska show that the new warmer temperature regime brings growth-limiting difficulties for this species when soils become drier or the plant tries to grow in the warmer spring air temperatures before there is liquid water available to the roots. Another species found close to the North American tree line is the white spruce. Once again, studies tell a complicated story. This tree seems to come in two types, both of which are present in about the same numbers. One type grows well if summer temperatures are higher than 16°C, but for the other type, growth is reduced. Before 1950, it was rare to experience 16°C near the tree line, but now it has become quite common. My reaction on hearing about these studies was to think that probably a genetic condition is involved, called a *polymorphism*, that maintains two different varieties in a population. It enables a quick response to changing conditions. Perhaps this is the case, but more recent studies show that in most tree line study areas, the "warm tolerant" types are also beginning to show less enthusiasm for the changing conditions. It is quite likely that grazing from herbivores, especially caribou and reindeer, is also influencing the dynamics of vegetation change. Another complicating factor is forest fires. Historically, these were rare events on the tundra, partly because of temperature and the late disappearance of snow but also because of the scarcity of potential fuel. Now we have accelerating shrub growth and a corresponding increase in the frequency and severity of fires, particularly in Alaska.

Confronted with this convoluted picture, some botanists are proposing that the data fit better with a theory that suggests ecosystems tolerate new conditions until critical thresholds are passed. A quick change then occurs, perhaps leading to a new succession of a different type of ecosystem.

I am not a botanist, but I wonder if what we are seeing is simply the slow way in which a new climax community will be established on the southern tundra. Here on the west coast of Canada, the climax community is dominated by cedar and Douglas fir. When logging strips the forest, it is initially replaced by a mixed bag of shrubs, including alder. After about several decades, the alder gives way to a community dominated by tall coniferous species, such as hemlock. Finally, after the passage of more decades, the cedar and Douglas fir reassert their dominance.

Arctic vegetation is an obviously vital element for the habitat of Arctic terrestrial animals. How are they impacted by change? Lemmings exist as several species in North America and Eurasia. To a large degree, they shape the tundra through their grazing of vegetation and by being the main food for predators, such as the snowy owl and Arctic fox. Typically, their populations undergo cycles. At the cycle peak (which usually lasts for only one year), they are most numerous. The population then crashes and reaches another peak after about three to five years. Recently, a number of studies suggest a trend towards lower peaks and a slower frequency of the cycle. Both trends may be caused by winter snow conditions being less than ideal for reproduction (which occurs in winter). In the sub-Arctic of Norway, recurrent snow-thaw cycles and winter rain in a warming climate create icy layers that make it difficult for lemmings to reach their food plants. We will see the same problem with caribou and reindeer.

Caribou and reindeer also undergo population cycles that in this case are quite slow. Although it varies between herds, a cycle may take 30–40 years to complete. This makes it difficult to detect population trends that may be associated with climate change over the last few decades. For example, since 1970, the number of caribou and wild reindeer in 23 surveyed circumpolar herds has declined from about 5.5 million to 2.7 million. There are now some indications that in just the last two years, the declining phase may have ended and that at least for the herds occupying the western and central parts of the coastal regions of North America, an increasing trend may be setting in.

Phenology is the study of the relationship between the life cycles of animal and plant species and the periodicity of cycles in their natural environment. When climate changes rapidly, plant and animal species and communities can suddenly find that such relationships are disrupted. For example, several climate-related stresses have been observed to impact caribou and reindeer in the North American and Eurasian herds. We have already encountered how intermittent winter freeze and thaw cycles can create a strong crust of ice over vegetation. This makes it very difficult for caribou and reindeer to reach their winter food plants. Robert Corell has told me that the Sami reindeer herders refer to this type of snow/weather condition as *goavvi*. In these circumstances, herd losses of upwards of 50–75% have been observed.

Another worrying group of observations by indigenous peoples and scientists is called *trophic mismatch*. Herds have evolved traditional migratory behaviours and pathways to move from winter-feeding areas to calving grounds and to summer areas. The prime time for feeding is when vigorous new growth is appearing on the shrubs. However, if snow conditions prolong the first part of a migration pathway that involves crossing a river, the herd may arrive at the river after ice breakup. This could lead to the significant mortality of calves as they try to cross. Exactly this scenario has been observed occurring with the Porcupine herd as they cross the Porcupine River (north-western Canada). This river is progressively breaking up earlier in response to regional warming. The compendium of Krupnik and Jolly and the 2005 ACIA provides a comprehensive description of these types of impacts. Caribou and reindeer are a vital source of food and clothing for Arctic circumpolar indigenous peoples.

One of the big unknowns concerning potential changes in some terrestrial Arctic ecosystems relates to the long-term availability of water. Much of the Arctic has little precipitation, particularly the Canadian Arctic, which comes as a surprise to some visitors when they see the vast expanse of lake and muskeg. It is all a question of evaporation. In the past, little snow or rain evaporated and permafrost inhibited drainage. But what will happen when a critical temperature is reached where evaporation and runoff exceed precipitation? This may be one of the most important questions being faced by the future tundra and northern boreal forest ecosystems.

In the oceans, cryospheric change is also the proximate cause of the most significant ecosystem and socioeconomic impacts. We have already discussed the layer of very low salinity (almost fresh) water that has been accumulating in (and extending beyond) the Beaufort Sea and its potential to enhance acidification of the Arctic Ocean. According to the 2011 SWIPA team, this reservoir (which has only appeared in recent years) now amounts to 7,700 km^3 of freshwater – equivalent to 1 metre of water over the entire continent of Australia. Oceanographers and climate scientists are watching the fate of this water with interest – partly because it inhibits mixing with underlying waters, partly because of its impact on upper ocean pH and partly because of events that occurred during the last interglacial period. Very sudden injections of freshwater resulting from the collapse of the Laurentide ice sheet are thought to have very significantly weakened the Atlantic meridional overturning circulation (AMOC) and hence also weakened the Gulf Stream.[19] This led to major climate repercussions around the world, especially in Europe, which experienced a dramatic cooling. However, the amount of freshwater involved in the Laurentide event vastly exceeds that likely to be

---

[19] The movement of surface waters to depth in the Arctic and Antarctic is called *meridional overturning* and is an example of thermohaline circulation. Without meridional overturning, the ocean's capacity to efficiently transport CO_2 and heat from surface waters to the deep would be dramatically reduced. The North Atlantic Arctic region and some parts of the Antarctic contain virtually the only areas where surface waters are sufficiently cold and subjected to such large-scale salt rejection from the creation of sea ice to result in the creation of huge quantities of heavy (high density) cold and saline water. The deep water formed in the Arctic and Antarctic flows towards the equator. The Atlantic deep water penetrates far into the Southern Hemisphere.

suddenly available in the present-day Arctic. Nevertheless, IPCC AR5 considers that "[i]t is *very likely* that the AMOC will weaken over the 21st century. Best estimates and ranges for the reduction are 11% (1 to 24%) in RCP2.6 and 34% (12 to 54%) in RCP8.5."[20] You will recall that at the moment, some researchers fear that RCP6.0 or even RCP8.5 appears to be the GHG emission pathway that world economies are following. There is a little good news in that most scientists consider it is very unlikely that the AMOC will undergo a rapid collapse in the present century.

Sea ice supports a unique (epontic) ecosystem founded on algae and amphipods that occupy the undersurface of the ice. It is a major element of a food web that leads via Arctic cod to seabirds (especially the ivory gull), marine mammals and, ultimately, the polar bear. Clearly, the epontic food web itself will become rare as the surface area of available ice is diminished. However, it appears that in some areas, such as the Beaufort Sea, the epontic algae and amphipods have been in decline for some time. This is probably because they are intolerant to the low-salinity surface water that has been accumulating there over the last decade. Ringed seals birth and nurse their pups in snow caves they build on continuous sea ice and from which they exploit the marine foods associated with the ice and the ice edge. These marine foods include Arctic cod. Bearded seals, a benthic (bottom-feeding) species, birth and nurse their pups on shallow water pack ice. Both species also use the ice for long periods during moulting. They are (particularly the ringed seal) the polar bear's primary prey. The long-term outlook for all three species suggests a risk of extinction. Polar bear populations in some areas are already showing signs of malnutrition. Remember that unlike the omnivorous grizzly and brown bears, the polar bear is exclusively carnivorous. It has no opportunity for adaptation to a marine life without ice. There are 19 subpopulations of polar bear, seven of which are in decline. The results of two studies linked subpopulation decline (survival) to changes in the seasonal availability of sea ice (western Hudson Bay and southern Beaufort Sea populations).

The walrus is a large benthic-feeding marine mammal. By drifting with the moving pack ice edge, it has historically avoided

---

[20] IPCC, 2013. "Summary for Policymakers." In Stocker et al. (eds.), *Climate Change 2013: The Physical Science Basis*, p. 24. Cambridge, UK, and New York: Cambridge University Press.

overgrazing its feeding grounds. In response to the ice edge no longer being over shallow waters, some populations settle on rocky islands or headlands and are already tending to deplete their summer food supply. Seals and walrus provide a very large proportion of the total diet of Inuit communities. Like most indigenous peoples who maintain a traditional way of life, there is no acceptable replacement for these foods because their entire culture is based on them. They are (as already noted in the chapter on POPs and heavy metals) also much more healthy to eat than store-bought foods available in northern towns or settlements. A dietary transition to store-bought foods also imposes economic stress in many Arctic communities.

Finally, we have already examined in the POPs and heavy metals chapter how climate warming is already beginning to remobilize certain toxic, semivolatile pollutants, which had previously been immobilized in Arctic soil, sediments, snow and ice. Once remobilized, they then become available for biological uptake and have the potential to lead to wildlife and human health implications.

This is the end of my threadbare attempt at ecological and societal impacts. It falls a very long way from being comprehensive, but I hope it has given an idea of the scope of change that can be expected even if those changes may at present be only in their infancy. At present, what is clear is that major elements of the Arctic marine ecosystem based on sea ice, such as polar bears and several species of seal, are possibly facing extinction within the next few decades. With this come the most important socioeconomic and cultural impacts imaginable for the people whose lives depend on these resources. What the future holds for terrestrial ecosystems is at present less clear, but it is safe to predict that major ecosystem-wide impacts lie ahead.

Indeed, the Arctic is not alone. It is generally accepted that early signs of climate-driven, large-scale regime shifts in terrestrial, marine and freshwater ecosystems can now be detected in many regions of the globe – from polar regions to tropical coral reef systems. To find out more on how widespread these changes are, I recommend chapter 18 of the IPCC *AR5 Working Group II Report*.

The Long and the Short of It

This brings our very brief review of Arctic climate change to an end. However, there is much more that should be said, for example, on ecological, health and socioeconomic ramifications. This is especially so in terms of the potential development of hydrocarbon and mineral resources, but I think we can now claim we have a sufficiently robust understanding of how:

- Self-regulation of the world climate system is being severely impacted by our ever-increasing injection of more and more GHGs into the atmosphere.
- The result has been a global mean 100-year surface warming temperature trend from 1906 to 2005 of 0.74°C.
- Computer models project that the future will unfold to eventually settle on a new and warmer global climate and empirical observations around the world all tend to support the models.
- The geological record and climate models agree that even if we could reach zero GHG emissions in this century, temperatures would hardly drop in the next 1,000 years due to the huge stockpile of CO_2 we have added to the atmosphere.
- The annual average Arctic temperature has increased at about twice the rate of the rest of the world. Alaska, western Arctic Canada and central Arctic Russia are warming more than other regions. In Alaska and western Arctic Canada, average winter temperatures have risen by between 3 and 4°C.
- Arctic snow, glacial and ice sheet ice and permafrost are disappearing much faster than had been expected. Mean global sea levels are projected to rise by 0.9–1.6 metres by 2100, but regional variability will give significantly higher levels in some areas. The sea level rise will not be geographically uniform.
- Summer Arctic sea ice is melting much faster and earlier than expected. We now expect to see an ice-free Arctic ocean in summer somewhere between 2030 and 2050.
- Increased levels of CO_2 in the atmosphere and its subsequent uptake by the ocean is leading to a lowering of ocean pH (ocean acidification). The Arctic Ocean is particularly sensitive to the chemistry that leads to acidification. In some parts of the Arctic, levels of acidification have already been recorded that have the potential to cause significant ecosystem impacts.
- There are signs that a warming Arctic that lacks the insulation of summer ice between sea and air may be beginning to significantly influence mid-latitude weather.
- Arctic marine, terrestrial and freshwater ecosystems are all beginning to show signs of being impacted by Arctic warming.

- For better or for worse, Arctic indigenous and nonindigenous peoples face an uncertain future.
- We and our political leaders are running out of time to effectively control or alleviate the progression of these events.

With these facts in mind, what about that question posed near the beginning of the ecosystem impact section? Are we watching the initial collapse of an Arctic ecosystem (physical and biological) that has existed for at least the past several thousand years? All the evidence indicates we have embarked on a course towards this outcome. We have learned in this chapter of the moderating role that the Arctic plays in our global climate system. Could the Arctic Messenger therefore be telling us something foreboding about the possible future of ecosystems elsewhere once this moderating role has weakened?

PART IV

What Does All This Mean?

In this section, we take stock of where we are. Our knowledge of the state of the Arctic environment ultimately depends on universities producing scientists capable of conducting focused interdisciplinary research, on Arctic peoples being willing to enrich such work with their own indigenous cultural knowledge and on governments providing adequate funding and infrastructure. In the chapter entitled "The Long and the Short of It," we will revisit our six environmental stressors by asking the following three questions: (1) When was the stressor anticipated, (2) when was it perceived (detected) and (3) what did we as a regional or global community do about it? This exercise enables us to evaluate how successful governments have been in organising international cooperative remedial actions and to identify future needs. Climate change is the most foreboding issue we face, but it remains the only stressor lacking any promising international cooperative action founded on a science-based approach. In the Epilogue, the possibility of more direct involvement by the Arctic Council in addressing climate change remedial actions is discussed.

11

Thoughts on Education, the Training of Arctic Scientists and Arctic Research

"The great aim of education is not knowledge but action."
Herbert Spencer, English philosopher

Before starting this chapter, I have to admit I have no educational expertise. To share my thoughts on education is presumptuous or foolhardy or hazardous. It is probably all three. However, I am going to take the risk. I firmly believe that the importance of environmental education cannot be overestimated. Three aspects really surprise me when talking to nonspecialists about the state of the Arctic environment and of its implications to the environments in which most of us live. The first is the general public's high level of interest in the Arctic. The second is how little many people know about the Arctic and environmental sciences in comparison with their understanding of topics that are more demanding on the mind. The third is the widespread reach of many misconceptions about the state of the Arctic and global environments.

As global society moves deeper into the twenty-first century, it is clear that socioeconomic models of the last two centuries are unsustainable. Our globe has limited resources and limited capacity to deal with our persistent wastes. Developing countries are determined to gain their fair share of global resources and attain lifestyles presently only enjoyed in developed countries. Our unrelenting use of hydrocarbons to power our economies is fundamentally altering Earth's climate system in a way that will be difficult to reverse. At the same time, the global human population is increasing. It grew from 1 billion to 7 billion between 1800 and 2011 and it will hit 8 billion by 2025. It will pass 10 billion in 2040. We are entering unknown territory. We do not know what all this will mean. However, it is clear we are exceeding the capacity of our planet to support the implied future utopian global economy and its ability to recycle our wastes. There will be uncertainty ahead for many alive today and for all our descendants. What we do know is that nowhere is being impacted more than the Arctic and its indigenous communities.

What can be done? The usual response is to call for political action. However, the necessary actions needed to secure our common future carry the risk of short-term pain and lifestyle adjustment. Most readers will come from countries led by democratically elected governments. A political party with a policy aimed at implementing such actions can expect swift slaughter at the polls because our electorate can quickly be persuaded there is nothing to be concerned about. They will use their vote elsewhere. In the middle of the first decade of this century, a Canadian mainstream political party (that has frequently formed the government) fought a federal election on a sustainable development platform. It has still not recovered.

How can we respond democratically to this situation? I see no antidote to the malaise other than education and research. We need to equip our general population with a sound knowledge of what we are doing to the Arctic and to our planet, an understanding of why the ways of the past are unsustainable and a willingness to make genuine changes to current lifestyles. Only then can we expect scientific knowledge to foster meaningful national and international political action.

I think there are at least five elements to this educative process:

1. **Education throughout the school system – from kindergarten to high school graduation:** This is the most important element of all. It is during our childhood and teenage years that our minds are moulded. It is when we learn how to behave in our society and our culture. We learn about our responsibilities to others and form our expectations for the future. My generation was born between about 1940 and 1960. Many of our world leaders received their school education in this period. We learned and absorbed the then-prevailing paradigm for humankind's relationship with our environment. It went something like this: "It is ours to use and its resources are without limit." For those of us with a Judeo-Christian background, it was a philosophy that can be traced back to an interpretation or translation of the first chapter of Genesis (verse 26) and it has served us well. I learned absolutely nothing to challenge this paradigm at any of my schools. No teacher suggested that we, along with all biota, live in an interrelated state of balance that also includes the geophysical world. It was a worldview far removed from

that of many other societies where all individuals lived more closely with their environment. Is it any wonder that the people of my generation have such difficulty in relating the limits of sustainability of our planet's environment to our personal goals of wealth generation?

An alternative view of humankind's relationship to its environment was made famous by the words attributed to Chief Si'ahl (Seattle) in a letter to the U.S. government in the early nineteenth century: "Will you teach your children what we have taught our children? That the earth is our mother? What befalls the earth befalls all the sons of the earth."

In the mid-1970s, school curricula at junior and high school levels began to slowly address these shortcomings. However, the impact varied greatly from school to school depending in large part on whether individual teachers had been prepared to donate their own time to extracurricular activities. Of course, I have no data, but as I mentioned in the first chapter, I cannot believe that the children enriched by these teachers have not questioned the old exploitation paradigm into their adult lives. It must be far easier for them to absorb, for example, ideas on sustainable economics and the relationship between our lifestyle and the looming spectres of global warming and population growth.

If our minds are indeed moulded in our childhood and youth (and neuroscience tells us that they are), then our long-term future depends on the worldview of children now still in school. In the meantime, we continue to make decisions that compound the problems to be faced by these children in their adult lives. We are severely reducing the options that will be available to them when they are forced by circumstances to address the need for global sustainability. It is unconscionable that everything is not done to provide them with an education to prepare them for the task ahead.

2. **Adult education:**   Adult education is crucial if we have any hopes of seeing the concept of sustainability being promoted by our present-day electorate. The ultimate long-term goal of an educational strategy in the context of a sustainable economy should be to enable the general population to make informed decisions on the way they live in relation to their environment. This would enable them to make educated choices on the types of governmental policies they would support at election time.

Ongoing adult education using mass media, such as television, magazines, newspapers and the Internet, are all readily available.

It is bewildering how much high-quality information is available on such websites as those of NASA, NOAA and TED. The Internet is also now being used to make available "massive open online courses" (MOOCs). Another recent example is the web-based Snowy Owl Talks, a collaboration between the University of the Arctic (see the forthcoming discussion) and GRID Arendal (a Norwegian foundation that works with UNEP to support and facilitate environmental decision making). Could some mechanism be used to organise these fine resources into narrative themes aimed at maintaining and updating the level of environmental knowledge reached at the time of school or college graduation? A residual problem that remains is finding ways to entice and engage the public in using these resources.

3. **Primary and high school needs of Arctic indigenous communities**: In much (but not all) of the Arctic, specialists who deliver regional programmes (including the conduct of scientific research) are rarely indigenous to the Arctic. This is particularly evident in Canada. When the Arctic Council was being negotiated in the mid-1990s, I spent some time supporting Mary Simon, who was at the time Canada's ambassador for Denmark and for circumpolar affairs. Mary is an Inuit leader from Northern Québec. She has spent most of her life working for organisations concerned with the rights and well-being of Inuit in Canada and in the circumpolar Arctic. It was from Mary that I first learned a sobering statistic: Only about 25% of Inuit schoolchildren in Canada graduate from high school.

Mary mentioned that statistic several times to me during those years, particularly whenever I was carried away by the needs of Arctic science. It was something that troubled her a great deal. About 10 years later, Mary was leading a national Canadian Inuit organisation (the Inuit Tapiriit Kanatami). In 2006, she launched an initiative that resulted in an accord being signed in 2009 between Inuit organisations and regional educational agencies (those responsible for delivering education in Nunavut). It set up a National Committee on Inuit Education (chaired by Mary) that was tasked with developing a national strategy on Inuit education. The strategy was published in 2011. It includes a pragmatic set of recommendations that strike a good balance between initiatives that preserve Inuit uniqueness, such as language and culture, with a more typically mainstream Canadian education. The blend of these two aspects should appeal to young Inuit and better prepare them for occupying all levels of society in Arctic Canada. Particular attention is paid to very early childhood education. I think the strategy is a perfect example of the type of initiative that could dramatically improve the high school graduation rate. Of course, to succeed, it will require strong and sustained financial and political support.

There is no limit to the age at which schoolchildren can become absorbed with Arctic science. Most Arctic countries have programmes to address this early intellectual appetite, but the only country for which I have some knowledge is Canada. In the early to mid-1990s, my colleagues and I faced the problem of explaining to Inuit how Arctic foods had become contaminated as a result of radionuclide and contaminant releases at mid- and low latitudes. One of our actions was to fund indigenous peoples' organisations in the Arctic and territorial governments to develop and deliver school units explaining the risks and benefits of consuming or not consuming these otherwise very healthy traditional foods. The curriculum units encompassed many educational age groups and the children were eager participants. More recently ArcticNet, the main institution now funding Arctic science in Canada (under the care of Martin Fortier), has integrated northern educational outreach into its portfolio of activities. On a circumpolar scale, this theme was also prominent in IPY-4 and many of the initiatives born at that time survive today.

4. **The availability of university-level education in the Arctic region**: Although there are several well-established universities in Arctic Scandinavia, Alaska and Russia, there are none in Arctic Canada. Instead, higher education is available through community colleges (such as Yukon College, Aurora College and Nunavut College). These are excellent institutions, but they do not at present provide an education that can lead to and deliver postgraduate studies. They are not yet equipped to prepare Arctic indigenous peoples with the skills that would enable them to occupy all levels of society. The emotional stress on Arctic indigenous young people who are obliged to leave their cultural environment in order to obtain a university degree presents a difficult hurdle to their achievement of academic success. The lack of such institutions also presents a barrier for research into topics of particular interest to Canada's Arctic indigenous peoples (including the environmental sciences). It is not surprising that one of the recommendations from the national strategy on Inuit education in Canada is for the establishment of a university in Arctic Canada.

The creation of the circumpolar University of the Arctic is an interesting story. It also demonstrates what can be achieved if a handful of people with different vocations set to work on an idea. In January 1997, the Netherlands hosted a reception at an Arctic Monitoring Assessment Programme (AMAP) meeting in Groningen (the Netherlands). In casual conversation, Bill Heal (from Durham University in the United Kingdom) asked Lars-Erik Liljelund and me what we thought about the idea of a University of the Arctic. He explained the concept of a consortium of existing universities and institutions cooperating on a circumpolar scale and capitalizing on each other's strengths. It was a fascinating idea. I asked Bill to give me a brief proposal for the next meeting of senior Arctic officials (SAOs) that was only about six weeks away (March 1997 in Kautokeino, Norway). In the meantime, I set about finding a way to slip it into the agenda. It was too late to do it officially, so we slid it into my report on AMAP progress, calling it "A Concept Paper From Canada and Sweden for a Possible University of the Arctic". At the same time, I explained the idea to Terry Fenge of the Inuit Circumpolar Council in Ottawa and Terry offered to see if the indigenous peoples' organisations represented in the Arctic Council would support the proposal. Terry came back with their support within a week.

We used Bill's draft in the concept paper exactly as written but added a few paragraphs to give it Arctic Council context. We then made some off-the-cuff proposals on how to proceed as we watched the reactions of the SAOs at Kautokeino. There was interest (particularly from Finland), but the SAOs were not prepared to express support at that time. We therefore sought and were given a decision that asked Canada, Sweden and AMAP to further review the feasibility of the proposal by undertaking consultations and to propose options and plans for the next SAO session, scheduled for October 1997.

Canada and Finland, with the University of Lapland and the Circumpolar Universities Association based in Rovaniemi, agreed to support a small international task force chaired by Bill Heal to undertake the consultations and to prepare structural proposals for the university. It was at this point that I began working with Outi Snellman of the University of Lapland and Sally Ross of Yukon College (Canada). I met with Outi in Québec City prior to the 10–13 October meeting of the Circumpolar Ministers of Education to ensure the ministers could reassure the Arctic Council they were willing to back the initiative. I was able to locate supportive funding for the task force.

At the October 1997 SAO meeting in Ottawa, the task force report "A University of the Arctic: Turning Concept Into Reality" was well received by SAOs and a number of them began to express interest in a possible announcement on the topic at the autumn 1998 ministerial conference of the Arctic Council. They requested that Canada, Norway and Sweden ask the Circumpolar Universities Association to further develop the proposal in association with the circumpolar permanent participant indigenous peoples' organisations. I was charged with writing a "letter of invitation" to ask Outi Snellman and the Circumpolar Universities Association to prepare a feasibility study for the University of the Arctic. I was also requested to include guidelines on content. I sent the letter by e-mail on 16 October 1997 and on the following day received an e-mail from Outi accepting the new assignment. The task force was replaced by a working group of the Circumpolar Universities Association, chaired by Professor Asgeir Brekke of the University of Tromso (Norway). Most of the financial support I was able to organise from Canada went directly to Yukon College to support meetings of the working group in Whitehorse.

The next step came with the presentation of a progress report from the working group (made by Peter Johnson and Sally Ross) at a 9–11 May 1998 meeting of SAOs. At the close of the meeting, Canadian SAO chair Mary Simon (the then–Canadian ambassador for circumpolar affairs) remarked that the initiative could be a potential deliverable for the Ministerial Conference in Iqaluit (Canada) on 17–18 September 1998. To make this a reality, Mary asked Outi for another progress report. She also requested confirmation that the Circumpolar Universities Association would present the working group report to the Arctic Council and would welcome some form of endorsement of the initiative by the council ministers. All went well, and at a preparatory meeting in London (UK), I drafted language for the Arctic Council conference declaration concerning the creation of the university. The Arctic Council meeting took place from 14 to 16 September 1998 in Iqaluit. The report "With Shared Voices: Launching the University of the Arctic" was presented by Oran Young on behalf of the Circumpolar Universities Association. It was so well received that Oran and I were able to strengthen the text for the ministerial declaration. It finally read: We "welcome, and are pleased to announce, the establishment of a University of the Arctic, a university without walls, as proposed by a working group of the Circumpolar Universities Association. We note the kind offer of Finland to support the interim secretariat. We encourage the working group to continue its efforts to consult with northern educational and indigenous authorities and colleges. We look forward to further reports on this issue and to seeking ways to promote the success of this initiative". It was language we knew could be built on at subsequent ministerial meetings of the Arctic Council.

After the Iqaluit Ministerial Conference, I ceased to play any role in the development of the university, but I managed to arrange for support from my budget to assist with design of one of the first concrete deliverables: the curriculum for the bachelor of circumpolar studies. This was led by Aron Senkpiel and Sally Ross. An Interim Council of Universities and Colleges of Higher Education that wished to join the University of the Arctic was established, and in June 2001, the university was formally launched in Rovaniemi, Finland. At that time, there were 33 member institutions. Today, there are 157. The university has truly flourished and provided a new generation of graduates with a breadth of education and experience that would have been quite impossible to acquire from any single organisation. It is all the more remarkable because universities are normally competitive organisations. Asgeir Brekke, Bill Heal, Sally Ross, the late Aron Senkpiel, Outi Snellman and their colleagues should feel immensely proud of the foundation they built in the late 1990s. It could serve as a case study on what can be achieved when a group of individuals with a wide range of different responsibilities in academia can recognize a need, see an opportunity to bring it to fruition, translate that opportunity into a critical path and run with it.

My only disappointment is that it is not clear that the University of the Arctic has been able to significantly meet the needs of Arctic indigenous peoples in Canada. Perhaps this was too much to expect. As identified in the National Strategy on Inuit Education, Canada needs a university located in the Arctic that is truly designed to serve the needs of the region and its indigenous peoples. I think such an institution should have more than one campus. If an institution of this nature could be established, its dimensions could be extended through participation in the circumpolar University of the Arctic. However, Arctic Canadians may have a long wait for any of these needs to be addressed. In the last few years, even Canadian funding for the University of the Arctic has become insecure and we still have no Canadian university in the Arctic.

It must often be frustrating for Arctic peoples to constantly listen to expertise from the South. In the 1980s, a well-intentioned group of southern Canadians formed a committee to essentially function as an Arctic lobby group. The aim was to pressure the government in Ottawa about Arctic environmental protection in the face of economic development. I think their efforts were generally appreciated in the North. However, when I was listening to a presentation by one member of this committee in Inuvik, a respected Inuit leader turned to me and whispered: "David, up here, we are thinking of setting up a committee to advise on urban sprawl in the Toronto-Windsor corridor!"

To enable Arctic indigenous peoples to genuinely participate in debates concerning the future of their homelands and the utilization of their renewable and nonrenewable resources, they must have the same ease of access to higher education that is available elsewhere. Only then can they be relatively free from those southerners "who know best" – particularly those who would like to turn the Arctic into a wildlife park and those who would like the freedom of resource exploitation at any cost.

5. **Producing and maintaining a research community to serve the Arctic.** Historically, universities described their function as being to educate and conduct research. More recently, these two functions have been joined by or framed around addressing the needs of society. This evolution can be seen in the enormous changes that have taken place in the nature of Arctic science over the last 40 years. The proportion of work that is curiosity driven has dramatically declined and been replaced by work that is directly focused on national and international perceptions of the priorities of society. The latter span a wide range of issues that reflect different ways in which interested parties view the Arctic.

How have governments described the needs of society with respect to the Arctic? At the present time, Canada states that its vision for the Arctic is "a stable, rules-based region with clearly defined boundaries" (illustrating a concern about sovereignty), "dynamic economic growth and trade, vibrant northern communities, and healthy and productive ecosystems". At the same time, the EU has described its three main policy objectives for the Arctic this way: "[1.] protecting and preserving the Arctic in unison with its population, [2.] promoting the sustainable use of resources, and [3.] contributing to enhanced Arctic multilateral governance". While these two visions are not necessarily contradictory, the language suggests significant differences in relative balance. When the chair of the Arctic Council passed to Canada in 2013, Canada announced that economic cooperation would be on the top of the agenda. At this time, the council expressed its vision under seven headings as follows:

- "A peaceful Arctic"
- "The Arctic home", reflecting the view that the Arctic Council countries are responsible for the well-being of all Arctic people as the region develops
- "A prosperous Arctic", reflecting its view that the economic potential of the Arctic is enormous and its sustainable development is key to the region's resilience and prosperity
- "A safe Arctic"
- "A healthy Arctic environment", expressing its commitment to work within the Arctic and globally to address the environmental challenges facing the region: "We remain committed to managing the region with an ecosystem-based approach which balances conservation and sustainable use of the environment."
- "Arctic knowledge", including an intent to "continue to deepen the knowledge and understanding of the Arctic, both inside and outside the region, and to strengthen Arctic research and transdisciplinary science, encourage cooperation between higher education institutions and society and synergies between traditional knowledge and science"
- "A strong Arctic Council"

Whatever one may think of them, these three governmental interpretations of the present and future needs of knowledge and expertise of society and particularly of northern societies are very useful. They help identify the types of educational and research programmes that governments value.

To fulfill the Arctic Council's vision of the Arctic of tomorrow requires an eclectic spectrum of scientific and technological expertise. Science in the Arctic has long said farewell to the individual specialist focusing on a discrete topic with tenuous linkages to the needs of society. The science paradigm of today and of tomorrow is that Arctic issues can only be tackled and financed by interdisciplinary teams often representing quite disparate areas of expertise.

It may not have been true 15 or 20 years ago, but today, most academic, educational and research organisations have fully grasped the demands, opportunities and funding realities associated with the new Arctic research paradigm. They are producing graduates and postgraduates capable of taking up the challenge. However, I am not sure these new potential researchers will find employment in the Arctic because it remains to be seen how much Arctic governments have themselves understood the implications of the paradigm shift in relation to their aspirations. If one looks again at the 2013 Arctic Council's vision, many readers may think that the two objectives of unleashing the Arctic's economic potential with maintaining a healthy Arctic environment are mutually antagonistic. The catch is the word "sustainable" that is tagged onto "economic development".

Trying hard not to take sides in these paragraphs, I will just say that some Arctic resources can be carefully exploited in a sustainable way. However, depending on the activity, a great deal of environmental information and understanding is required to reach this conclusion with a reasonable level of confidence. At present, this body of knowledge is simply not yet available for a significant number of the economic development proposals currently on the table. Governments will not be able to honestly make environmentally sustainable decisions for Arctic resource development without a major new deployment of scientists. The reality is that even if this happens, the answers will not come quickly. This is presumably the reason for the "Arctic knowledge" component of the Arctic Council's vision. But will governments wait for and be guided by any emerging science consensus? The answer is probably that some will and some will not. A guide to what we can expect is to look at which governments have (and which have not) implemented actions that are already significantly reducing their national GHG emissions. Which governments are investing in expertise, such as Arctic marine and freshwater toxicology, climate science and the physics of long-range atmospheric transport of pollutants, and which are not? Which governments encourage unimpeded public access to government-funded environmental scientists and which do not? It will then be obvious which governments are truly committed to sustainable Arctic development and to the support of knowledge-based decision making.

To sum it all up, the future well-being of the Arctic and of its peoples depends on the wisdom of decisions to be taken now and in the next few decades. Wise decisions will come from listening to those who live in the Arctic and from assessing the findings of the Arctic environmental research community that hopefully will include more and more people from indigenous communities. It is implicit that a wise Arctic country is one that environmentally educates its youth and nurtures a vibrant Arctic research community. Fortunately, no Arctic country has to do all the work itself. There has been a long historical tradition of cooperation in polar research forced by the high cost of logistics and a general shortage of expertise. This tradition is still being maintained, as demonstrated by the activities of such organisations as the Arctic Council and AMAP, the IASC, the Barents Euro-Arctic Council, the SAON and the various legacy initiatives of the IPY and, most recently, by the powerful contributions of the European Framework Programme and its European Polar Consortium.

12

The Long and the Short of It: Has the Arctic Messenger Been Noticed? What Can Be Done?

Life is a strange thing indeed
-just when you are getting the hang of it,
a gibbet marks the skyline.
        Dáithí Ó hÓgáin, *"Too Late" from* Footsteps From Another World

When it is obvious that your goals cannot be reached, do not
adjust the goals.
Adjust the action steps.

                                                Confucius, The Analects

Earth scientists use the geological time scale to describe the passage of time in Earth's history. The scale consists of a hierarchy of units, each of which is characterized by a combination of different conditions that prevailed at the time of that unit. In the 1980s, ecologist Eugene F. Stoermer proposed the term "anthropocene" to describe the stage in which we now live. The rationale is that humankind is responsible for major and fundamental long-lasting geophysical changes in the atmosphere and oceans. They are at the same level of magnitude as changes that mark the boundaries between many past geological periods. Our impact on the global ecosystem will be the dominating feature that geologists thousands or millions of years into the future will use to describe the present time. The idea gained momentum with support in 2000 from Dutch Nobel Prize winner Paul Crutzen (who we met earlier in the chapter on stratospheric ozone depletion). This proposed designation is now being examined to see if it should be formally embraced in the geological time scale. A decision is planned for 2016. The idea that the term *anthropocene* has even been suggested as a geological epoch makes me pause. What are we doing to our planet?

What Drives Societies to Make Decisions That Prove to Be Fatal for the Sustainability of Their Environments?

As I began to work my way through writing this book, I became more and more apprehensive. How would I finish it? It was a task I felt ill equipped to tackle. However, I remembered Jared Diamond's 2005 book *Collapse: How Societies Choose to Fail or Succeed*. It is an excellent work that should be on the bookshelf of every head of state in the world and every executive of large corporations and on the curriculum for the final year of every school system. Come to think of it, perhaps it should be in every home.

Diamond describes the rise and fall of a number of prehistoric and historic societies that existed in very different ecological environments. Despite this diversity, he found eight self-inflicted common threads that eroded the ability of the environment to sustain each society. The eight traits were "[1.] deforestation and habitat destruction, [2.] soil problems (erosion, salinization, and soil fertility losses), [3.] water management issues, [4.] overhunting, [5.] overfishing, [6.] effects of introduced species on native species, [7.] human population growth, and [8.] increased per-capita impact of people". These will sound very familiar to anyone who has browsed IPCC reports. Consequential food shortages and starvation resulted when too many people competed for dwindling resources. This brought wars, the collapse of political control and the demise of each of the studied societies. Diamond considered that the eight traits still face us today, but to make them fully up to date, he added four more. They are (1) climate change brought on by the anthropogenic emission of Green House Gases (GHGs), (2) the environmental accumulation of toxic chemicals, (3) energy shortages and (4) "the full utilization of the Earth's photosynthetic capacity".

What is really intriguing for our purpose is that Diamond went on to search for explanations as to why each society undertook disastrous activities and made fateful decisions. He found four themes:

1. Sometimes, societies appeared to have failed to anticipate a problem before it arrived – perhaps because they had no prior experience of the problem or the seriousness of an earlier encounter with the problem was forgotten. A further possible explanation for this failure could be that the serious nature of the new and fateful problem was underestimated because a false analogy was drawn between it and familiar problems that were not too intimidating. In a modern context, we may believe our modern technological cornucopia will

always find a solution to environmental stress, resource availability and population growth because in our memory, it always has. Paul Sabin has reminded us that Paul Ehrlich, famous for his neo-Malthusian views on ecological limits to economic growth, lost his famous 1980 wager with the "cornucopian" economist Julian Simon on the 10-year increase in cost of five crucial metals.

2. Societies may have failed to perceive a problem even when it has arrived. In this case, the problem was usually a slow trend in an attribute that was obscured by irregular fluctuations.

3. Societies may have perceived a problem but either failed to address it or did so ineffectively.

4. Societies may have anticipated, perceived and even tried to solve a problem but still failed for a number of reasons. For example, the solution may have been beyond the extant technological capacity of the society. It may have been a matter of "too little, too late" or perhaps the attempted solution itself was disastrous.

You have probably already guessed where I am going here. Each of the six issues we studied earlier (radioactivity, acidification, stratospheric ozone depletion, persistent organic pollutants, mercury and climate change) has the potential to create major environmental and/or human health impacts. How these impacts would be mediated varies with each issue, but the end results would ultimately place significant pressures on regional and, eventually, global ecosystems, global economics and global societies.

Not all places in the world are equally sensitive to change as a result of impact from one or more of our studied issues. However, the scientific community and indigenous peoples who have provided the Arctic Messenger with a voice have taught us that the Arctic is in one way or another especially sensitive to all of them. Therefore, it is the place to watch closely because of concern for the Arctic and because of concern for ourselves. The impacts of each issue to Arctic ecosystems and Arctic societies have been well documented and assessed by AMAP. Furthermore, the Arctic Messenger has shown how Arctic impacts reach out to influence geophysical processes (such as weather patterns and the frequency of extreme events) in mid-latitudes of the Northern Hemisphere.

The potency of five of the six issues to change our world, our economies and our lives has been recognized. Governments have been galvanized to take cooperative and legally binding action to address them. One could argue about the level of effectiveness of each of these cooperative actions, but the world is much better off with them than without them. That leaves us with the sixth of our issues: the warming climate. On this one, we have no living global action plan to reduce GHG emissions that stands a hope of doing the job. This despite an overwhelming scientific consensus that identifies anthropogenic GHG emissions as the root cause of global warming. IPCC expectations for the future include the major geographical redistribution of rainfall leading to food and water shortages, mass migrations, coastal flooding of large areas and enormous adaptation costs. It is the most malevolent and intimidating of the six issues. Why are world governments reluctant to deal with it even when they have publicly recognized that even moderate warming scenarios carry major risks for significant environmental and socioeconomic consequences? To help answer this question, we will examine our six issues again, paying particular attention to each of the preceding "The Long and the Short of It" sections. This time, we will view them through the lens of Jared Diamond's analytical methodology, identifying how our modern society has or has not dealt with them.

Our first step will be to return to the five issues where national and international cooperation have been set up to deal with environmental hazards and risks. Before we go on, I probably need to clarify Jared Diamond's use of the words "anticipated" and "perceived" in the context of how I am using them here. Our present societies can *anticipate* an environmental issue if scientists have been able to predict it will happen. They have been able to *perceive* the issue when evidence begins to accumulate to support the predictions.

Radioactivity

Anticipation: The nuclear powers from 1945 until the early 1960s probably did not anticipate the scale and global scope of environmental and human health risks related to the atmospheric testing of nuclear weapons. If they did, they likely considered such effects as costs justified by the global politics of the Cold War. At the same time, I expect they believed that careful design and operation would prevent accidents at civilian reactor and waste storage sites.

Perception: The global risks resulting from atmospheric testing had certainly been perceived by the late 1950s, when high body burdens of radionuclides were discovered in Sami consumers of reindeer meat. It was perhaps the first time the Arctic Messenger told the world of the unusual sensitivity of the Arctic environment

and also warned of global environmental and human health implications attributable to human activity in distant locations.

Reaction: By 1963, the United Kingdom, the United States, the Soviet
Union and a large number of nonnuclear states had signed the
Limited Test Ban Treaty. This banned the testing of nuclear
weapons in the atmosphere, underwater or in outer space but
allowed underground tests. France and China never signed the
treaty and continued atmospheric testing until 1974 and 1980,
respectively. More recently, the Comprehensive Test Ban Treaty
(that came into force in 1996) also included the prohibition of
underground tests. These actions brought an end to the injection
of radionuclides into the atmosphere from weapons testing.

Is there a lesson we can learn from this that may be useful to
remember when we look at the other issues? I think there is if we
frame the question differently. Why were countries able to react quite
quickly and cooperatively once the global risk from radioactive fallout
had been perceived? I think the first reason was that the number of
countries holding nuclear weapons was small. This always simplifies
negotiations.

However, a second driving force was the convergence of scientific
evidence for the risks to human health and the environment and the
fact that there could be no clear winner even in the case of a limited
exchange of nuclear weapons. A glimpse of Hiroshima and Nagasaki
told everyone what must be avoided. At the same time, there was a
convergence between this acceptance and a growing unease on both
sides of the Iron Curtain with the scale and speed of the 1950s–1960s
arms race. The 1983 paper published in the journal *Science* by Richard
Turco and colleagues added a totally new way of looking at the impact of
a moderate to large-scale nuclear war in the Northern Hemisphere.
Their paper is today known as the TTAPS study (after the names of the
authors) and it introduced the world to the phrase "nuclear winter".
Instead of concentrating on the then-traditional approach of dealing
with the effects of blast, heat and radiation, it focused on the scale
and impacts resulting from smoke and small particulates injected high
into the atmosphere by the intense heat generated by nuclear explosions. Once in the atmosphere, the resulting massive and dense cloud
would rapidly spread over much of the Northern Hemisphere, which
would be plunged into darkness (the nuclear targets were assumed
to largely be in the Northern Hemisphere). Most sunlight would be
blocked for weeks to months and surface temperatures would plunge.

When combined with that of high radiation levels, these impacts would destroy much of the Northern Hemisphere's flora (including agricultural capacity) and fauna (including human populations). The fate of the Southern Hemisphere was less clear.

Public opposition to nuclear weapons in Western countries became widespread and vociferous. I remember the chill of the 1962 Cuban crisis. In England, we felt little comfort from the four minutes of warning we were promised in the event of a nuclear attack. What would I do with those four minutes? Two years ago, Thérèse and I stood at ground zero in Nagasaki. Memories of the Cuban crisis flooded back. Possibly the third and most powerful reason that fuelled the comparatively swift progress on test ban treaties was the power of public opinion. We should remember these three drivers when we reach the end of our review on climate change.

The 1950s saw the development of civilian use of nuclear power to generate electricity. The first such power station to be connected to an electrical grid opened in 1954 at Obninsk about 110 kilometres from Moscow. It was followed in 1956 by the Calder Hall station in the United Kingdom near the present nuclear complex at Sellafield. It was clear that nuclear power would spread around the world. Along with the benefits of nuclear power came a problem: What should be done with radioactive waste? In 1957, the International Atomic Energy Agency (IAEA) was set up through the United Nations with a mandate that spans military (it is the depository of several of the weapons' agreements) and civilian issues. The statute of the organisation declares that its work has "three pillars: nuclear verification and security, safety and technology transfer". This resulted in a bewildering number of treaties, protocols and agreements concerning almost every aspect of the nuclear power industry, including the handling of nuclear waste.

The result has been a nuclear power industry that is generally safe but is also not perfect. Very serious avoidable accidents continue to happen and devastating consequences are avoided more by heroism and luck than by design. It always strikes me as a giant anachronism to contrast the technical sophistication of a nuclear power plant with the desperate reliance on such measures as dumping water and cement on the facilities at Fukushima and Chernobyl.

It is equally perplexing that very few arrangements have been made for the permanent storage of nuclear waste. The Eurasian Arctic continues to host a very high density of temporarily stored radioactive waste that requires assessment and remediation.

Future needs: Much progress has been made to improve nuclear safety. However, if nuclear power is to play a significant role in the plans of countries to approach carbon neutrality, then safety measures must be further improved and the problems of long-term storage of wastes solved, particularly in the Arctic.

Acidification

Anticipation: There seems to have been no general anticipation that industrial emissions of such substances as the dioxides of sulphur and nitrogen could acidify rain, which would lead to damaged freshwater and forest ecosystems.

Perception: Robert Smith first demonstrated the relationship between acid rain and atmospheric pollution in 1853. He was an English chemist working in the rapidly developing industrial heartland around Manchester. In 1872, Smith published a book entitled *Air and Rain: The Beginnings of a Chemical Climatology* in which he introduced the phrase "acid rain". The first concerns were related to the increased erosion of limestone building materials.

Despite this early perception, it was not until the late 1960s and early 1970s that scientists began to link damaged forest and freshwater ecosystems to the long-range transport of acidifying substances from sources in distant countries. Regions that were mountainous (encouraging rain to fall), possessed an underlying geology dominated by acidic rocks (such as granite) and downwind of industrial sources were particularly sensitive even if the distance between source and deposition was in the order of 2,000 kilometres.

Reaction: The first reaction to the observations of Smith was to build taller chimneys in order to help winds carry the acidifying pollutants away. As industrial plants became larger, smokestacks became taller. This approach was still alive and well in the late 1960s. At that time, I worked at Harwell in southern England. On the horizon, we watched a massive coal-fired power station being constructed at Didcot. The chimneys were 199.5 metres (655 feet) high. It was a perfect formula to spread acid rain over long distances because it released gases into the regional atmospheric circulation. This explained the large distances observed between emission sources in, for example, England and impacted ecosystems in Norway.

Lobby groups vigorously denied the scientific findings. However, in 1979, the Convention on Long-Range Transboundary Air Pollution

(CLRTAP) was established to enable parties to the agreement to cooperatively reduce acid rain. Several legally binding protocols have been added to the convention. The more recent protocols balance specified emission levels with knowledge about atmospheric transport and the sensitivity of the receiving deposition environments. Adequacy of the actions to protect ecosystems is measured by several different methods, including environmental monitoring.

Future needs: The CLRTAP is generally considered to have been a success. In 2004, sulphur emissions in Europe and North America were down 60% and 50%, respectively, in comparison to 1980, while ecosystems have been recovering accordingly. Further emission reductions are ongoing.

At the time when the Rovaniemi negotiations were creating the Arctic Environmental Protection Strategy in 1989–1991, little information was available on the extent of acidification in the Arctic. Since then, two assessments from Arctic Monitoring Assessment Programme (AMAP) have found little evidence for chronic and widespread ecosystem impacts attributable to long-range atmospheric transport. However, there are several areas, mainly in the Barents Region, where severe localized forest and freshwater ecosystem degradation has occurred in association with Russian smelters. Although acidification is now generally not regarded to be a pressing issue in the Arctic, I have included it in the Arctic Messenger's story because lessons learned in the CLRTAP acidification programmes shaped the approach taken later when countries turned to the control of POPs and heavy metals.

Stratospheric Ozone

Anticipation: In the late 1960s and early 1970s, atmospheric scientists warned that the ozone layer could be eroded by such substances as nitrogen oxides (NOx) released within the stratosphere by supersonic aircraft and by the then-planned space shuttle. By 1973, it was known that chlorofluorocarbon (CFC) gases (widely used as refrigerants and as spray propellants) were ubiquitous in the atmosphere at surprisingly high concentrations. Almost immediately, laboratory studies showed that CFCs could be very effective at eroding the ozone layer under the conditions that exist in the stratosphere. Environmental agencies began to consider regulatory actions despite vigorous activity from lobby groups.

Perception: In 1982, a massive springtime depletion of stratospheric ozone above Antarctica was recorded, covering a geographical area larger than that of the United States. In subsequent years, the spring Antarctic ozone hole grew to up to 27 million square kilometres. Further studies revealed widespread seasonal thinning of the ozone layer, particularly over the Arctic.

Reaction: This was very unusual because the first regulations to control CFCs as spray can propellants (not as refrigerants) came into effect in the United States in 1979. In other words, this was before it had been proven that the ozone layer was being significantly depleted. Similar regulations quickly followed in other countries. Other uses of CFCs remained unregulated, but international political action was swift. In 1985, the Vienna Convention for the Protection of the Ozone Layer (VCPOL) was negotiated and followed in 1987 by the Montreal Protocol on Substances That Deplete the Ozone Layer. As of this writing, 197 countries have ratified it.

The Montreal Protocol includes a mechanism that requires parties to regularly review monitoring information on the concentrations of substances that deplete the ozone layer, including CFCs. These studies showed that as a result of actions taken by countries to comply with the protocol, the atmospheric concentrations of these long-lived chemicals are no longer increasing. However, substances that were introduced into the market to replace CFCs and that were thought to be relatively benign have now also appeared in the stratosphere. Fortunately, this tendency has been "caught" through the monitoring review process and the "new" ozone-depleting substances are now also controlled under the protocol.

Future needs: It is with good justification that the Montreal Protocol is widely considered to be the most successful international environmental agreement. It is expected that a return to pre-1980 ozone levels at Northern Hemisphere mid-latitudes could occur by (or before) 2050 and even rise above 1980 levels by the latter part of the present century. The Arctic will recover a little more slowly. Although it does seem that everything that should be done is being done, there are two major caveats. The widespread reduction in emissions of earlier ozone-depleting substances was achieved by replacing them with other substances, particularly HFCs. HFCs are benign to stratospheric ozone but are very powerful GHGs. Second, the work of Johannes Laube and colleagues, published in 2014, shows that we should never assume we have identified all the ozone-depleting substances present in the atmosphere.

Persistent Organic Pollutants

Anticipation: It is unlikely that the chronic environmental and human health effects of persistent organic pollutants (POPs) were ever anticipated.

Perception: Awareness began in the 1950s, when biologists started to report problems in wildlife (especially birds) that seemed to be linked to the use of pesticides. With the publication of her famous book *Silent Spring* in 1962, Rachel Carson awakened the general public overnight to the impact that certain pesticides could have in the environment. We have already seen that most early pesticides were members of the category of substances we now call POPs. By the late 1960s, David Peakall measured the DDT metabolite DDE in peregrine falcon eggs from Alaska and showed a clear inverse relationship between DDE content and eggshell thickness. He was able to demonstrate this relationship back to the mid-1940s in England and California.

Reaction: By 1970, countries began to take national regulatory actions against DDT, such as prohibiting its use in agriculture. However, the regulatory actions were piecemeal and directed against individual substances. It was not until the mid- to late 1980s that it was realized that most of these substances share properties that can cause them to be transported in the environment far from their source regions. Here, they can aggressively biomagnify up the food chain to top predators and people. Once the perception of POPs as a long-range transboundary issue was brought to the attention of governments in 1989, it was not until 1998 and 2001, respectively, that the CLRTAP POPs protocol and the Stockholm Convention were signed. It was a further three years before the Stockholm Convention entered into force. Ramon Guardans published a short paper in *Atmospheric Pollution Research* in 2012 in which he estimated that it has historically taken about 20–30 years from the time that there is a solid case built for international regulation of a substance to the time when there is general implementation of those regulations. POPs were more or less on schedule.

Mercury

Anticipation: Mercury is a strange story because the neurological effects of chronic exposure were common knowledge in the days of "mad hatters" in the nineteenth century. Therefore, impacts of chronic environmental exposure should have been anticipated (but were not) long before the tragedies of Minamata and southern Ontario occurred.

Perception: Even after the sad lessons from Minamata and Ontario were perceived, there was still a failure to recognize the impacts of chronic low-level environmental exposure and to realize that mercury can also travel far from source regions and biomagnify to levels capable of causing physiological and behavioural impacts in wildlife and people.

Reaction: It was not until the early 1990s that countries began to seek international controls. This initially took place under the CLRTAP. However, the CLRTAP is not a global forum. Although AMAP information in 2000 spurred the Arctic Council to ask the Governing Council of United Nations Environment Programme (UNEP) to consider a global convention devoted to the control of mercury, it would not be until 2013 that the Minamata Convention was signed.

Future needs for POPs and mercury: It is well established that persistent biomagnifying substances are reaping their chronic toxicological toll inside and outside the Arctic. But there is little evidence outside the Registration, Evaluation, Authorisation and Restriction of Chemicals (REACH) initiative in the European Union to show that agencies responsible for evaluating the safety and regulating the use and disposal of new substances entering commerce (including pharmaceuticals) are considering the full suite of properties of POPs and of endocrine disruption. At the time of writing, an innovative EU attempt to deal with the environmental disposal of pharmaceuticals is running into very determined opposition.

What has our society learned from the POPs and mercury stories? The Arctic Messenger would conclude that our societies have not learned too much. They have *not* learned that:

- Humankind, with an ever-increasing industrialized population, cannot continue to use its finite environment as a sewer for persistent, chronically toxic and biomagnifying substances.

- Continuing to apply regulatory controls only after a huge persistent inventory has accumulated in the environment is a strategy that is unsustainable.

- To pursue this strategy will continue to place us in the situation where we wait, for example, for evidence to accumulate from studies

of cohorts of children showing that a particular substance is endocrine disrupting or neurotoxic or that it affects behaviour at very low
doses of exposure before birth.

- A new paradigm is essential to efficiently bridge the stages of scientific anticipation and the perception of an issue to an appropriate
regulatory response. Without it, how can we avoid the tragic human
and environmental costs implied by the growing evidence of long-
term and even intergenerational impacts from exposure to endocrine-
disrupting POPs?

We have been so slow to learn because powerful elements of our
society promote the belief that a precautionary approach to chemical
regulation is not financially feasible. What this really means is that they
are not prepared to accept any short-term pains that are necessary
to ensure long-term environmental and economic sustainability.
Biological, social and commercial history is littered with the carcasses
of species, political regimes and commercial enterprises that were
unable to change. Some important industrial lobby groups that perpetu-
ate this state of affairs in, for example, North America strongly influence political decision making.

It is a mistake to believe that commercial lobby groups are always
against environmental regulation. For example, I was strongly supported by the Canadian Chemical Manufacturers Association (CCMA)
when trying to set a progressive Canadian negotiating position for the
CLRTAP POPs protocol in the 1990s. Their rationale was that international controls on hazardous substances would give them a competitive, level playing field for marketing their newer and safer products. At
that time, some of the federal Canadian economic departments opposed
several key actions being proposed on the grounds that they would
severely injure Canada's chemical industry. I told them that the proposed measures were supported by the CCMA. This brought an edifying
outburst from my opponents: "David, the CCMA is not responsible for
deciding what is good for the Canadian chemical industry!"

Some lobby groups are motivated by libertarian ideals rather
than by being linked to a particular industry. They are just utterly
opposed to any government intervention in the fabric of society.
They can also be very intimidating. In those early CLRTAP days, I was
lucky to have an early desk telephone model that identified the
number of the incoming call. I received so many calls from lobbyists
based in the United States that I eventually stopped picking up the
receiver unless I recognized the number. Unfortunately, lobby groups

are a generic feature of the Arctic Messenger's story, so we will come back to them again when we close in on climate. Although their power is greatest in the United States, their success in influencing U.S. policy affects us all. The U.S. market still dominates the world. The only ways in which their impact has been reduced took place when public opinion was overwhelming. This is a very important point to remember.

Climate

Anticipation: Using our interpretation of Jared Diamond's analytical scheme, when could humankind have anticipated climate warming? The basic physics of the greenhouse effect were worked out in 1827 by Jean Baptiste Joseph Fourier, one of Napoleon's savants. In 1896, Svante Arrhenius calculated how much our global temperature would change from a given rise in CO_2. In other words, he had made an estimate of climate sensitivity. His calculation of 5°C for the tropics and 6°C for high latitudes remains close to those estimated today by the Intergovernmental Panel on Climate Change (IPCC), which puts climate sensitivity (for a doubling of atmospheric CO_2) as most likely lying between 1.5°C and 4.5°C. Thus, we have no excuse – we should have seen it coming since at least 1900.

Perception: By 1938, Guy Stewart Callendar had shown that CO_2 levels and temperature in the atmosphere had been rising over the previous 50 years. In 1960, Charles Keeling confirmed that CO_2 levels in the atmosphere were rising and produced what is now called the Keeling Curve to describe it. In 1979, the U.S. National Research Council concluded: "When it is assumed that the CO_2 content of the atmosphere is doubled and statistical thermal equilibrium is achieved, the more realistic of the modelling efforts predict a global surface warming of between 2°C and 3.5°C with greater increases at high latitudes." Notice the anticipation of greater temperature increases in the Arctic and in Antarctica. In 1988, the WMO organised a World Conference on the Changing Atmosphere: Implications for Global Security. The conference concluded that changes in the atmosphere due to human activity "represent a major threat to international security and are already having harmful consequences over many parts of the globe". They recommended that by 2005, world global emissions of CO_2 should be 18% lower than in 1988.

Therefore, the date at which humankind's scientific community perceived that climate warming was under way and attributable to anthropogenic GHG emissions could be said to be somewhere after 1938 and before 1988. For world political leaders, I think it would be reasonable to pick 1992, when the United Nations Framework Convention on Climate Change (UNFCCC) was opened for signature.

With every IPCC assessment and with every new issue of the most influential scientific journals, the reality of climate change and its attribution to GHG emissions become more clearly defined. Here is an example. As I was completing the penultimate draft of this book in December 2013, James Overland and colleagues published a paper in the journal *Earth's Future*. It took a new look at predicting the scale of temperature projections for the Arctic in the second half of the twenty-first century using RCP scenarios. They chose the RCP8.5 scenario to represent rising emissions, resulting in 8.5 W/m^2 (about 1,370 ppm CO_2) by 2100 and chose RCP4.5 to represent an emission trajectory that (through the application of mitigation) achieves stabilization at 4.5 W/m^2 (about 650 ppm CO_2) by about 2060. Using global circulation models, the authors found that the RCP8.5 scenario (business as usual) leads in 2100 to an Arctic-wide "increase of +13°C in late fall and +5°C in late spring". In contrast, the RCP4.5 mitigation scenario leads in 2100 to an Arctic-wide "increase of +7°C in late fall and +3°C in late spring". The projected increases are relative to a 1981–2005 baseline. This sounds alarmingly similar to the personal communication from Bob Corell mentioned in the climate chapter under the section "Sea Level Rise in the Arctic and Around the Planet." Even if we can succeed in ignoring the impacts of an Arctic-wide temperature increase on the scale indicated by these studies, can we ignore the impact on sea level elsewhere? However one looks at it, results such as these plead for swift and effective mitigation action.

Reaction: The two classes of reaction we may expect governments to take are actions to alleviate the expected impacts of climate change over a given time frame. An example of the first would be to increase the protection of vulnerable coastal areas and cities to rising sea levels and storm surges. This is called *adaptation* and its implementation is mainly a national or subnational responsibility. The second is to attack the root cause of the issue, which is our unabated cumulative GHG emissions. This is called *mitigation*. The Kyoto Protocol was an attempt to organise global mitigation commitments. However, history will show that even as I write, humankind has not perceived global warming to the point that it is prepared to work cooperatively to undertake what really needs to be done. No globally coherent plan is being implemented to meaningfully reduce total global GHG emissions towards a goal of carbon neutrality over a time frame of at least the next 100 years. Instead, following the virtual abandonment of the Kyoto Protocol, we wait until a meeting scheduled for 2015 to see if a more workable agreement can provide robust arrangements to achieve the global reduction in GHG emission that is required. In the meantime, although some countries are clearly trying very hard to reduce their total GHG emissions, others are just as clearly clouding their inaction with obfuscation.

In the studies undertaken by Jared Diamond, historical societies that collapsed were those that failed to anticipate or failed to perceive (notice) or failed to act sufficiently to counteract the circumstances that led to their end. Why are so many governments not reacting to the biblical writing on the wall? When we analyse the situation more carefully, it is going to sound as if I am targeting the United States. This is far from my intent. The United States and U.S. government agencies have made the largest contributions to understanding climate science. I could have told the tale equally well from the perspective of any one of the countries that sits in the upper part of the list of per capita emissions of CO_2. However, the United States is still the world's economic powerhouse. A global cooperative plan to address climate change is very unlikely to be effective without leadership from Washington, DC.

Can we expect such leadership? In 2011, the House of Representatives voted in favour of a bill to amend the Clean Air Act to prohibit the administrator of the Environmental Protection Agency (EPA) from promulgating any regulation related to or taking into consideration the emission of a GHG to address climate change. It excluded GHGs from the definition of "air pollutant" for purposes of addressing climate change. This bill passed the House. If this does not sound incredible, there is more to come. During debate on the bill, Democrats tried to introduce the following words: "Congress accepts the scientific findings of the EPA that climate change is occurring, is caused largely by human activities and poses significant risks for public health and welfare." This proposed amendment to the bill failed. In fact, 240 representatives voted against it.

Here is a sample of some of the arguments presented by some of the representatives. (You can watch part of the debate at http://my .barackobama.com/Climate-Change.):

- "Anthropogenic global warming is an issue that the scientists are still debating – and you know it and I know it."
- "I do not think CO_2 is a problem. Therefore, I do not think it needs to be regulated."
- "It is almost comical. Every time we exhale, we exhale carbon dioxide." (This was from one of the most senior and influential Republicans in Washington, DC.)
- "We all breathe CO_2. Climate changes, but there is no evidence at all that it is man-made CO_2 that causes the climate to change."
- "The ice caps are melting, which we see over and over again. Yeah, they're melting on Mars too."
- "The idea of human-induced global climate change is one of the greatest hoaxes perpetrated out of the scientific community. It is a hoax."

How is it possible for such powerful lawmakers to have such unfounded convictions?

Perhaps they understand very well the message coming from the science community but are not prepared to sacrifice a portion of our present wealth in favour of a secure and healthy environment for future generations.

Or perhaps they do not understand or do not believe the unequivocal consensus from the scientific community. If this is so, then their opinions have probably been influenced by others, so here we return again to the power of well-funded lobby groups that can undermine the basic tenets of democracy. The excellent 2010 book by Naomi Oreskes and Erik Conway (see the bibliography) argues that the campaigns of the lobbyists have historically given up trying to build a scientifically credible alternative explanation to counter a strong scientific consensus. Instead, they favour a different approach that needs no scientific justification and no support from peer-reviewed science. This more useful method works by delaying effective actions to address the issue in question. To do this, they focus on convincing lawmakers that scientists are still debating the root cause of the issue or that the entire scientific community is unethically creating an issue. It is an approach that can be traced from fighting efforts to reduce tobacco smoking to those of toxic chemical regulation.

If we cannot win the battle of influence within our governments, our only hope for a change in political direction lies in the electorate empowering our governments to act on established truth. Back in the ozone and POPs chapters, we saw how important public opinion was in bringing about change in the use of chemicals. To help our governments become more accountable, we need to:

- Succeed in explaining that present-day climate warming is real, serious, caused by anthropogenic GHG emissions and is going to be much more pronounced if we ignore the cause. This is the fundamental cumulative consensus of essentially the entire scientific community. Contrary to rumour, there is no significant scientific debate on these basic facts.
- Ensure that government climate scientists are allowed to present and explain their results to the public without regulation by their employer.

In the words of Henry David Thoreau: "It takes two to speak the truth – one to speak and the other to hear."

Future needs: science-based frameworks for international action on climate:
How can a world of competing economies set about dealing with cumulative GHG emissions – the cause of our present climate warming? This would never be easy, but at the present time, it is a particularly difficult question as countries pull themselves out of the economic problems of the first decade of the twenty-first century.

There is no shortage of proposals and many are not easy to understand. Clearly, the world needs an overall framework for GHG emission reduction – one that is "fair" to all countries regardless of their state of economic development. The underlying paradigm that has (and continues to be) the basis of approaches taken under the UNFCCC is the concept of "contract and convergence". Under this simple framework (developed in the United Kingdom in the 1990s), countries would agree on mitigation measures to collectively achieve a contraction path for their GHG emissions. Ultimately and ideally, all the national contraction pathways would converge on a similar per capita GHG emission level for all countries. The nature of the pathways would be based primarily on scientific projections as to what would be needed to prevent the global mean temperature exceeding an agreed level (say, +2°). A person in Kenya, for example, would then have an identical CO_2 allocation as a person from the United States, China or the European Union. In other words, we would have achieved carbon equity by contracting and converging our CO_2 emissions. At some later negotiable date, a country's allocation would become fixed and not adjusted to reflect population density. Finally, the total trajectory would continue downwards until we have all reached or almost reached carbon neutrality.[1]

Is this achievable? The 2012 Global Energy Assessment (GEA), coordinated by the International Institute of Applied System Analysis (IIASA), concluded that indeed it is, but it requires firm dedication to bring it about. The GEA argued that it is a fatal hallucination to believe we have any long-term sustainable policy options other than to urgently move in the direction of carbon neutrality. Furthermore, it shows that with political and commercial commitment, we have the capacity to change our energy use and sources to provide enough energy for us all wherever we live. All that is required is to work together and to accept change. Finally, remember that the GEA is not a product of physicists, chemists, earth scientists or biologists. It is largely a product of economists and the global business community.

So far, so good. However, when attempting to implement the approach, countries have immediately come up against short-term implications that they have found hard to accept or address. Most developing countries with low per capita GHG emissions are reluctant to undertake any actions that will threaten to inhibit their economic growth. At the same time, most developed economies find it very difficult to contemplate any actions that risk a short-term contraction of their economies.

To have hope of success, any proposed plan of action must recognize some highly inconvenient global economic and political realities.

First, the biggest proportion of the cumulative GHG emissions since the Industrial Revolution has come from developed nations representing less than 20% of the global population. The world population is rapidly growing, mainly in countries with developing economies. They have every right to expect their own fair share of Earth's energy resources and to anticipate a lifestyle similar to that which I enjoy.

Second, a global GDP growth of 2% or more means that the global economies will double in about 35 years with a concomitant growth in the demand for energy, water and food. It is simply unsustainable to expect that global society can fairly and indefinitely power its economic future on fossil fuels regardless of whatever we do to our environment.

---

[1] Andrew Weaver's book *Keeping Our Cool* includes a more complete summary of the concept of "contract and converge".

It was against this background that the Kyoto Protocol emission targets slumped to cover only 15% of global emissions by 2013.

The Kyoto Protocol is fading and the 196 parties to the UNFCCC have set 2015 as the year in which they hope to have developed a new "replacement" global agreement. At the time of writing in late 2013, this date seems to be very close, and for a good short overview of how things stand at this time, I recommend the paper by Elliot Diringer (see the bibliography). Diringer points out that another "top-down" approach (such as Kyoto) based on negotiated legally binding emission targets is unlikely to emerge in 2015. The prospects are poor for a model based on obligations that are restricted to countries with developed economies. In other words, a modification of the Kyoto design is not expected. At the opposite end of the spectrum is a "bottom-up" approach, which grew from the 2009 UNFCCC conference in Copenhagen and which led to the Cancun Agreements a year later. Basically, what happened here is that countries have set their own voluntary pledges for emission targets to be realized by 2020. The trouble with this approach is that some of the Cancun pledges will not even take us to 50% of the emission reductions needed to keep global warming below 2°C. It is also depressing that some of the countries with the highest per capita emissions look as if they will be unable to meet their own voluntary reduction targets.

Where does all this leave us? The United States has been proposing a way to impose more rigour and ambition into the bottom-up model. Countries would decide their own emission targets, but through reporting mechanisms, they would be able to review each other's targets and progress. Diringer reports that these ideas have been garnering substantial interest, although predictably, there is a wide range of views on how to balance the stimulation of ambition without infringing on national sovereignty in setting emission reduction targets.

Perhaps this does represent the way forward – but only if countries take actions that are truly designed to aggressively bring down cumulative GHG emissions. This is where the need for an impartial science framework comes in. It is very unfortunate, but only one statistic matters in terms of climate mitigation. It is the single volume of total GHG emissions to the atmosphere. Therefore, if we are serious about reducing those cumulative GHG emissions to levels that will limit global warming, those of us who are contributing most to the problem on a per capita basis must be prepared for big changes in the way we fuel our economies. This is such a salient fact that it deserves a paragraph by itself.

A mitigation plan that is expressed as per capita reductions of CO_2 or as a reduction in carbon intensity of an economic activity is meaningless to the global climate system if total emissions have not also gone down. A country that is not reducing its total carbon emissions is doing nothing but exacerbating global warming. Total emissions must go down very significantly because we are working against the huge inventory of anthropogenic CO_2 already injected into the system. Remember, it is the amount of water in the proverbial bathtub that matters, so even if we turn down the tap, the water is still going to rise. In fact, if we wish our grandchildren to live in a world that resembles that which we inherited from our grandparents, we need to approach (if not actually reach) a carbon-neutral economy very quickly.

Every week or so, the evening news seems to contain an item that drives home the urgent need for some sort of framework to guide global emissions. By spring 2013, the climate station on Mauna Loa in Hawaii hit a CO_2 concentration of 400 ppm. (See the 2013 paper in *Nature* by Monastersky). The atmosphere has not been at such a level since at least 3–5 million years ago. That is a time in the geological past before our species evolved.

With that thought, we can prepare to move on to the last chapter, but first, I want to come back to our mentor: the Arctic Messenger. After all, the power of so much of this story comes from what is happening in the Arctic and on the implications to the rest of the world. Here is a little thought experiment similar to those popular with early twentieth-century nuclear physicists. Imagine that the Arctic Messenger has read Jared Diamond. In its wisdom, it ponders the effectiveness of the reaction of humankind to each of the six environmental issues we have discussed. What would it conclude on the prospects for a healthy future for the Arctic and for the whole of humankind?

Here is my guess: It could conclude that in each case of environmental degradation that has been examined in this book, humankind came to a fork in the road. We had a choice of going in one direction to a dismal future and in the other to a healthy one. For stratospheric ozone depletion and acid rain, the Arctic Messenger could feel quite comfortable that we have found our way and are heading in the right direction at a reasonable speed. For POPs, heavy metals and radioactivity, the judgement could be that we have found our way and are heading in a healthy direction but too slowly. We need to move faster and we have far to go. What our Arctic sage would find utterly mystifying is our reaction to climate warming – the most serious of all the issues. We are on the track and heading as fast as we can in a wrong direction. We have our eyes wide open and fully understand the implications to our children, our environment, our long-term economic well-being and ourselves. We know that all we have to do is to turn around. But we refuse to do so.

What will the Arctic Messenger conclude from this? Well, the Arctic Council was set up to address such issues. Will it heed the call for help? This is the topic of the following epilogue.

13

Epilogue: Keeping the Rovaniemi Flame Alive

"A poisoned Inuk child, a poisoned Arctic and a poisoned planet are all one in the same. We are all in this together."
Sheila Watt-Cloutier, Inuit advocate on human, cultural and environmental rights

If we need a précis of the entire tale from the Arctic Messenger, this quote from Sheila Watt-Cloutier says it all.

In this final and short chapter, our tale will come to an end. Our last act will be to speculate about what the Arctic Messenger would (if it were able) ask Arctic Council governments to do at this moment in time. But as a prelude before we dive into the fray, I want to spend a little time refreshing our memories about the unique nature of the Arctic Council. At the same time, we can remind ourselves about some key individuals and developments that have made the Arctic Council into the capable circumpolar organisation in which the Arctic Messenger would expect to find a strong ally. They have taught us lessons we would be wise to keep in mind. Their work has succoured the "Rovaniemi flame" – that innervating spirit that gave the Arctic Messenger a voice. Over the last 20 years, the council has exposed the dilemma captured by Sheila-Watt Cloutier's words and we need it now more than ever before.

Lessons From Heroines and Heroes: From Rovaniemi to the Arctic Council

Until 1989–1991, no advocate existed that combined the capacity of putting its finger on the pulse of the Arctic environment with the responsibility to simultaneously advise Arctic governments on the status of the patient's health. No mechanism existed that would empower governments to consider responding to such advice with cooperative environmental protection measures. All that suddenly changed in 1989 with the bold initiative of Kalevi Sorsa, Kaj Barlund and the Finnish government. In a clever and practical suggestion, they showed how the environmental elements of Mr. Gorbachev's 1987 Murmansk speech could easily be put into action without any complicated legal agreement. The so-called Finnish Initiative, or Rovaniemi Process, bore fruit, and in 1991, the eight Arctic countries signed on to the Arctic Environmental Protection Strategy in Rovaniemi, Finland. The Rovaniemi flame was alight and it has served the Arctic well. It has motivated much of the cooperative scientific progress, understanding and environmental actions we have been following in this book.

We all owe a great debt of gratitude to Mikhail Gorbachev, Kalevi Sorsa, Kaj Barlund and the Finnish government. They taught us to never be afraid of proposing something we believe in and to always seek the most practical ways to achieve our goals. The subsequent transformation of the Arctic Environmental Protection Strategy (AEPS) into the Arctic Council now led by ministers of foreign affairs or ministers of state further strengthened the scope and responsibility of the organisation (to include economic development) until Canada, at the beginning of its two-year term as chair, decided to reverse the trend by appointing its environment minister to this role in 2013.

The Finnish proposal recognized that protection of the Arctic required circumpolar cooperative environmental monitoring. The Norwegian government took up this challenge and led the negotiations to create Arctic Monitoring and Assessment Programme (AMAP) as a fully integrated component of the Arctic Environmental Protection Strategy. They sought a way to collect and assess reliable circumpolar scientific information and to provide a path where the assessments would be presented to circumpolar governments and international organisations responsible for environmental protection. The result was an organisation with an amazing record of achievements. These include providing the lion's share of information that led to the negotiation of the Convention on Long-Range Transboundary Air Pollution (CLRTAP) persistent organic pollutants (POPs), and heavy metal protocols and to the global Stockholm and Minamata conventions. AMAP also played a significant role in supporting the international cooperative actions set up to deal with radioactive legacies of the Cold War in the Arctic region and to convince Arctic governments to examine the benefits of taking quick action on short-lived climate forcers.

Norway realized that the key to maintaining long-term vitality in AMAP lies in the existence of a secure and adequately funded coordination mechanism (the AMAP secretariat). Norway houses the secretariat and provides its operational budget through an independent foundation that Norway established specifically for this purpose. This stability enabled the secretariat to recruit an outstandingly gifted and dedicated team led by Lars-Otto Reiersen and Simon Wilson. Lars-Otto's vision and tenacity and Simon's skills at organising scientific assessments have been the key to every AMAP accomplishment.

It is the Norwegian government, Lars-Otto Reiersen, Simon Wilson and hundreds of Arctic scientists we must thank for providing us with a reliable, long-lasting tool to detect, assess and understand Arctic environmental change.

In the early 1990s, environmentalists and economists came to view our future in terms of "sustainable economic development", a paradigm developed by the World Commission on Economic Development (WCED), chaired by Gro Harlem Brundtland. It subsequently formed the foundation for the United Nations Conference on Environment and Development (UNCED) held in Rio de Janeiro in 1992. The WCED defined the concept as follows: "Sustainable development is development that meets the needs of the present without compromising the ability of future generations to meet their own needs. It contains within it two key concepts: the concept of needs, in particular the essential needs of the world's poor, to which overriding priority should be given; and the idea of limitations imposed by the state of technology and social organisation on the environment's ability to meet present and future needs." We should never forget that this is the definition of sustainable development.

Arctic indigenous peoples have produced a number of extremely dedicated people over the last 25 years whose passion for doing what is right rubbed off on us all. I was fortunate to learn from and work with some of these people, including Cindy Dixon, Rosemarie Kuptana, Aggaluk Lynge, Leif Halonen, Gunn-Britt Retter, Mary Simon, Sheila Watt-Cloutier and Pavel Sulyandziga. Of course, being from Canada, it was with Mary Simon and Sheila Watt-Cloutier that I spent the most time. In addition to forcefully advocating indigenous rights, Mary was convinced that circumpolar Arctic cooperation would be far more effective if such cooperation could operate by adopting the principles of sustainable development. By the early 1990s, she had moved from being the president of the Inuit Circumpolar Council (ICC) to being Canada's ambassador for circumpolar affairs. In this role, she chaired negotiations that led to the Arctic Environmental Protection Strategy being subsumed into a new organisation: the Arctic Council. This metamorphosis created a potentially more powerful organisation because the Arctic Council is composed of ministers responsible for foreign affairs of the participating countries. It is largely due to Mary Simon's tenacity that it came to pass. In theory, the Arctic Council has the clout to orchestrate great achievements.

Finally, once more, we come to Sheila Watt-Cloutier. I first met Sheila at the beginning of her 1995–2001 tenure as president of ICC Canada, a role she then continued as the international president of the ICC. She is a passionate and uncompromising advocate for the fundamental rights of Inuit people and of their "oneness" with the Arctic environment. But her passion for the well-being of Inuit never prevents her from being concerned about the well-being of us all – wherever we live. Time and again, I have been fascinated as she has captured the sympathy of a room full of sceptics. You will have seen glimpses of Sheila in action from the quotes and anecdotes of her work in the sections on the Stockholm Convention. It is from Sheila that we should learn the wisdom of negotiating an agreement that gets the job done while leaving everyone proud of the achievement. For readers interested in game theory, she is a natural master of the non-zero-sum game where one player's gain need not be bad news for the other players.

This list could go on and on. However, I believe that the noted individuals and countries deserve special recognition. They played seminal roles in creating an organisation capable of taking stock of the health of the Arctic homeland and of promoting international actions to restore that health when necessary. They ensured that the Rovaniemi flame came to life and flourished within the Arctic Council. Collectively, they epitomize the heart of the Arctic Messenger.

In the preceding chapter, we decided that the Arctic Messenger probably feels fairly comfortable with the actions taken to deal with stratospheric ozone depletion and acid rain. For POPs, heavy metals and radioactivity, it could judge that we are heading slowly in the right direction but at the same time question why a precautionary approach to chemical regulation has not become universal in at least all the Arctic countries. But with climate change, our Arctic Messenger would justifiably be puzzled. The Arctic environment, as it was known by Arctic indigenous peoples born before about 1980, is vanishing before our eyes. The fact that "we are all in this together" wherever we live should be blatantly obvious to anyone who gives it a moment of thought. But the world still fails to agree on what to do.

I believe that climate change is the challenge that the Arctic Council must address head-on. It is clearly within its mandate to do so. The 1996 founding declaration of the council affirmed:

- Its "commitment to the well-being of the inhabitants of the Arctic, including recognition of the special relationship and unique contributions to the Arctic of indigenous peoples and their communities".
- Its "commitment to sustainable development in the Arctic region, including economic and social development, improved health conditions and cultural well-being".
- Its concurrent "commitment to the protection of the Arctic environment, including the health of Arctic ecosystems, maintenance of biodiversity in the Arctic region and conservation and sustainable use of natural resources".

Over the last decade, Arctic Council ministers have received several comprehensive reports from AMAP and others on Arctic climate change. The council has progressively responded to these reports with stronger and stronger language, urging, for example, parties to the United Nations Framework Convention on Climate Change (UNFCCC) to continue to take urgent measures "to meet the long-term goal aimed at limiting increase in global average temperature to below 2 degrees Celsius above pre-industrial levels" (Kiruna Declaration, May 2013). Other actions by the Council have ranged from asking for reports on Arctic resilience to climate change to exploring how to achieve reductions in short-lived climate-forcing agents. These are all very worthy activities. However, lurking and ignored in the background is the true monster: that cumulative, ever-growing inventory of atmospheric carbon dioxide that is overwhelming the natural capacity of our planet to deal with it. The important first step is for the Arctic Council to bring its voice into the emission reduction debate. How may the Arctic Council grasp this initiative?

I believe it is time for one or more of the Arctic countries to step forward and make a bold proposal, as Finland did in 1989. The proposal could be something along the following lines:

The Arctic region is experiencing climate warming that is twice the global average. Arctic ecosystems are changing or vanishing before our eyes.

We recognize that the Arctic Council member states are collectively responsible for about 30% of global CO_2 emissions.

We believe that the Arctic Council must show leadership in addressing the root cause of the issue: the anthropogenic emissions of greenhouse gases

(GHGs) (primarily carbon dioxide). We suggest that Arctic Council states consider seizing the opportunity before the UNFCCC meeting in 2015 to explore and describe cooperative actions or strategies aimed to guide or influence their individual national cumulative emission reduction targets. This could take the form of a common regional framework for GHG emission reduction by Arctic Council countries that recognizes and reinforces the Arctic Council's long-term goal to keep the global average temperature increase below 2°C.

We propose that member states consider setting up a task force or some other group charged with developing options for such a common framework. The options should be graded according to their anticipated ability to meet the 2°C objective and their prospects for gaining political backing. The development of options for a framework would not imply at this stage any future legal commitment, but the next step could be for Arctic Council countries to develop GHG emission reduction plans based on the common framework. The proposal could suggest the consideration of the development of other regional agreements under the UNFCCC. The Arctic Council could provide a "demonstration project" by developing the structure and content of such an agreement between its member states.

The proposed task force could be formal or informal.

Why do I think such an exercise would be useful, possible and quite simply the correct thing to do?

It would be useful because:

- Although the task would not be easy, it would be much easier than trying to develop an accord with more than 150 countries within the UNFCCC. It is always simpler to reach agreement in small groups, especially when they share a common vulnerable ecosystem, the Arctic and its peoples.
- The Arctic Council contains large, small, and medium-sized countries. Some have economies that include a large hydrocarbon (including coal) production and export sector, while others rely on imports for all their hydrocarbon needs. Some are heavily industrialized and some are not. This breadth of different economic backgrounds should result in options that would have appeal from countries well beyond Arctic countries. If it wished, the Arctic Council could consider whether to ask some of its observer states to join the exercise.
- Completion of the exercise should improve the ability of Arctic Council member states to achieve mutually satisfactory results at the UNFCCC.

It would be possible because:

- Although the Arctic Council will not meet again until spring 2015, it has created a way for opportunities to be seized between meetings. The May 2013 Kiruna Declaration empowered senior Arctic officials to adjust mandates and work plans of the council and of its working groups and to establish new ones.

It would be correct because:

- The three elements from the Arctic Council founding declaration quoted earlier seem to require the council to take a lead role in addressing the root cause of Arctic warming and emission reduction. The founding declaration certainly does not imply that Arctic countries would address a response to issues of this magnitude without the benefit of circumpolar cooperation.
- If the Arctic Council is not willing to take actions aimed at addressing climate change from a circumpolar as well as a global perspective, who will?

Are these thoughts too ambitious? Perhaps they are. However, I do believe that the founders of the Rovaniemi Process and of the Arctic Council would have suggested actions to achieve objectives such as those I have described. It would be a major and assertive step forward for the Arctic Council. Climate policy is seen by some as being so closely related to economic competitiveness that doing anything cooperatively could be viewed as being too risky. But the Arctic Council is an organisation that embraces sustainable economic development as well as environmental protection. The possession of a common framework for actions to prevent crossing the 2°C line could present substantial benefits for the Arctic states at the UNFCCC. It could also offer the best chances that the climate concerns of the Arctic Messenger will be heard and addressed. The legacy of Rovaniemi would be alive and well. It would be a fitting and promising point for us to leave the Arctic Messenger as it continues its journey through environmental change.

At the end of the preceding chapter, I asked whether the Arctic Council would heed the call from the Arctic Messenger. Has it listened? Has it understood? Will it act? These questions remain unanswered at this time, but the legacy of the Rovaniemi Process and of the Arctic Council gives cause for some optimism.

We have reached the end of the Arctic Messenger's story – for now. I hope it will ultimately have a happy ending.

Thank you for persevering with me to the end!

The Intergovernmental Panel
on Climate Change

The background related to global climate change will come indirectly from the work of the Intergovernmental Panel on Climate Change (IPCC). The World Meteorological Organization (WMO) and the United Nations Environment Programme (UNEP) set up this panel through a resolution of the UN General Assembly in 1988. Its purpose is to scientifically support the United Nations Framework Convention on Climate Change (UNFCCC) process and to provide governments with a clear view of what is happening to the world's climate. To do this, the IPCC has produced periodic assessments of the present state of knowledge from three topic-specific working groups:

- Working Group I reports on understanding changes in the physical climate systems.
- Working Group II assesses the vulnerability of socioeconomic and natural systems to climate change and options for adaptation (increasing our resilience to adverse impacts resulting from climate change).
- Working Group III looks at options for the mitigation of climate change emissions.

IPCC reports are packed solid with information but are not easy reading. You need to read a couple of pages and then ruminate for a day. They remind me of the Christmas puddings of my childhood in southeast England – a rich concoction of suet, preserved fruits and stout all glued together in a matrix of cholesterol and served up under a crown of rum butter and cognac. One portion had the weight of an ingot of lead. A couple of mouthfuls and you had taken on board enough calories to make it from Christmas Day to New Year's Eve.

The first IPCC assessment provided the technical rationale for the UNFCC in 1992, designed to "reduce global warming resulting from human activities and to cope with the consequences of climate change". This apparently swift progress on the political front was possible because, like the Vienna Convention on Stratospheric Ozone and the Convention on Long-Range Transboundary Air Pollution (CLRTAP), the treaty itself is quite general with no specific commitments required of the parties (national governments who have signed on to the UNFCCC). It sets out such things as broad goals and objectives, together with systems of governance, to build a regime in which the real and difficult actions can be prescribed by legally binding protocols. Such protocols can then be independently ratified by the parties. The first such protocol under the convention was the Kyoto Protocol, which is essentially an action plan aimed at achieving the objectives of the UNFCCC. As of December 2012, 191 countries are parties to the protocol. This means they have either ratified or acceded to it. Canada and the United States – two of the countries with the highest per capita rate of emissions of carbon dioxide – are not parties. (Canada originally signed but subsequently abandoned the protocol on the 15 December 2012). Sadly, the protocol is generally judged by the scientific community as being unable to prevent global warming surpassing an increase of +2°C (the level that many countries have informally agreed as a boundary not to be passed) and it is probably now consigned to the history books. But its demise had more to do with a lack of commitment to its policy intent of reducing greenhouse gases (GHGs) emissions than to any dissatisfaction about the protocol's scientific foundations.

The fourth IPCC assessment (AR4) was published in 2007 and the fifth assessment (AR5) appeared over the period 2013–2014.

What Will Happen in the Future If We Do Nothing or If We Try Very Hard to Aggressively Reduce GHG Emissions: Projected Change Under Different Emission Scenarios

The ultimate cause of our present climate warming is the massive input of greenhouse gases (GHGs) to the atmosphere due to human activity over the last 200 or more years. Intergovernmental Panel on Climate Change (IPCC) AR4 and AR5 confirm the high level of confidence of the global scientific community in making this statement. However, the future is still largely in our hands. We have already injected so much carbon dioxide into the atmosphere to commit the globe to a certain degree of warming. Nevertheless, it is still just possible that actions taken today could limit the amount of warming to be not far from the range within which our civilizations have developed. What will happen next depends entirely on how our energy-driven world economies evolve. Therefore, to look into the future, we have to combine scenarios for global economic development and then calculate what their individual climate impact may be. The first step has been to quantify the relationship between carbon dioxide emission and variables indicative of how economies and societies may evolve over the next 100 years. The following are examples of these variables (which for simplicity only consider CO_2):

- *Global population level (GPL):* The higher the population, the greater the global demand is for energy.
- *Global prosperity (GP):* A convergence of global prosperity towards that enjoyed by the more developed world economies will be driven by energy.
- *The level of energy intensity (EI):* This is the rate of energy use per unit of economic activity. Economic activities vary in their demand for energy.

- *The emission level (EL) of carbon dioxide per watt of energy:* Some energy sources vary considerably in their emission of carbon dioxide and some produce no GHG emissions. Even within the fossil fuel category, there is considerable variation. For example, natural gas releases far less carbon dioxide than coal or tar (oil) sands. It is sad and ironic that at present, Canada is vigorously promoting the latter energy source.

The relationship between emissions and these types of variables is sometimes simplified as:

$$CO_2 \text{ emission} = GPL \times GP \times EI \times EL$$

In 2000, the Intergovernmental Panel on Climate Change (IPCC) published a *Special Report on Emission Scenarios (SRES)* in which it explored this type of relationship by including different demographic, economic and technological paths for the types of variables described for the twenty-first century. See: www.ipcc.ch/pdf/special-reports/spm/sres-en. pdf for the *Summary for Policymakers.* The resulting SRES scenarios are grouped according to a storyline and associated families.

The A1 storyline and family explore a future of very rapid economic growth, a global population peaking in mid-century and then a decline and a rapid introduction of new and more efficient technologies. There is a general global convergence in levels of economic development and of per capita income. There are three families within the A1 scenario. Each family describes different possible paths of energy production and use: fossil intensive (A1FI); nonfossil energy sources (A1T); or a balance across all sources (A1B).

The A2 storyline and family explore a future with pronounced differences in regional economic development, lower per capita economic growth and a continuously growing global population.

The B1 storyline and scenario family portray a similar future as the A1 storyline but with rapid changes towards less energy-dependent economies and intensification of clean and resource-efficient technologies and environmental sustainability.

The B2 storyline and scenario family portray a future characterized by local solutions to economic, social and environmental sustainability. Economic development and technological change are less rapid and more diverse than in the B1 and A1 storylines, while the global population continues to increase but at a rate lower than in A2.

Altogether, 40 scenarios were developed under the SRES framework. The GHG emissions anticipated from these scenarios have been fed into the atmosphere-ocean global climate models (AOGCMs) we met earlier when looking at why climate scientists are able to confidently attribute present global warming to the increase in anthropogenic GHG emissions. The model runs usually begin in preindustrial times using observed natural forcing agents (such as solar activity) and observed values for GHG and aerosol concentrations. They then proceed into the future, "feeding" on the GHG and aerosol atmospheric concentrations computed for whichever of the SRES scenarios is under study (such as A1FI). At least 20 different climate models have been used to examine these scenarios.

The SRES scenarios do not consider the possibility of active global mitigation by world governments to reduce their GHG emissions. A new approach has been used in the fifth IPCC assessment. This consists of the development of a set of four scenarios containing emission, concentration and land-use trajectories referred to as *representative concentration pathways* (RCPs) for the main forcing agents of climate change. Four emission trajectories have been selected based on how much forcing they would produce at the end of the century: 2.6, 4.5, 6 and 8.5 watts per square metre. Integrated assessment models (IAMs) can then be used to look at different technological, socioeconomic and policy options for the present and the future that could lead to a particular magnitude of climate change. This approach holds promise for the evaluation of the costs and benefits of long-term climate goals and for the investigation of the possible roles of adaptation. The results should be directly supportive of international policy development aimed at the cooperative control of global warming through GHG emission reduction.

Nevertheless, since the global efforts of mitigation are as yet barely evident in terms of total global emissions, the existing SRES scenarios are still useful to capture the spectrum of the currently most plausible global economic development for the twenty-first century. In addition, you will find (if you read further) that much of the scientific literature prior to 2013 used the SRES approach. This is why I have explained their development in some detail.

What do the modelling studies using the SRES scenarios indicate for the future global climate? There is a very quick, rough and lazy way to answer this question. Essentially, it is that the climate trends we have been witnessing over the last 50 years will continue – except they will become more and more pronounced and the rate of change will generally increase.

By the end of the twenty-first century, the average global temperature will have increased by between 1.8°C (range 1.1 to 2.9) and 4.0°C (range 2.4 to 6.4) since 1995. Remember that we have already experienced a 0.6°C increase between the beginning of modern industrialization and 1995. Therefore, we should really add 0.6°C to these numbers to appreciate the full picture. The low estimate is for the lowest emission scenario (B1), while the high estimate comes from the highest emission scenario (A1FI). The A2 scenario is quite similar to A1FI (3.4°C within a range of 2.0 to 5.4). These latter two scenarios sadly come closest to describing the emerging economic world for at least the next 25 years, as shown by the most recent global CO_2 emission data we examined in the climate chapter. In the absence of action to aggressively reduce our GHG emissions (not just a reduction of the carbon intensity of various elements of national economies), global warming will continue for centuries beyond the year 2100. Unfortunately, very few governments appear to be prepared for the necessary action. Regardless as to how we look at it, by the year 2100, we can expect that our children and grandchildren will be living in a world that has an average global temperature somewhere between 2°C and 7°C warmer than that known by our parents. It is becoming increasingly probable that the often-stated desire to restrict warming to less than 2°C is not going to be achieved.

To put the significance of this into context, a global mean temperature decrease of about 3.5 to 5.0°C is all that separates us from the last glaciation and we will be moving into territory that has not been encountered during the evolution of the world's civilizations.

The uncertainties that show up as the ranges of projected temperature increase are not just a product of us being unable to guess how the world's energy-driven economies will develop over the next 100 years. For example, another important element of uncertainty is the behaviour of the various positive and negative feedback mechanisms that influence Earth's radiation budget and their resulting potential impact on climate sensitivity (roughly how much the climate will change for a doubling of carbon dioxide concentration).

The projected warming is also not globally uniform. It will continue to be most intense in the Arctic. Under the more moderate B1 SRES scenario, half the Arctic summer sea ice would be lost by 2100. However, we appear on track to probably having a summer ice-free Arctic Ocean by or before 2050.

One practical point needs heavy emphasis: If our governments have any true aspiration to restrict global warming to a level that is below 2°C, you will already have noticed that it will not suffice to simply stabilize emissions. Total emissions must be very considerably reduced to the point where atmospheric concentrations substantially decline in the first part of this century. This will not be easy, but we no longer have time to rely on the actions of others in the future. When politicians congratulate themselves that national emissions have plateaued, even though gross domestic product has risen, you know this is exactly what they are doing: relying with the faith of Micawber that someone somewhere will come up with an easy solution. Meanwhile, our children and grandchildren are looking towards a future that is very different from that which was bequeathed to us by our forebears. Past and future anthropogenic carbon dioxide emissions will continue to contribute to warming, sea level rise and acidifying oceans for more than a millennium even if emissions could be immediately stabilized today.

APPENDIX III

Some Geophysical Background Notes Related to Climate and Weather

The troposphere, the stratosphere and the polar vortex:

There are only two layers of the atmosphere we need to think about in this discussion: the troposphere and the stratosphere. The troposphere extends upwards from Earth's surface to an altitude of about 18 kilometres over equatorial regions and to about 8 kilometres over the High Arctic and Antarctic. The upper boundary is known as the *tropopause* and is the altitude at which air ceases to cool with height. Above the tropopause lies the stratosphere, where temperature increases with altitude. This is caused by the presence of the ozone maximum layer at an altitude of between 15 and 35 kilometres. Stratospheric ozone is formed naturally at these altitudes, where solar ultraviolet radiation breaks oxygen (O_2) molecules into free oxygen atoms that are then able to combine with intact oxygen molecules to produce ozone (O_3). Ozone in the stratosphere absorbs solar UV radiation. Therefore, the stratospheric temperature gradient runs in the opposite direction to that seen in the troposphere, which is warmest close to Earth. This temperature trend is part of the definition of the troposphere (cooling with height) and the stratosphere (warming with height). The upper boundary of the stratosphere (the stratopause) lies at an altitude of about 50 kilometres. Above this level, temperature decreases with altitude out into space.

The stratospheric polar vortex is a large-scale region of low pressure air that is constrained by a strong west-to-east jet stream that circles the polar region. It usually has two centres: one over Baffin Island and the other over north-east Siberia. The polar vortex extends from the upper troposphere through the stratosphere. Low values of ozone and cold temperatures are associated with the air inside the vortex, which acts as a barrier to prevent the movement of air from the South.

Coriolis effect:

Earth rotates towards the east, but the velocity of rotation varies according to latitude on Earth's surface. If I stand at the equator, where Earth's circumference is 40,075 kilometres, I will travel eastwards at 1,669 kilometres an hour. If at the same time my wife were standing at the Arctic Circle, where the circumference is only 17,662 kilometres, she would lazily amble eastwards at a sedate 736 kilometres an hour. If she were standing exactly at the pole, she would not have moved at all (only rotated). Anything that moves north from the equator that is not attached to the ground, such as a parcel of air, will conserve (retain) the eastward speed that it acquired at the equator. Therefore, as it moves north, it will be moving ever faster than the ground immediately beneath because it is moving closer to the axis of Earth's rotation. The result is that air (or water) travelling northwards away from the equator will seem to be deflected towards the right (east). This deflection to the right of the direction of movement in the Northern Hemisphere is said to be caused by the Coriolis effect. In the Southern Hemisphere, air moving away from the equator is again deflected to the east, but this time, the deflection is to the left. The Coriolis effect is zero at the equator.

In reality, the faster the air flows down a pressure gradient (from an area of high pressure to an area of low pressure), the harder it is for it to be pushed to the right. Eventually, a balance is reached between the pressure gradient force and the Coriolis effect, resulting in an airflow that is parallel to the isobars (lines on a chart showing areas of equal pressure). Winds that are in such a state of balance are called *geostrophic winds*. The result is the characteristic anticlockwise (cyclonic) circulation around cyclones (low-pressure systems) in the Northern Hemisphere.

Air being displaced out of a high-pressure system (anticyclone) in the Northern Hemisphere will move to the right, producing a clockwise (anticyclonic) circulation.

The jet stream:

The jet streams are narrow, fast-flowing rivers of air in the upper troposphere. The four strongest jet streams flowing from west to east are the two subpolar jets and the two less powerful subtropical jets of the Northern and

Southern hemispheres, respectively. Both types are located close to the tropopause, where temperature gradients are very intense. Being a high-flow core in a "fluid atmosphere", the jet stream is dynamically unstable and small perturbations can make it wiggle strongly north and south. These wiggles, or loops, are Rossby waves that normally move from west to east at a slower speed than that of the air within the jet stream itself. Because the Coriolis effect varies with latitude, the waves start tilting south-west to north-east in the Northern Hemisphere and break in mid-latitude storms at their polar tips. Normal waves (as in the open ocean) do not transfer momentum (do not move anything), but breaking waves do transfer momentum (and can move things, as on a beach). Therefore, as the Rossby waves break, they transfer momentum (and, hence, air) from the tropics to their breaking tips in the mid-latitudes.

The jets are usually continuous over long distances and follow the seasons, with, for example, the northern jet moving south during the northern winter. The subpolar (or polar) jet only really exists during the winter. At that point, there is a strong temperature gradient between areas in sunlight (getting energy from storms spinning off the poleward edge of the subtropical jet) and areas without any heat source (such as those areas experiencing the 24-hour darkness of Arctic winter). This leads to a strong pressure gradient and, through the Coriolis effect, a strong perpendicular wind. Although present in the summer, the subtropical jet is again mostly a winter phenomenon. This is because, in summer, Earth's inclination is such that at the time of the northern summer solstice, the top of the atmosphere at the pole is receiving slightly more sunlight than is the equator. At Earth's polar surface, less of this solar radiation can be absorbed because of the higher albedo of ice and snow. Of course, the solstice conditions do not last long, so there is not really time for a "more tropical" equilibrium to develop. But the point is that during the summer, there is very little north-south temperature gradient, thus not much to drive winds. Hence, the jet streams are primarily a winter phenomenon.

The jet streams are of great interest to climate scientists because they influence and are influenced by atmospheric processes taking place lower in the troposphere, including the formation of mid-latitude depressions (or storms) described earlier. They can also act as the mediators of teleconnections. This term is used here to describe a situation where an atmospheric event in one geographic locality is related to events perhaps thousands of kilometres away. For example, the El Niño Southern Oscillation (ENSO) influences the location of the jet streams, resulting in temperature and rainfall events as far away as North and South America and Western Europe.

The North Atlantic Oscillation (NAO):

This is the name given to year-to-year variations in the relative strengths and locations of the Azores High and the Icelandic Low. A large difference in pressure between the two systems (known as an index of NAO⁺) leads to increased moist westerly winds, giving cool summers and mild wet winters in Europe. Alternatively, when the pressure difference is low (NAO⁻), the moist westerly winds weaken, resulting in cold winters in Europe. Cyclones then track further south into the Mediterranean and over North Africa.

Under NAO⁺ conditions, northern Canada and Greenland are colder and drier, but this situation is reversed during a NAO⁻ phase. The NAO also exerts an influence on winter weather over much of eastern North America. When the index is high (NAO⁺), the Icelandic Low reflects a strong south-westerly circulation of air over this region at the expense of Arctic air moving south. Particularly during El Niño years, this results in warmer winters over the north-eastern United States and south-eastern Canada. During NAO⁻ years, cold air is able to penetrate further south and the eastern and south-eastern North American continent experiences winter cold outbreaks as far south as Florida.

The Arctic Oscillation (AO):

This is closely related to the NAO and describes two contrasting atmospheric circulation patterns over the Arctic. When the AO index is positive (AO⁺), surface pressure is low in the polar region. A strong jet stream supports a strong polar vortex and cold Arctic air is trapped in the polar region. When the AO index is negative (AO⁻), atmospheric pressure is higher over the polar region and the west to east winds and the polar vortex are weakened. This allows cold Arctic air to break out in winter into North America and Europe.

A word of caution: It is important to remember, however, that AO/NAO are just indices. The AO is just a measure of the strength of the polar vortex, so it does not really make sense to say it causes the strength of the vortex to change. Similarly, the NAO is an index of the latitudinal position of the storm track in the North Atlantic.

Orbital Forcing

There are three main orbital cycles that are known as the Milankovitch cycles and which concern how Earth moves around the sun. They are:

1.Orbital eccentricity:

The shape of Earth's orbit varies over a cycle of 100,000 years – from being nearly circular to mildly elliptical. At present, we are near the minimum of the cycle (almost circular), with only about a 6% difference in the amount of solar energy reaching Earth between perihelion (smallest Earth-sun distance) and aphelion (largest distance). However, when Earth's orbit is at its most elliptical, this solar energy difference is between 20 and 30%, which significantly increases or reduces the amount of energy received by Earth during different seasons, even though the net energy gain will be more or less constant.

2.Obliquity (axial tilt):

The tilt or inclination of Earth's axis of rotation in relation to the plane of the orbit around the sun oscillates between 21.5 and 24.5° and with a periodicity of about 41,000 years. The greater the tilt, the greater is the difference between our seasons. At present, we are located roughly in the middle of the obliquity cycle.

3.Precession of the equinoxes:

To understand this cycle, I find it easier to temporarily forget about the variation in axial tilt. Earth has a slow wobble as it spins on its axis. Although the angle of tilt remains the same, the direction of tilt will vary with time around Earth's orbit (wobble) over a 23,000-year cycle. At the present time, the direction of tilt is such that the North Pole is most inclined away from the sun (winter solstice) at perihelion (when Earth is closest to the sun) and most inclined towards the sun (summer solstice) at aphelion (when Earth is furthest from the sun). However, in 11,000 years, the slow wobble will mean Earth will be most inclined away from the sun at aphelion and most inclined towards the sun at perihelion.

The role of orbital forcing has been very important in the geological past. It is studied today in large part because paleoclimate processes help us understand how Earth has responded to climate forcing. However, they do not operate on time scales that are relevant to the climate warming that has been experienced over the last 50 or so years.

The Concept of Commitment

Climate scientists are confident that warming and sea level change will continue for centuries because of their knowledge of the carbon cycle in the case of carbon dioxide and of the lifetime of other Green House Gases (GHGs) in the atmosphere. For example, the lifetimes for methane and hydrochlorofluorocarbon-22 (as a representative of these types of chemicals) is about 12 years and for nitrous oxide about 110 years. The concentration of any GHG in the atmosphere depends on how much of the gas is being added to the atmosphere in comparison to how much is being removed. Therefore, for these substances (excluding carbon dioxide), it is not difficult to calculate the atmospheric concentration that would result from a reduction in the emissions of the gas. Stabilization and even a return to preindustrial levels (if emissions are eliminated) are possible within decades or a few centuries for substances with a short atmospheric lifetime.

However, this is not the case with carbon dioxide, for which an atmospheric lifetime cannot be calculated. Unfortunately, carbon dioxide is by far the most important GHG. Until the age of industrialisation, the concentration of carbon dioxide in the atmosphere was maintained over the long term at a roughly constant level (approximately 280 ppm) through the carbon cycle, which involves gas exchange between the atmosphere, the ocean and the lithosphere and which is made up of slow and fast components. At present, the oceans are absorbing much of the "excess" carbon dioxide we are adding to the atmosphere – but not all. About 20% will remain in the atmosphere for many millennia.

What about that 375 billion tonnes of carbon that humankind has injected into the atmosphere through our CO_2 emissions since the beginning of the Industrial Revolution? The oceans are a very significant component of the carbon cycle. In fact, the oceans are thought to have already absorbed about 48% of the carbon emitted to the atmosphere from anthropogenic sources between 1800 and 1994. However, carbon dioxide absorption by the oceans is a slow process. There are major bottlenecks in the route to the deep ocean (meridional overturning circulations), a warm ocean can hold less carbon dioxide than a cold ocean, and the depletion of carbonate in the surface waters will slow the conversion of carbon dioxide into bicarbonate. Although it makes up only 15% of the total world ocean surface, the North Atlantic contains nearly 25% of the carbon released by human activity since 1800. The net result of all this bad news is that over time, the ocean will become saturated with CO_2, beginning first with its surface waters. As this state is approached, the ocean's ability to absorb "excess" carbon dioxide will decline and atmospheric concentrations will increase. Some studies have already reported regional saturation of CO_2 in the ocean.

The stark implication is that even if we could stabilize our carbon dioxide emissions at current levels, we will simply be adding to the CO_2 stockpile we have already dumped into the atmosphere. Therefore, we are "committed" to seeing a continuous increase of atmospheric carbon dioxide concentrations for centuries to come (and of increasing temperatures). To make matters worse, we all know we are far from achieving even a stabilization of current levels. The rate at which we are adding carbon dioxide to the atmosphere greatly exceeds its rate of removal. Combine this with our knowledge about the slow atmospheric removal processes of the carbon cycle and we can foresee that even moderate reductions in global emissions will only reduce the rate of increase of atmospheric concentrations in coming decades. The practical importance in terms of climate policy of this huge "stockpile" of carbon dioxide that now resides in our atmosphere/ocean system was shown in a 2007 study by Montenegro and colleagues. They estimated how long anthropogenically produced carbon dioxide will stay in the atmosphere following one of the emission scenarios used by the Intergovernmental Panel on Climate Change (IPCC). The scenario chosen was the A2 scenario until the end of the twenty-first century, followed by a linear decline in emissions until zero emissions are reached in 2300. It was found that 75% of all anthropogenic carbon dioxide will remain in the atmosphere for an average of 1,800 years and that 25% will still be there after 5,000 years. In terms of everyone alive today and as far as we can possibly look into the future of our own personal kin, this means "forever".

The AR4 states that a 50% emission reduction (a scenario not contemplated by most of our world politicians) would only manage to stabilise atmospheric carbon dioxide concentrations for about a decade. After this, atmospheric CO_2 would rise again as the absorption capacity of land and ocean sinks declines. Complete elimination of CO_2 emissions was estimated in AR4 to lead to a slow decrease in atmospheric CO_2 of only about 40 ppm over the course of the twenty-first century. It is indeed something to ponder.

Bibliography

Chapter 2: The Arctic Messenger

Arctic Council Indigenous Peoples Secretariat. Fram Centre, N-9296 Tromso, Norway. www.arcticpeoples.org.

United Nations Permanent Forum on Indigenous Issues. http://undesadspd.org/ indigenouspeoples.aspx.

Chapter 3: The Arctic Messenger Gains a Voice: The Arctic Monitoring and Assessment Programme

The primary sources for this chapter have been the ministerial reports of the Arctic Environmental Protection Strategy (AEPS) and the Arctic Council. They are presented chronologically.

The Rovaniemi Declaration on the Protection of the Arctic Environment and the Establishment of the Arctic Environmental Protection Strategy, 14 June 1991. Rovaniemi, Finland.

The Nuuk Declaration on Environment and Development in the Arctic, 1993. Report of the Second Ministerial Conference on the Arctic Environment. Copenhagen, Denmark: Ministry of Foreign Affairs.

The Inuvik Declaration on Environmental Protection and Sustainable Development in the Arctic, 1996. Report of the Third Ministerial Conference on the Protection of the Arctic Environment. Ottawa, Canada: Department of Indian and Northern Affairs.

The Ottawa Declaration on the Establishment of the Arctic Council, 19 September 1996. Ottawa, Canada.

The Alta Declaration on the Arctic Environmental Protection Strategy, 1997. The Fourth Ministerial Meeting Under the Arctic Environmental Protection Strategy. Alta, Norway.

The Iqaluit Declaration on the Occasion of the First Ministerial Meeting of the Arctic Council, 17–18 September 1998. Ottawa, Canada.

The Barrow Declaration on the Occasion of the Second Ministerial Meeting of the Arctic Council, 13 October 2000. Barrow, Alaska.

The Inari Declaration on the Occasion of the Third Ministerial Meeting of the Arctic Council, 10 October 2002. Inari, Finland.

The Reykjavik Declaration on the Occasion of the Fourth Ministerial Meeting of the Arctic Council, 24 November 2004. Reykjavik, Iceland.

The Salekhard Declaration on the Occasion of the Fifth Ministerial Meeting of the Arctic Council, 26 October 2006. Salekhard, Russia.

The Tromso Declaration on the Occasion of the Sixth Ministerial Meeting of the Arctic Council, 29 April 2009. Tromso, Norway.

The Nuuk Declaration on the Occasion of the Seventh Ministerial Meeting of the Arctic Council, 12 May 2011. Nuuk, Greenland.

The Kiruna Declaration on the Occasion of the Eighth Ministerial Meeting of the Arctic Council, 15 May 2013. Kiruna, Sweden.

Suggested Further Reading

Arctic Governance Project, 2010. *Arctic Governance in an Era of Transformative Change: Critical Questions, Governance Principles, Ways Forward.* www.arcticgovernance.org.

Axworthy, T. S., T. Koivurova and W. Hasanat, 2012. *The Arctic Council: Its Place in the Future of Arctic Governance.* Collection of papers originally presented 17–18 January 2012 at a conference of the same name – a collaboration between the Munk-Gordon Arctic Security Program and the University of Lapland. Munk-Gordon Arctic Security Program.

English, J., 2013. *Ice and Water: Politics, Peoples, and the Arctic Council.* Toronto: Allen Lane and Penguin Canada.

Gorbachev, Mikhail, 1 October 1987. *Speech Given in Murmansk at the Ceremonial Meeting on the Occasion of the Presentation of the Order of Lenin and the Gold Star to the City of Murmansk.* www.lecerclepolaire.com/en/documentation-uk/news-a-views-uk/454-meeting-with-the-former-president-mikhail-gorbachev. Murmansk, USSR.

IASC Bulletin 2013. International Arctic Science Committee. www.iasc.info.

Kankaanpaa, P. and O. R. Young, 2012. "The Effectiveness of the Arctic Council." *Polar Research,* 31. http://dx.doi.org/10.3402/polar.v31i0.17176.

Keskitalo, E. C. H., 2004. *Negotiating the Arctic: The Construction of an International Region.* New York and London: Routledge.

Young, O. R., 2005. "Governing the Arctic: From Cold War Theatre to Mosaic of Cooperation." *Global Governance,* 11:9–15.

Chapter 4: Radioactivity

The primary sources for this chapter have been:

AMAP, 1998. *AMAP Assessment Report: Arctic Pollution Issues.* Arctic Monitoring and Assessment Programme. Oslo, Norway.

AMAP, 2002. *AMAP Assessment Report: Radioactivity in the Arctic.* Arctic Monitoring and Assessment Programme. Oslo, Norway.

AMAP, 2009. *AMAP Assessment Report: Radioactivity in the Arctic.* Arctic Monitoring and Assessment Programme. Oslo, Norway.

Sjoblom, K.-L. and G. Linsley, 1998. International Arctic Seas Assessment Project (IASAP): "Summary." *IAEA Bulletin,* 40/4/1998.

Strand, P. and A. Cooke (eds.), 1995. *Environmental Radioactivity in the Arctic.* Proceedings of the Second International Conference on Environmental Radioactivity in the Arctic. C/O Norwegian Radiation Protection Authority. Østorås, Norway.

Strand, P. and E. Holm (eds.), 1993. *Environmental Radioactivity in the Arctic and Antarctic.* Scientific Committee of the International Conference on Environmental Radioactivity in the Arctic and Antarctic. C/O Norwegian Radiation Protection Authority. Østorås, Norway.

Suggested Further Reading

AMAP/Nordic Environment Finance Corporation (NEFCO), 1995. *Barents Region Environmental Programme: Proposals for Environmentally Sound Investment Projects in the Russian Part of the Barents Region.* Volume II: Radioactive Contamination.

IAEA, 1993. *International Meeting on Assessment of Actual and Potential Consequences of Dumping of Radioactive Waste Into Arctic Seas* (Oslo, Norway, 1–5 February 1993). Working Material of the IAEA.

IAEA, 1998. *The Radiological Accident in the Reprocessing Plant at Tomsk.* International Atomic Energy Agency. Vienna, Austria.

IAEA, 2008. Contact Expert Group (CEG). *Current Developments in the Nuclear Legacy Programmes of the CEG Members and Partners by September 2008.* Newsletter No. 2. Contact Expert Group for International Radwaste Projects in the Russian Federation.

Joint Norwegian-Russian Expert Group for Investigation of Radioactive Contamination in Northern Areas, 1997. *Sources Contributing to Radioactive Contamination of the Techa River and Areas Surrounding the Mayak Production Association, Urals, Russia: Programme of Investigation of Possible Impacts of Mayak PA Activities on Radioactive Contamination of the Barents and Kara Seas.* Norwegian Secretariat. Norwegian Radiation Protection Authority. Østorås, Norway.

Layton, D., R. Edson, M. Varela and B. Napier, 1997. *Radionuclides in the Arctic Seas From the Former Soviet Union: Potential Health and Ecological Risks.* Arctic Nuclear Waste Assessment Program (ANWAP): Office of Naval Research.

Lind, O.C., D.H. Oughton, B. Salbu, I. Skipperud, M.A. Sickel et al., 2006. "Transport of Low ^{240}Pu/^{239}Pu Atom Ratio Plutonium-Species in the Ob and Yenisei Rivers to the Kara Sea." *Earth and Planetary Science Letters*, 251:33–43.

Lidén, K., 1961. "Caesium 137 Burdens in Swedish Laplanders and Reindeer." *Acta Radiologica*, 56:237–240.

Luoto, J. and H. Haapala, 2005. *Environmental Cooperation in the Barents Region.* Ten-Year Review. Ministry of Environment. Finland.

NEFCO, 2013. Nordic Environment Finance Corporation. *Annual Review: 2012.* NEFCO. Helsinki, Finland.

Office of Technology Assessment, 1995. *Nuclear Wastes in the Arctic: An Analysis of Arctic and Other Regional Impacts From Soviet Nuclear Contamination.* OTA-ENV-623. Washington, DC.

Porfiriev, B. N., 1996. "Environmental Aftermath of the Radiation Accident at Tomsk-7." *Environmental Management*, 20(1):25–33.

Sivintsev, Y. and O. Kiknadze, 1998. Components of Nuclear Reactors Dumped in the Kara Sea and the Sea of Japan, and Assessment of Radionuclides Contained in them. In the Afterword to the *White Book* (Yablokov Commission), 172–181. Proceedings of the International Seminar, 19–21 January 1998. International Science and Technology Centre: CDB "Lazurit". Nizhny Novgorod, Russia.

Skipperud, L., D. H. Oughton, L. K. Fifield, O. C. Lind et al., 2004. "Plutonium Isotope Ratios in the Yenisei and Ob Estuaries." *Applied Radiation and Isotopes*, 60:588–93.

Smith, J. N., K. M. Ellis and L. R. Kilius, 1998. "^{129}I and ^{137}Cs Tracer Measurements in the Arctic Ocean." *Deep Sea Research*, 45:959–84.

Smith, J. N., K. M. Ellis and T. Boyd, 1999. "Circulation Features in the Central Arctic Ocean Revealed by Nuclear Fuel Reprocessing Tracers From Scientific Ice Expeditions, 1995–1996." *Journal of Geophysical Research*, 104:29663–67.

Strand, P., A. I. Nikitin, B. Lind, B. Salbu and G. C. Christensen, 1997. *Dumping of Radioactive Waste and Investigation of Radioactive Contamination in the Kara Sea. Results from three years of investigations (1992–1994) in the Kara Sea.* Joint Norwegian-Russian Expert Group for investigation of radioactive contamination in the northern areas. Norwegian Radiation Protection Authority. Østorås, Norway.

UNSCEAR, 2000. *Report of the Scientific Committee on the Effects of Atomic Radiation (UNSCEAR) to the United Nations General Assembly.* Volume 1.

Yablokov, A. (ed.), 1992–1993. *Facts and Problems Related to Radioactive Waste Disposal in the Seas Adjacent to the Territory of the Russian Federation: Materials for a Report by the Governmental Commission on Matters Related to Radioactive Waste Disposal at Sea.* Established by Decree No. 613 of the Russian Federation President, Moscow, 24 October 1992 (in Russian). English translation by the International Maritime Organization: LC16/INF.2. 1993.

Chapter 6: Acidification and Arctic Haze

The primary sources for this chapter have been:

AMAP, 1998. *AMAP Assessment Report: Arctic Pollution Issues.* Arctic Monitoring and Assessment Programme. Oslo, Norway.

AMAP, 2006. *AMAP Assessment Report: Acidifying Pollutants, Arctic Haze, and Acidification in the Arctic.* Arctic Monitoring and Assessment Programme. Oslo, Norway.

Nordberg, L., 2010. *Air Pollution: Promoting Regional Cooperation.* United Nations Environment Programme. Nairobi, Kenya.

Symon, Carolyn, 2006. *Arctic Pollution 2006: Acidification and Arctic Haze.* Arctic Monitoring and Assessment Programme. Oslo, Norway.

Suggested Further Reading

Oreskes, N. and E. M. Conway 2010. *Merchants of Doubt.* New York: Bloomsbury Press.

Sliggers, J. and W. Kakebeeke, 2004. *Clearing the Air: 25 Years of the Convention on Long-Range Transboundary Air Pollution.* United Nations Economic Commission for Europe. Geneva, Switzerland.

UNECE, 1984. *The Geneva Protocol on Long-Term Financing of the Cooperative Programme for Monitoring and Evaluation of the Long-Range Transmission of Air Pollutants in Europe (EMEP).* United Nations Economic Commission for Europe. Geneva, Switzerland.

UNECE, 2004. *Handbook for the 1979 Convention on Long-Range Transboundary Air Pollution and Its Protocols.* United Nations Economic Commission for Europe. Geneva, Switzerland.

UNECE, 2004. *Protocol to the 1979 Convention on Long-Range Transboundary Air Pollution to Abate Acidification, Eutrophication and Ground-Level Ozone – Consolidated Text of the Amended Protocol.* United Nations Economic Commission for Europe. Geneva, Switzerland.

UNECE, 2010. *Hemispheric Transport of Air Pollution 2010: Part A, Ozone and Particulate Matter.* Air Pollution Studies No. 17. United Nations Economic Commission for Europe. Geneva, Switzerland.

UNECE, 2010. *Hemispheric Transport of Air Pollution 2010: Part D, Answers to Policy Relevant Questions.* Air Pollution Studies No. 20. United Nations Economic Commission for Europe. Geneva, Switzerland.

Victor, D. G., K. Raustiala and E. B. Skolnikoff (eds.), 1998. *The Implementation and Effectiveness of International Environmental Commitments: Theory and Practice.* Cambridge and London: MIT Press.

Chapter 7: Stratospheric Ozone Depletion

The primary sources for this chapter have been:

ACIA, 2004. *Impacts of a Warming Arctic.* Arctic Climate Impact Assessment. Cambridge, UK: Cambridge University Press.

ACIA, 2005. *Ozone and Ultraviolet Radiation.* Arctic Climate Impact Assessment. Cambridge, UK: Cambridge University Press.

Newman, P. A., M. Rex et al., 2006. Polar Ozone: Past and Present. Chapter 4 in *Scientific Assessment of Ozone Depletion: 2006.* Global Ozone Research and Monitoring Project – Report No. 50. World Meteorological Organization. Geneva, Switzerland.

UNEP, 2010 Assessment. *Environmental Effects of Ozone Depletion.* United Nations Environment Programme.

WMO, 2007. *Scientific Assessment of Ozone Depletion: 2006.* Global Ozone Research and Monitoring Project – Report No. 50. World Meteorological Organization. Geneva, Switzerland.

WMO, 2011. *Scientific Assessment of Ozone Depletion: 2010.* Global Ozone Research and Monitoring Project – Report No. 52. World Meteorological Organization. Geneva, Switzerland.

Suggested Further Reading

AMAP, 1998. "Chapter 11: Climate Change, Ozone, and Ultraviolet Radiation." In *AMAP Assessment Report: Arctic Pollution Issues.* Arctic Monitoring and Assessment Programme. Oslo, Norway.

Balter, M., 2013. "Archaeologists Say the 'Anthropocene' Is Here – But It Began Long Ago." *Science,* 340(6130):261–62.

Bernhard, G., G. Manney, V. Fioletov, J.-U. Grooß, A. Heikkilä, B. Johnsen, T. Koskela, K. Lakkala, R. Müller, C. L. Myhre and M. Rex, 2012. "Arctic Ozone and UV Radiation" (in "State of the Climate in 2011"). *Bulletin of the American Meteorology Society,* 93(7):S129–S132.

Crutzen, P. J., 1970. "The Influence of Nitrogen Oxides on the Atmospheric Ozone Content." *Quarterly Journal of the Royal Meteorological Society,* 96:320–25.

Dotto, L. and H. I. Schiff, 1978. *The Ozone War.* Garden City, NY: Doubleday.

Farman, J., B. Gardiner and J. Shanklin, 1985. "Large Losses of Total Ozone in Antarctica Reveal Seasonal ClOx/NOx Interaction." *Nature,* 315:207–10.

Gao, K., Z. Ruan, V. E. Villafañe, J.-P. Gattuso and E. W. Helbling, 2009. "Ocean Acidification Exacerbates the Effect of UV Radiation on the Calcifying Phytoplankter *Emiliania huxleyi.*" *Limnology and Oceanography,* 54(6):1855–62.

Johnston, H. 1971. "Reduction of Stratospheric Ozone by Nitrogen Oxide Catalysts From Supersonic Transport Exhaust." *Science,* 173:517–22.

Lange J. R., B. E. Palis, D. C. Chang, S. J. Soong and C. M. Balch, 2007. "Melanoma in Children and Teenagers: An Analysis of Patients From the National Cancer Data Base." *Journal of Clinical Oncology,* 25:1363–68.

Laube, J. C., M. J. Newland and C. Hogan et al., 2014. "Newly Detected Ozone-Depleting Substances in the Atmosphere." *Nature Geoscience*. www.nature.com/ngeo/journal/v7/n4/full/ngeo2109.html.

Lovelock, J. E., 1971. "Atmospheric Fluorine Compounds as Indicators of Air Movements." *Nature*, 230(5293):379.

Lovelock, J. E., R. J. Maggs and R. J. Wade, 1973. "Halogenated Hydrocarbons in and Over the Atlantic." *Nature*, 241(5386):194–96.

Manney, G. L., M. L. Santee, M. Rex., N. J. Livesey et al., 2011. "Unprecedented Arctic Ozone Loss in 2011." *Nature*, 478:469–75.

Molina, M. J. and F. S. Rowland, 1974. "Stratospheric Sink for Chlorofluoro-methanes: Chlorine Atom Catalysed Destruction of Ozone." *Nature*, 249:810–12.

Newsham, K. K. and S. A. Robinson, 2009. "Responses of Plants in Polar Regions to UVB Exposure: A Meta-Analysis." *Global Change Biology*, 15:2574–89.

Oreskes, N. and E. M. Conway, 2010. *Merchants of Doubt*. New York: Bloomsbury Press.

Rex, M., R. J. Salawitch, P. von der Gathen, N. R. P. Harris, M. P. Chipperfield and B. Naujokat, 2004. "Arctic Ozone Loss and Climate Change." *Geophysical Research Letters*, 31:L04116.

Rowland, F. S. and M. J. Molina, 1975. "Chlorofluoromethanes in the Environ-ment." *Reviews of Geophysics and Space Physics*, 13:1–35.

Solomon, S., 1999. "Stratospheric Ozone Depletion: Review of Concepts and History." *Review of Geophysics*, 37:275–316.

Stolarski, R. and R. Cicerone, 1974. "Stratospheric Chlorine: A Possible Sink for Ozone." *Canadian Journal of Chemistry*, 52:1610–15.

Strahan, S. E., A. R. Douglass and P. A. Newman, 2013. "The Contributions of Chemistry and Transport to Low Arctic Ozone in March 2011 Derived From Aura MLS Observations." *Journal of Geophysical Research: Atmospheres*, 118 (3):1563–76.

UNEP, 2012. *Handbook for the Montreal Protocol on Substances That Deplete the Ozone Layer*. Ozone Secretariat. United Nations Environment Programme. Nairobi, Kenya.

UNEP, 2012. *Handbook for the Vienna Convention for the Protection of the Ozone Layer (1985)*. Ninth edition (2012). Ozone Secretariat. United Nations Environment Programme. Nairobi, Kenya.

WMO/NASA/UNEP, 1988. *Report of the International Ozone Trends Panel* - Report No. 18, Vols. 1 and 2.

Zavala, J. A., C. L. Casteel, E. H. DeLucia and M. R. Berenbaum, 2008. "Anthropo-genic Increase in Carbon Dioxide Compromises Plant Defense Against Invasive Insects." *Proceedings of the National Academy of Sciences of the United States*. 105:5129–33.

Zavala, J. A. and D. A. Ravetta, 2002. "The Effect of Solar UV-B Radiation on Terpenes and Biomass Production in *Grindelia chiloensis* (Asteraceae), a Woody Perennial of Patagonia, Argentina." *Plant Ecology*, 161:185–91.

Chapter 8: Persistent Organic Pollutants and Heavy Metals (Including Mercury)

The primary sources for this chapter have been:

AMAP, 1998. *AMAP Assessment Report: Arctic Pollution Issues*. Arctic Monitoring and Assessment Programme. Oslo, Norway.

AMAP, 2002. *AMAP Assessment Report: Persistent Organic Pollutants in the Arctic*. Arctic Monitoring and Assessment Programme. Oslo, Norway.

AMAP, 2002. *AMAP Assessment Report: The Influence of Global Change on Contaminant Pathways to, Within, and From the Arctic.* Arctic Monitoring and Assessment Programme. Oslo, Norway.

AMAP, 2004. *AMAP Assessment Report: Persistent Toxic Substances, Food Security and Indigenous Peoples of the Russian North.* Final Report. Arctic Monitoring and Assessment Programme. Oslo, Norway.

AMAP, 2009. *AMAP Assessment Report: Human Health in the Arctic.* Arctic Monitoring and Assessment Programme. Oslo, Norway.

AMAP, 2009. *AMAP Assessment Report: Persistent Organic Pollutants in the Arctic.* Arctic Monitoring and Assessment Programme. Oslo, Norway.

AMAP, 2011. *AMAP Assessment Report: Mercury in the Arctic.* Arctic Monitoring and Assessment Programme. Oslo, Norway.

Kallenborn, R. et al., 2011. *Combined Effects of Selected Pollutants and Climate Change in the Arctic Environment.* AMAP Technical Report No. 5. Arctic Monitoring and Assessment Programme. Oslo, Norway.

UNECE, 2010. *Hemispheric Transport of Air Pollution 2010: Part B, Mercury.* Air Pollution Studies No. 18. United Nations Economic Commission for Europe. Geneva, Switzerland.

UNECE, 2010. *Hemispheric Transport of Air Pollution 2010: Part C, Persistent Organic Pollutants.* Air Pollution Studies No. 19. United Nations Economic Commission for Europe. Geneva, Switzerland.

UNECE, 2010. *Hemispheric Transport of Air Pollution 2010: Part D, Answers to Policy Relevant Questions.* Air Pollution Studies No. 20. United Nations Economic Commission for Europe. Geneva, Switzerland.

UNEP, 2013. *Global Mercury Assessment 2013: Sources, Emissions, Releases and Environmental Transport.* UNEP Chemicals Branch: Geneva, Switzerland. Technical background report available at: www.amap.no/documents/doc/technical-background-report-for-the-global-mercury-assessment-2013/848.

Suggested Further Reading

Addison, R. F. and P. F. Brodie, 1973. "Occurrence of DDT Residues in Beluga Whales (Delpinapterus leucas) From the Mackenzie Delta, N.W.T." *Journal of the Fisheries Research Board of Canada*, 30:1733–36.

Addison, R. F. and T. G. Smith, 1974. "Organochlorine Residue Levels in Arctic Ringed Seals: Variation With Age and Sex." *Oikos*, 25:335–37.

Arctic Council, 2000. *The Barrow Declaration on the Occasion of the Second Ministerial Meeting of the Arctic Council.* Barrow, Alaska.

Aubail, A., R. Dietz, F. Riget, C. Sonne et. al., 2012. "Temporal Trend of Mercury in Polar Bears (Ursus maritimus) From Svalbard Using Teeth as a Biomonitoring Tissue." *Journal of Environmental Monitoring*, 14:56–63.

Ayotte P., G. Muckle, J. L. Jacobson, S. W. Jacobson and É. Dewailly, 2003. "Assessment of Pre- and Postnatal Exposure to Polychlorinated Biphenyls: Lessons From the Inuit Cohort Study." *Environmental Health Perspectives*, 111:1253–58.

Balmford, A., 2013. "Pollution, Politics, and Vultures." *Science*, 339:653–54.

Barr D. B., P. Weihe, M. D. Davis, L. L. Needham and P. Grandjean, 2006. "Serum Polychlorinated Biphenyl and Organochlorine Insecticide Concentrations in a Faroese Birth Cohort." *Chemosphere*, 62(7):1167–82.

Becker, S. et al., 2008. "Long-Term Trends in Atmospheric Concentrations of α- and γ-HCH in the Arctic Provide Insight Into the Effects of Legislation and Climatic Fluctuations on Contaminants Levels." *Atmospheric Environment*, 42:8225–33.

Bellinger, D. C., 2012. "A Strategy for Comparing the Contributions of Environmental Chemicals and Other Risk Factors to Neurodevelopment of Children." *Environmental Health Perspectives*, 120(4):501–07.

Benskin, J. P., D. Muir, B. F. Scott, C. Spencer, A. O. De Silva, H. Kylin, J. W. Martin, A. Morris, R. Lohmann, G. Tomy, B. Rosenberg, S. Taniyasu and N. Yamashita, 2012. "Perfluorinated Compounds in the Arctic and Atlantic Oceans." *Environmental Science and Technology*, 46(11):5815–23.

Bergman, Å., J. Heindel, S. Jobling, K. A. Kidd and R. T. Zoeller (eds.), 2012. *State of the Science of Endocrine Disrupting Chemicals*. World Health Organization and United Nations Environment Programme. World Health Organization. Geneva: WHO Press.

Bidleman, T. F., G. W. Patton, D. A. Hinckley, M. D. Walla, W. E. Cotham and B. T. Hargrave, 1990. "Chlorinated Pesticides and Polychlorinated Biphenyls in the Atmosphere of the Canadian Arctic." In D. A. Kurtz (ed.), *Long-Range Transport of Pesticides*. Chelsea, MI: Lewis Publishers.

Bidleman, T. F., P. A. Helm, B. M. Braune and G. W. Gabrielsen, 2010. "Polychlorinated Naphthalenes in Polar Environments – A Review." *Science of the Total Environment*, 408(15):2919–35.

Bogdal, C., E. Abad, M. Abalos, B. van Bavel, J. Hagberg, H. Fiedler and M. Scheringer, 2012. "Worldwide Distribution of Persistent Organic Pollutants in Air, Including Results of Air Monitoring by Passive Air Sampling in Five Continents." *Trends in Analytical Chemistry*. http://dx.doi.org/10.1016/j.trac.2012.05.011.

Boucher O., C. H. Bastien, D. Saint-Amour, É. Dewailly, P. Ayotte, J. L. Jacobson et al., 2010. "Prenatal Exposure to Methylmercury and PCBs Affects Distinct Stages of Information Processing: An Event-Related Potential Study With Inuit Children." *Neurotoxicology*, 31:373–84.

Boucher O., M. J. Burden, G. Muckle, D. Saint-Amour, P. Ayotte, É. Dewailly et al., 2012. "Response Inhibition and Error Monitoring during a Visual Go/No Go Task in Inuit Children Exposed to Lead, Polychlorinated Biphenyls, and Methylmercury." *Environmental Health Perspectives*, 120(4):608–15.

Boucher, O., S. W. Jacobson, P. Plusquellec, É. Dewailly, P. Ayotte, N. Forget-Dubois, J. L. Jacobson and G. Muckle, 2012. "Prenatal Methylmercury, Postnatal Lead Exposure, and Evidence of Attention Deficit/Hyperactivity Disorder Among Inuit Children in Arctic Québec." *Environmental Health Perspectives*, 120 (10):1456–61.

Braune, B. M., 2007. "Temporal Trends of Organochlorines and Mercury in Seabird Eggs From the Canadian Arctic, 1975–2003." *Environmental Pollution*, 148:599–613.

Braune, B. M., P. M. Outridge, A. T. Fisk, D. C. G. Muir et al., 2005. "Persistent Organic Pollutants and Mercury in Marine Biota of the Canadian Arctic: An Overview of Spatial and Temporal Trends." *Science of the Total Environment*, 351:4–56.

Buck Louis, G. M., R. Sundaram, E. F. Schisterman, A. M. Sweeney et al., 2013. "Persistent Environmental Pollutants and Couple Fecundity: The LIFE Study." *Environmental Health Perspectives*, 121(2):231–36.

Butt, C. M., D. C. G. Muir and S. A. Mabury, 2010. "Elucidating the Pathways of Poly- and Perfluorinated Acid Formation in Rainbow Trout." *Environmental Science and Technology*, 44(13):4973–80.

Butt, C. M., U. Berger, R. Bossi, G. T. Tomy, 2010. "Levels and Trends of Poly- and Perfluorinated Compounds in the Arctic Environment." *Science of the Total Environment*, 408(15):2936–65.

Colborn, T., D. Dumanoski and J. P. Myers, 1996. *Our Stolen Future: Are We Threatening Our Fertility, Intelligence, and Survival? A Scientific Detective Story*. New York: Dutton.

Cuklev, F., E. Kristiansson, J. Fick et al., 2011. "Diclofenac in Fish: Blood Plasma Levels Similar to Human Therapeutic Levels Affect Global Hepatic Gene Expression." *Environmental Toxicology and Chemistry*, 30(9):2126–34.

Debes, F., E. Budtz-Jorgensen, P. Weihe, R.F. White and P. Grandjean, 2006. "Impact of Prenatal Methylmercury Exposure on Neurobehavioural Function at Age 14 Years." *Neurotoxicology and Teratology*, 28(5):536–47.

Dietz, R., C. Sonne, N. Basu, B. Braune, T. O'Hara et al., 2013. "What Are the Toxicological Effects of Mercury in Arctic Biota?" *Science of the Total Environment*, 443:775–90.

Dietz, R., C. O. Nielsen, M. M. Hansen and C. T. Hansen, 1990. "Organic Mercury in Greenland Birds and Mammals." *Science of the Total Environment*, 95:41–51.

Dietz, R., E. W. Born, F. Riget, A. Aubail et al., 2011. "Temporal Trends and Future Predictions of Mercury Concentrations in Northwest Greenland Polar Bear (Ursus maritimus) Hair." *Environmental Science and Technology*, 45:1458–65.

Dietz, R., F. Riget, D. Boertmann et al., 2006. "Time Trends of Mercury in Feathers of West Greenland Birds of Prey During 1851–2003." *Environmental Science and Technology*, 40:5911–16.

Dietz, R., F. Riget, E. W. Born et al., 2006. "Trends in Mercury in Hair of Greenlandic Polar Bears (Ursus maritimus) During 1892–2001." *Environmental Science and Technology*, 40:1120–25.

Dietz, R., P. M. Outridge and K. A. Hobson, 2009. "Anthropogenic Contributions to Mercury Levels in Present-Day Arctic Animals: A Review." *Science of the Total Environment*, 407(24):6120–31.

Downie, D. L. and T. Fenge (eds.), 2003. *Northern Lights Against POPs: Combatting Toxic Threats in the Arctic*. Montreal and Kingston: McGill-Queen's University Press.

Dudarev, A. A., 2012. "Dietary Exposure to Persistent Organic Pollutants and Metals Among Inuit and Chukchi in Russian Arctic Chukotka." *Journal of Circumpolar Health*, 71:1–12.

Eckley, N. and H. Selin, 2003. "The Arctic at Risk: Arctic Pollution 2002." *Environment* 45(7):37–40.

Eskenazi, B., J. Chevrier, S. A. Rauch et al., 2012. "In Utero and Childhood Polybrominated Diphenyl Ether (PBDE) Exposures and Neurodevelopment in the CHAMACOS Study." *Environmental Health Perspectives*. doi:10.1289/ehp.1205597.

Genualdi, S., S. C. Lee, M. Shoeib, A. Gawor, L. Ahrens and T. Harner, 2010. "Global Pilot Study of Legacy and Emerging Persistent Organic Pollutants Using Sorbent-Impregnated Polyurethane Foam Disk Passive Air Samplers." *Environmental Science and Technology*, 44:5534–39.

Genualdi, S., T. Harner, Y. Cheng, M. MacLeod et al., 2011. "Global Distribution of Linear and Cyclic Volatile Methyl Siloxanes in Air." *Environmental Science and Technology*, 45:3349–54.

Gilbert, N., 2012. "Drug Pollution Law All Washed Up: EU Initiative to Clean Up Waterways Faces Tough Opposition." *Nature*, 491(7425):503–04.

Gouin, T., J. M. Armitage, I. T. Cousins, D. C. G. Muir, C. A. Ng, L. Reid and S. Tao, 2013. "Influence of Global Climate Change on Chemical Fate and Bioaccumulation: The Role of Multimedia Models." *Environmental Toxicology and Chemistry*, 32 (1):20–31.

Gouteux, B., M. Alaee, S. Mabury, G. Pacepavicius and D. Muir, 2008. "Polymeric Brominated Flame Retardants: Are They a Relevant Source of Emerging Brominated Aromatic Compounds in the Environment?" *Environmental Science and Technology*, 42:9039–44.

Grandjean, P. et al., 1994. "Human Milk as a Source of Methylmercury Exposure in Infants." *Environmental Health Perspectives*, 102(1):74–77.

Grandjean, P. et al., 1997. "Cognitive Deficit in 7-Year-Old Children With Prenatal Exposure to Methylmercury." *Neurotoxicology and Teratology*, 19 (6):417–28.

Grandjean, P., P. Weihe, P. J. Jordensen et al., 1992. "Impact of Maternal Seafood Diet on Fetal Exposure to Mercury, Selenium, and Lead." *Archives of Environmental Health*, 47(3):185–95.

Grandjean, P., P. Weihe, R. F. White and F. Debes, 1998. "Cognitive Performance of Children Prenatally Exposed to 'Safe' Levels of Methylmercury." *Environmental Research*, 77(2): 165–72.

Grandjean P., P. Weihe, V. W. Burse, L. L. Needham, E. Storr-Hansen et al., 2001. "Neurobehavioural Deficits Associated With PCB in 7-Year-Old Children Prenatally Exposed to Seafood Neurotoxicants." *Neurotoxicology and Teratology*, 23:305–17.

Gregor, D. J., 1990. "Deposition and Accumulation of Selected Agricultural Pesticides in Canadian Arctic Snow." In D. A. Kurtz (ed.), *Long-Range Transport of Pesticides*. Chelsea, MI: Lewis Publishers.

Hansen, J. C., B. Deutch and J. O. Odland, 2008(2). *Dietary Transition and Contaminants in the Arctic: Emphasis on Greenland: Circumpolar Health Supplements*. International Association of Circumpolar Health Publishers. Aapistie, Finland.

Hoff, R. M., D. C. Muir and N. P. Grift, 1992. "Annual Cycle of Polychlorinated Biphenyls and Organohalogen Pesticides in Air in Southern Ontario: Atmospheric Transport and Sources." *Environmental Science and Technology*, 26:276–83.

Houde, M., D. C. G. Muir, K. Kidd, S. Guildford, K. Drouillard, M. Evans, X. Wang, M. Whittle, D. Haffner and H. Kling, 2008. "Influence of Lake Characteristics on the Biomagnification of Persistent Organic Pollutants in Lake Trout Food Webs." *Environmental Toxicology and Chemistry*, 27:2169–78.

Howard, P. H. and D. C. G. Muir, 2010. "Identifying New Persistent and Bioaccumulative Organics Among Chemicals in Commerce." *Environmental Science and Technology*, 44:2277–85.

Hung, H., P. Blanchard, C. J. Halsall, T. F. Bidlemam et al., 2005. "Temporal and Spatial Variabilities of Atmospheric POPs in the Canadian Arctic: Results From a Decade of Monitoring." *Science of the Total Environment*, 342:119–44.

Hung, H. et al., 2010. "Atmospheric Monitoring of Organic Pollutants in the Arctic Under the Arctic Monitoring and Assessment Programme (AMAP): 1993–2006." *Science of the Total Environment*, 408:2854–73.

Jacobson, J. L. and S. W. Jacobson, 1996. "Intellectual Impairment in Children Exposed to Polychlorinated Biphenyls in Utero." *New England Journal of Medicine*, 335(11):783–89.

Jacobson, J. L. and S. W. Jacobson, 2003. "Prenatal Exposure to Polychlorinated Biphenyls and Attention at School Age." *Journal of Pediatrics*, 143:780–88.

Jacobson, J. L, S. W. Jacobson, G. Muckle, M. Kaplan-Estrin, P. Ayotte and É. Dewailly, 2008. "Beneficial Effects of a Polyunsaturated Fatty Acid on Infant Development: Evidence From the Inuit of Arctic Québec." *Journal of Pediatrics*, 152(3):356–64.

Jacobson, J. L., S. W. Jacobson and H. E. B. Humphrey, 1990. "Effects of Exposure to PCBs and Related Compounds on Growth and Activity in Children." *Neurotoxicology and Teratology*, 12(4):319–26.

Jacobson, J. L., S. W. Jacobson, R. J. Padgett, G. A. Brumitt and R. L. Billings, 1992. "Effects of Prenatal PCB Exposure on Cognitive Processing Efficiency and Sustained Attention." *Developmental Psychology*, 28(2):297–306.

Jantunen, L. M. and T. F. Bidleman, 1995. "Reversal of the Air-Water Gas Exchange Direction of Hexachlorocyclohexanes in the Bering and Chukchi Seas: 1993 Versus 1988." *Environmental Science and Technology*, 29:1081–89.

Jobling, S., R. Williams, A. Johnson et al., 2006. "Predicted Exposures to Steroid Estrogens in U.K. Rivers Correlate With Widespread Sexual Disruption in Wild Fish Populations." *Environmental Health Perspectives*, 114(S-1):32–39.

Jorissen, J., 2007. "Literature Review: Outcomes Associated With Postnatal Exposure to Polychlorinated Biphenyls (PCBs) via Breast Milk." *Advances in Neonatal Care*, 7(5):230–37.

Kelly, E. N., D. W. Schindler, P. V. Hodson, J. W. Short, R. Radmanovich and C. C. Nielsen, 2010. "Oil Sands Development Contributes Elements Toxic at Low Concentrations to the Athabasca River and Its Tributaries." *Proceedings of the National Academy of Sciences of the United States*, 107(37):16178–183.

Kelly, E. N., J. W. Short, D. W. Schindler, P. V. Hodson, M. Ma, A. K. Kwan and B. L. Fortin, 2009. "Oil Sands Development Contributes Polycyclic Aromatic Compounds to the Athabasca River and Its Tributaries." *Proceedings of the National Academy of Sciences of the United States*, 106(52):22346–351.

Kidd, K. A., R. H. Hesslein, R. J. P. Fudge and K. A. Hallard, 1995. "The Influence of Trophic Level as Measured by $\delta^{15}N$ on the Mercury Concentrations in Freshwater Organisms." *Water, Air and Soil Pollution*, 80:1011–15.

Kirk, J. L., D. Muir, X. Wang, D. Antoniades, M. Douglas, M. Evans, T. Jackson, H. Kling, S. Lamoureux, D. S. S. Lim, R. Pienitz, J. Smol, K. Stewart and F. Yang, 2011. "Climate Change and Mercury Accumulation Rates in Canadian High and - Sub-Arctic Lake Sediments." *Environmental Science and Technology*, 45(3):964–70.

Klánová, J., P. Čupr, I. Holoubek, J. Borůvková, P. Přibylová et al., 2008. *Application of Passive Sampler for Monitoring of POPs in Ambient Air – Part VI: Pilot Study for Development of the Monitoring Network in the African Continent*. Brno, Czech Republic: Masaryk University.

Lamon, L. et al., 2009. "Modelling the Global Levels and Distribution of Polychlorinated Biphenyls in Air Under a Climate Change Scenario." *Environmental Science and Technology*, 43:5818–24.

Landrigan, P., L. Lambertini and L. S. Birnbaum, 2012. "A Research Strategy to Discover the Environmental Causes of Autism and Neurodevelopmental Disabilities." *Environmental Health Perspectives*, 120(7):258–59.

Letcher, R. J., J. O. Bustnes, R. Dietz, B. M. Jenssen et al., 2010. "Exposure and Effects Assessment of Persistent Organohalogen Contaminants in Arctic Wildlife and Fish." *Science of the Total Environment*, 408(15):2995–3043.

Li, C. S., J. Cornett and K. Ungar, 2003. "Long-Term Decrease of Cadmium Concentration in Canadian Arctic Air." *Geophysical Research Letters*, 30:1256–59.

Li, C. S, J. Cornett, S. Willie and J. Lam, 2009. "Mercury in Arctic Air: The Long-Term Trend." *Science of the Total Environment*, 407:2756–59.

Li, Y. F. et al., 2010. "Polychlorinated Biphenyls in Global Air and Surface Soil: Distributions, Air-Soil Exchange, and Fractionation Effect." *Environmental Science and Technology*, 44:2784–90.

Lindh, C. et al., 2012. "Blood Serum Concentrations of Perfluorinated Compounds in Men from Greenlandic Inuit and European Populations." *Chemosphere*, 88(11):1269–75.

Lu, J. Y., W. H. Schroeder, L. A. Barrie, A. Steffen et al., 2001. "Magnification of Atmospheric Mercury Deposition to Polar Regions in Springtime: The Link to Tropospheric Ozone Depletion Chemistry." *Geophysical Research Letters*, 28:3219–22.

Ma, J., H. Hung, C. Tian and R. Kallenborn, 2011. "Revolatilization of Persistent Organic Pollutants in the Arctic Induced by Climate Change." *Nature Climate Change*, 1:255–60.

Ma, J. and Z. Cao, 2010. "Quantifying the Perturbations of Persistent Organic Pollutants Induced by Climate Change." *Environmental Science and Technology*, 44:8567–73.

Macdonald, R. W., D. Mackay and B. Hickie, 2002. "Contaminant Amplification in the Environment: Revealing the Fundamental Mechanisms." *Environmental Science and Technology*, 36:456–62.

Macdonald, R. W., D. Mackay, Y. F. Li and B. Hickie, 2003. "How Will Global Climate Change Affect Risks From Long-Range Transport of Persistent Organic Pollutants?" *Human and Ecological Risk Assessment*, 9:643–60.

Macdonald, R. W. and L. L. Loseto, 2010. "Are Arctic Ocean Ecosystems Exceptionally Vulnerable to Global Emissions of Mercury? A Call for Emphasized Research on Methylation and the Consequences of Climate Change." *Environmental Chemistry*, 7:133–38.

Macdonald, R. W. and M. J. Bewers, 1996. "Contaminants in the Arctic Marine Environment: Priorities for Protection." *ICES Journal of Marine Science*, 53:537–63.

Macdonald, R. W., T. Harner and J. Fyfe, 2005. "Recent Climate Change in the Arctic and Its Impact on Contaminant Pathways and Interpretation of Temporal Trend Data." *Science of the Total Environment*, 342:5–86.

Mackay, D. and S. Paterson, 1981. "Calculating Fugacity." *Environmental Science and Technology*, 15:1006–14.

Mackay, D. and S. Paterson, 1982. "Fugacity Revisited." *Environmental Science and Technology*, 16:654A-60A.

Meyer, T. and F. Wania, 2008. "Organic Contaminant Amplification During Snowmelt." *Water Research*, 42:1847–65.

Meyer, T., Y. D Lei, I. Muradi and F. Wania, 2009. "Organic Contaminant Release From Melting Snow. 1. Influence of Chemical Partitioning." *Environmental Science and Technology*, 43:657–62.

Mozaffarian, D., 2009. "Fish, Mercury, Selenium, and Cardiovascular Risk: Current Evidence and Unanswered Questions." *International Journal of Environmental Research and Public Health*, 6:1894–1916.

Muckle, G., P. Ayotte, É. Dewailly, S. W. Jacobson and J. L. Jacobson, 2001. "Prenatal Exposure of the Northern Québec Inuit Infants to Environmental Contaminants." *Environmental Health Perspectives*, 109:1291–99.

Muir, D. C. G. and C. A. de Wit, 2010. "Trends of Legacy and New Persistent Organic Pollutants in the Circumpolar Arctic: Overview, Conclusions and Recommendations." *Science of the Total Environment*, 408(15):3044–51.

Muir, D. and R. Lohmann, 2013. "Water as a New Matrix for Global Assessment of Hydrophilic POPs." *Trends in Analytical Chemistry*, 46:162–72.

Muir, D. C. G. and P. H. Howard, 2006. "Are There Other Persistent Organic Pollutants? A Challenge for Environmental Chemists." *Environmental Science and Technology*, 40:7157–66.

Muir, D. C. G., X. Wang, F. Yang, N. Nguyen, T. A. Jackson, M. S. Evans, M. Douglas, G. Köck, S. Lamoureux, R. Pienitz, J. Smol, W. F. Vincent and A. P. Dastoor, 2009. "Spatial Trends and Historical Deposition of Mercury in Eastern and Northern Canada Inferred From Lake Sediment Cores." *Environmental Science and Technology*, 43:4802–09.

Müller, C. E., A. O. De Silva, J. Small, M. Williamson, X. Wang, A. Morris, S. Katz, M. Gamberg and D. C. G. Muir, 2011. "Biomagnification of Perfluorinated Compounds in a Remote Terrestrial Food Chain: Lichen-Caribou-Wolf." *Environmental Science and Technology*, 45:8665–73.

National Academy of Sciences of the United States, 2000. *Science Frontiers in Developmental Toxicity and Risk Assessment*. National Academy of Sciences. Washington, DC.

Nizzetto, L. et al., 2010. "Atlantic Ocean Surface Waters Buffer Declining Atmospheric Concentrations of Persistent Organic Pollutants." *Environmental Science and Technology*, 44:6978–84.

Noyes, P. D., M. McElwee, H. D. Miller et al., 2009. "The Toxicology of Climate Change: Environmental Contaminants in a Warming World." *Environment International*, 35(6): 971–86.

O'Driscoll, K., B. Mayer, T. Ilyina and T. Pohlmann, 2013. "Modelling the Cycling of Persistent Organic Pollutants (POPs) in the North Sea System: Fluxes, Loading, Seasonality, Trends." *Journal of Marine Systems*, 111–12:69–82.

Oehme, M., 1991. "Further Evidence for Long-Range Transport of Polychlorinated Aromates and Pesticides: North America and Eurasia to the Arctic." *Ambio*, 20 (7):293–97.

Oehme, M., P. Fürst, Chr. Krüger, H. A. Meemken and W. Groebel, 1988. "Presence of Polychlorinated Dibenzo-P-Dioxins, Dibenzofurans and Pesticides in Arctic Seal From Spitzbergen." *Chemosphere*, 17:1291–1300.

Oreskes, N. and E. M. Conway, 2010. *Merchants of Doubt*. New York: Bloomsbury Press.

Ottar, B., 1981. "The Transfer of Airborne Pollutants to the Arctic Region." *Atmospheric Environment*, 15:1439–45.

Outridge, P. M., K. A. Hobson and J. Savelle, 2009. "Long-Term Changes of Mercury Levels in Ringed Seal (Phoca hispida) From Amundsen Gulf and Beluga (Delphinapterus leucas) From the Beaufort Sea, Western Canadian Arctic." *Science of the Total Environment*, 407:6044–51.

Outridge, P. M., K. A. Hobson, R. McNeely and A. Dyke, 2002. "A Comparison of Modern and Preindustrial Levels of Mercury in the Teeth of Beluga in the Mackenzie Delta, Northwest Territories and Walrus at Igloolik, Nunavut, Canada." *Arctic*, 55:123–32.

Outridge, P. M., R. W. Macdonald, F. Wang, G. A. Stern and A. P. Dastoor, 2008. "A Mass Balance Inventory of Mercury in the Arctic Ocean." *Environmental Chemistry*, 5:89–111.

Pacyna, J. M. and M. Oehme, 1988. "Long-Range Transport of Some Organic Compounds to the Norwegian Arctic." *Atmospheric Environment*, 22(2):243–57.

Pozo, K., T. Harner, F. Wania, D. Muir, K. C. Jones and L. A. Barrie, 2006. "Toward a Global Network for Persistent Organic Pollutants in Air: Results From the GAPS Study." *Environmental Science and Technology*, 40:4867–73.

Pozo, K., T. Harner, S. C. Lee, F. Wania, D. Muir and K. C. Jones, 2009. "Seasonally Resolved Concentrations of Persistent Organic Pollutants in the Global Atmosphere From the First Year of the GAPS Study." *Environmental Science and Technology*, 43:796–803.

Riget, F., A. Bignert, B. Braune et al., 2010. "Temporal Trends of Legacy POPs in Arctic Biota: An Update." *Science of the Total Environment*, 408(15):2874–84.

Rogan, W. J., B. C. Gladen, J. D. McKinney, N. Carreras, P. Hardy, J. Thullen et al., 1987. "Polychlorinated Biphenyls (PCBs) and Dichlorodiphenyl Dichloroethene (DDE) in Human Milk: Effects on Growth, Morbidity, and Duration of Lactation." *American Journal of Public Health*, 77(10):1294–97.

Rooney, R. C., S. E. Bayley and D. W. Schindler, 2012. "Oil Sands Mining and Reclamation Cause Massive Loss of Peatland and Stored Carbon." *Proceedings of the National Academy of Sciences of the United States*, 109(13):4933–37.

Scheringer. M., S. Strempel, S. Hukari, C. A. Ng, M. Blepp and K. Hungerbuhler, 2012. "How Many Persistent Organic Pollutants Should We Expect?" *Atmospheric Pollution Research*, 3:383–91.

Schroeder, W. H., K. G. Anlauf, L. A. Barrie, J. Y. Lu, A. Steffen, D. R. Schneeberger and T. Berg, 1998. "Arctic Springtime Depletion of Mercury." *Nature*, 394:331–32.

Selin, H., 2000. *Towards International Chemical Safety: Taking Action on Persistent Organic Pollutants (POPs)*. Department of Water and Environmental Studies. Linkoping University, Sweden.

Selin, H., 2009. *Managing Hazardous Chemicals: Longer-Range Challenges*. The Pardee Papers/No. 5/March 2009. Boston University, Boston, Massachusetts.

Selin, H., 2010. *Global Governance of Hazardous Chemicals: Challenges of Multilevel Management*. Massachusetts Institute of Technology (MIT). Cambridge, Massachusetts.

Shoeib, M. and T. Harner, 2002. "Characterization and Comparison of Three Passive Air Samplers for Persistent Organic Pollutants." *Environmental Science and Technology*, 36:4142–51.

Shoeib, M., T. Harner and P. Vlahos, 2006. "Perfluorinated Chemicals in the Arctic Atmosphere." *Environmental Science and Technology*, 40:7577–83.

Sliggers, J. and W. Kakebeeke (eds.), 2004. *Clearing the Air: 25 Years of the Convention on Long-Range Transboundary Air Pollution*. United Nations Economic Commission for Europe (UNECE). Geneva, Switzerland.

Sonne, C., 2010. "Health Effects From Long-Range Transported Contaminants in Arctic Top Predators: An Integrated Review Based on Studies of Polar Bears and Relevant Model Species." *Environment International*, 36(5):461–91.

Sorensen, N., K. Murata, E. Budtz-Jorgensen, P. Weihe and P. Grandjean, 1999. "Prenatal Methylmercury Exposure as a Cardiovascular Risk Factor at Seven Years of Age." *Epidemiology*, 10:370–75.

Steffen, A., T. Douglas and M. Amyot, 2008. "A Synthesis of Atmospheric Mercury Depletion Event Chemistry in the Atmosphere and Snow." *Atmospheric Chemistry and Physics*, 8:1445–48.

Stern, G. A., R. W. Macdonald, P. M. Outridge, S. Wilson et al., 2012. "How Does Climate Change Influence Arctic Mercury?" *Science of the Total Environment*, 414:22–42.

Strachan, W. M. J. and S. J. Eisenreich, 1990. "Mass Balance Accounting of Chemicals in the Great Lakes." In D. A. Kurtz (ed.), *Long-Range Transport of Pesticides*. Chelsea, MI: Lewis Publishers.

Su, Y. et al., 2006. "Spatial and Seasonal Variations of Hexachlorocyclohexanes (HCHs) and Hexachlorobenzene (HCB) in the Arctic Atmosphere." *Environmental Science and Technology*, 40:6601–07.

Sunderland, E. M. and R. P. Mason, 2007. "Human Impacts on Open Ocean Mercury Concentrations." *Global Biogeochemical Cycles*, 21:GB4022.

Tian, W., G. M. Egeland, I. Sobol and H. M. Chan, 2011. "Mercury Hair Concentrations and Dietary Exposure Among Inuit Preschool Children in Nunavut, Canada." *Environment International*, 37:42–48.

Toose, L., D. G. Woodfine, M. Macleod, D. Mackay and J. Gouin, 2004. "BETR-World: A Geographically Explicit Model of Chemical Fate: Application to Transport of Alpha-HCH to the Arctic." *Environmental Pollution*, 128:223–40.

Tysklind, M., I. Faengmark, S. Marklund, A. Lindskog et al., 1993. "Atmospheric Transport and Transformation of Polychlorinated Dibenzo-P-Dioxins and Dibenzofurans." *Environmental Science and Technology*, 27(10):2190–97.

UNECE, 1994. *State of Knowledge Report of the UNECE Task Force on Persistent Organic Pollutants*. Geneva, Switzerland.

UNECE, 2010. *Hemispheric Transport of Air Pollution 2010: Part C, Persistent Organic Pollutants*. Air Pollution Studies No. 19. United Nations Economic Commission for Europe. Geneva, Switzerland.

UNECE, 2010. *Hemispheric Transport of Air Pollution 2010: Part D, Answers to Policy-Relevant Science Questions*. Air Pollution Studies No. 20. United Nations Economic Commission for Europe. Geneva, Switzerland.

UNEP, 2013. *Global Mercury Assessment*. United Nations Environment Programme Chemicals Branch. Geneva, Switzerland.

UNEP, 2006. *Strategic Approach to International Chemicals Management: Comprising the Dubai Declaration on International Chemicals Management, the Overarching Policy Strategy and the Global Plan of Action*. Resolutions of the International Conference on Chemicals Management. United Nations Environment Programme. Geneva, Switzerland.

UNEP/AMAP, 2011. *Climate Change and POPs: Predicting the Impacts*. Report of the UNEP/AMAP Expert Group. Secretariat of the Stockholm Convention. Geneva, Switzerland.

University of Lapland Arctic Centre,1991. *The State of the Arctic Environment*. Arctic Centre University of Lapland. Rovaniemi, Finland.

Vandenberg, L. N., T. Colborn, T. B. Hayes, J. J. Heindel et al., 2012. "Hormones and Endocrine Disrupting Chemicals: Low-Dose Effects and Nonmonotonic Dose Responses." *Endocrine Reviews*. doi:10.1210/er.2011-1050.

Verner, M. A., P. Plusquellec, G. Muckle, P. Ayotte, É. Dewailly, S. W. Jacobson et al., 2010. "Alteration of Infant Attention and Activity by Polychlorinated Biphenyls: Unravelling Critical Windows of Susceptibility Using Physiologically Based Pharmacokinetic Modelling." *Neurotoxicology*, 31(5):424–31.

Wania, F. and D. Mackay, 1993. "Global Fractionation and Cold Condensation of Low Volatility Organochlorine Compounds in Polar Regions." *Ambio*, 22:10–18.

Weber, J., C. J. Halsall, D. Muir, C. Teixeira et al., 2010. "Endosulfan, a Global Pesticide: A Review of Its Fate in the Environment and Occurrence in the Arctic." *Science of the Total Environment*, 408(15):2966–84.

Weihe, P. and P. Grandjean, 2012. "Cohort Studies of Faroese Children Concerning Potential Adverse Health Effects After the Mothers' Exposure to Marine Contaminants During Pregnancy." *Acta Veterinaria Scandinavica*, 54(Suppl. 1):S7.

WHO/LRTAP Convention, 2003. *Health Risks of Persistent Organic Pollutants From Long-Range Transboundary Air Pollution*. Joint WHO/Convention Task Force on the Health Aspects of Air Pollution. World Health Organization. Copenhagen, Denmark.

de Wit, C. A., D. Herzke and K. Vorkamp, 2010. "Brominated Flame Retardants in the Arctic Environment: Trends and New Candidates." *Science of the Total Environment*, 408(15):2885–2918.

de Wit, C. A. and D. Muir, 2010. "Levels and Trends of New Contaminants, Temporal Trends of Legacy Contaminants and the Effects of Contaminants in the Arctic." *Science of the Total Environment*, 408(15):2852–53.

Wöhrnschimmel, H., M. MacLeod and K. Hungerbuhler, 2013. "Emissions, Fate and Transport of Persistent Organic Pollutants to the Arctic in a Changing Global Climate." *Environmental Science and Technology*, 47:2323–30.

Wong, F. et al., 2011. "Air-Water Exchange of Anthropogenic and Natural Organohalogens on International Polar Year (IPY) Expeditions in the Canadian Arctic." *Environmental Science and Technology*, 45:876–81.

Wong, M. P., 1985. *Chemical Residues in Fish and Wildlife Harvested in Northern Canada*. Environmental Studies Program. Department of Indian and Northern Affairs. Ottawa, Canada.

Woodruff, T. J., A. R. Zota and J. M. Schwartz, 2011. "Environmental Chemicals in Pregnant Women in the United States: NHANES 2003–2004." *Environmental Health Perspectives*, 119:878–85.

Wu, H., K. A. Bertrand, A. L. Choi, F. B. Hu, F. Laden, P. Grandjean and Q. Sun, 2013. "Persistent Organic Pollutants and Type 2 Diabetes: A Prospective Analysis in the Nurses' Health Study and Meta-Analysis." *Environmental Health Perspectives*, 121:153–61.

Yao, Y., T. Harner, K. Su, K. A. Brice et al., 2010. "A Captured Episode of γ-Hexachlorocyclohexane Air Pollution in the Toronto Area After the Canadian Lindane Ban." *Atmospheric Pollution Research*, 1:168–76.

Chapter 9: Conducting Marine Science in the Arctic

Suggested Further Reading

Dawkins, R., 1976. *The Selfish Gene*. New York: Oxford University Press.
Hardy, A., 1967. *Great Waters*. London: Collins.
Kintisch, E., 2013. "A Sea of Change for U.S. Oceanography." *Science*, 339:1138–43.
Wright, R.,1995. *The Moral Animal*. New York: Vintage Books.

Chapter 10: Climate Change in the Arctic

The primary sources for this chapter have been:

ACIA, 2004. *Impacts of a Warming Arctic*. Arctic Climate Impact Assessment. Cambridge: Cambridge University Press.
ACIA, 2005. *Arctic Climate Impact Assessment*. New York: Cambridge University Press.
AMAP, 1998. "AMAP Assessment Report, Chapter 11. Climate Change, Ozone, and Ultraviolet Radiation." In *AMAP Assessment Report: Arctic Pollution Issues*. Arctic Monitoring and Assessment Programme. Oslo, Norway.
AMAP, 2002. *AMAP Assessment Report: The Influence of Global Change on Contaminant Pathways to, Within, and From the Arctic*. Arctic Monitoring and Assessment Programme. Oslo, Norway.
AMAP, 2011. *Snow, Water, Ice and Permafrost in the Arctic (SWIPA): Climate Change and the Cryosphere*. Arctic Monitoring and Assessment Programme. Oslo, Norway.
Jeffries, M. O. and J. Richter-Menge, eds., 2011. "The Arctic." *Bulletin of the American Meteorological Society*, 93(7):S127-45.
Jeffries, M. O., J. A. Richter-Menge and J. E. Overland, eds., 2012. *Arctic Report Card 2012*. www.arctic.noaa.gov/report12.
Kallenborn, R. et al., 2011. "Combined Effects of Selected Pollutants and Climate Change in the Arctic Environment." *AMAP Technical Report No. 5*. Arctic Monitoring and Assessment Programme. Oslo, Norway.
Krupnik, I. and D. Jolly (eds.), 2002. *The Earth Is Faster Now: Indigenous Observations of Arctic Environmental Change*. Arctic Research Consortium of the United States. Fairbanks, Alaska.
Richter-Menge, J., M. O. Jeffries and J. E. Overland, eds., 2011. *Arctic Report Card 2011*. www.arctic.noaa.gov/report11.

Suggested Further Reading

AMAP, 2008. "The Impact of Short-Lived Pollutants on Arctic Climate." By P. K. Quinn et al. *AMAP Technical Report No. 1*. Arctic Monitoring and Assessment Programme. Oslo, Norway.
AMAP, 2008. "Sources and Mitigation Opportunities to Reduce Emissions of Short-Term Arctic Climate Forcers." By J. Bluestein, J. Rackley and E. Baum. *AMAP Technical Report No. 2*. Arctic Monitoring and Assessment Programme. Oslo, Norway.

AMAP, 2011. "The Impact of Black Carbon on Arctic Climate." By P. K. Quinn et al. *AMAP Technical Report No. 4*. Arctic Monitoring and Assessment Programme. Oslo, Norway.

AMAP, 2011. "The Impact of Black Carbon on Arctic Climate." *AMAP Expert Group on Short-Lived Climate Forcers.*" By P. K. Quinn, A. Stohl, A. Arneth, J. F. Burkhart et al. Arctic Monitoring and Assessment Programme. Oslo, Norway.

AMAP, 2013. *AMAP Assessment: Arctic Ocean Acidification*. Arctic Monitoring and Assessment Programme. Oslo, Norway.

Angerbjörn, A., D. Berteaux and R. Ims, 2012. "Arctic Fox." In *Arctic Report Card 2012*. www.arctic.noaa.gov/report12/arctic_fox.html.

Archer, D., 2009. *The Long Thaw: How Humans Are Changing the Next 100,000 Years of Earth's Climate*. Princeton and Oxford, UK: Princeton University Press.

Archer, D. and S. Rahmstorf, 2010. *The Climate Crisis*. Cambridge, UK: Cambridge University Press.

Arctic Council, 2011. "An Assessment of Emissions and Mitigation Options for Black Carbon for the Arctic Council." *Technical Report of the Arctic Council Task Force on Short-Lived Climate Forcers*. http://arctic-council.org.

Bamber, J. L. and W.P. Aspinall, 2013. "An Expert Judgement Assessment of Future Sea Level Rise from the Ice Sheets." *Nature Climate Change*. doi:10.1038/nclimate1778.

Barber, D. G., R. Galley, M. G. Asplin et al., 2009. "Perennial Pack Ice in the Southern Beaufort Sea Was Not as It Appeared in the Summer of 2009." *Geophysical Research Letters*, 36:L24501.

Bates, N. R. and J. T. Mathis, 2009. "The Arctic Ocean Marine Carbon Cycle: Evaluation of Air-Sea CO_2 Exchanges, Ocean Acidification Impacts and Potential Feedbacks." *Biogeosciences*, 6:2433–59.

Bates, N. R., J. T. Mathis and L. W. Cooper, 2009. "Ocean Acidification and Biologically Induced Seasonality of Carbonate Mineral Saturation States in the Western Arctic Ocean." *Journal of Geophysical Research: Oceans*, 114(C11): C1107.

Becker, S. et al., 2008. "Long-Term Trends in Atmospheric Concentrations of α- and γ-HCH in the Arctic Provide Insight Into the Effects of Legislation and Climatic Fluctuations on Contaminants Levels." *Atmospheric Environment*, 42:8225–33.

Berger, A. and M. F. Loutre, 2002. "An Exceptionally Long Interglacial Ahead?" *Science*, 297:1287–88.

Bhatt, U. S., D. A. Walker, M. K. Raynolds, J. C. Comiso et al., 2010. "Circumpolar Arctic Tundra Vegetation Change Is Linked to Sea-Ice Decline." *Earth Interactions*, 14:1–20.

Bintanja, R. and E. C. van der Linden, 2013. "The Changing Seasonal Climate in the Arctic." *Nature Scientific Reports*, 3(1556):1–8.

Bond, T. C., 2007. "Can Warming Particles Enter Global Climate Discussions?" *Environmental Research Letters*, 2(4). doi:10.1088/1748-9326/2/4/045030.

Bond, T. C. and H. L. Sun, 2005. "Can Reducing BC Emissions Counteract Global Warming?" *Environmental Science and Technology*, 39(16):5921–26.

Bond, T. C., S. J. Doherty, D. W. Fahey et al., 2013. "Bounding the Role of Black Carbon in the Climate System: A Scientific Assessment." *Journal of Geophysical Research: Atmosphere*, 118(11):5380–552.

Box, J. E. and D. T. Decker, 2011. "Greenland Marine-Terminating Glacier Area Changes: 2000–2010." *Annals of Glaciology*, 52(59):91–98.

Brown, R., C. Derksen and L. Wang, 2010. "A Multi-Data Set Analysis of Variability and Change in Arctic Spring Snow Cover Extent, 1967–2008." *Journal of Geophysical Research*, 115:D16111.

Brown, R. and D. Robinson, 2011. "Northern Hemisphere Spring Snow Cover Variability and Change Over 1922–2010 Including an Assessment of Uncertainty." *The Cryosphere*, 5:219–29.

CAFF, 2010. *Arctic Biodiversity Trends 2010. Selected Indicators of Trends.* Conservation of Arctic Flora and Fauna International Secretariat. Akureyri, Iceland.

Callaghan, T., M. Johansson, R. Brown, P. Groisman et al., 2011. "The Changing Face of Arctic Snow Cover: A Synthesis of Observed and Projected Changes." *Ambio*, 40:17–31.

Callaghan, T. V. et al., 2011. "Multi-Decadal Changes in Tundra Environments and Ecosystems: Synthesis of the International Polar Year Back to the Future Project (IPY-BTF)." *Ambio*, 40:705–16.

Callaghan, T. V., F. Bergholm, T. R. Christensen, C. Jonasson, U. Kokfelt and M. Johansson, 2010. "A New Climate Era in the Sub-Arctic: Accelerating Climate Changes and Multiple Impacts." *Geophysical Research Letters*, 37:L14705.

Cappa, C. D., T. B. Onasch, P. Massoli et al., 2012. "Radiative Absorption Enhancements Due to the Mixing State of Atmospheric Black Carbon." *Science*, 337:1078–81.

Cattiaux, J., R. Vautard, C. Cassou, P. Yiou, V. Masson-Delmotte and F. Codron, 2010. "Winter 2010 in Europe: A Cold Extreme in a Warming Climate." *Geophysical Research Letters*, 37:L20704.

Chierici, M. and A. Fransson, 2009. "Calcium Carbonate Saturation in the Surface Water of the Arctic Ocean: Undersaturation in Freshwater-Influenced Shelves." *Biogeosciences*, 6:2421–32.

Cramer, W., G. Yohe, M. Auffhammer, C. Huggel, U. Molau, M. A. F. Silva Dias, A. Solow, D. Stone, L. Tibig et al., 2014. "Detection and Attribution of Observed Impacts." In C. Field, V. Barros et al. (eds.), *Climate Change 2014: Impacts, Adaptation, and Vulnerability.* Contribution of Working Group II to the Fifth Assessment Report of the Intergovernmental Panel on Climate Change. In press.

Derksen, C. and R. Brown, 2012. "Snow." *Arctic Report Card: Update for 2012.* National Oceanic and Atmospheric Administration.

Derksen, C. and R. Brown, 2012. "Spring Snow Cover Extent Reductions in the 2008–2012 Period Exceeding Climate Model Projections." *Geophysical Research Letters*, 39:L19504.

Deser, C., R. Tomas, M. Alexander and D. Lawrence, 2010. "The Seasonal Atmospheric Response to Projected Arctic Sea Ice Loss in the Late Twenty-First Century." *Journal of Climate*, 23:333–51.

Elmendorf, S. C., G. H. R. Henry, R. D. Hollister et al., 2011. "Plot-Scale Evidence of Tundra Vegetation Change and Links to Recent Summer Warming." *Nature Climate Change*, 2:453–57.

Englander, J., 2013. *High Tide on Main Street: Rising Sea Level and the Coming Coastal Crisis* (2nd ed.). Boca Raton, FL: The Science Bookshelf.

Epstein, H. E., D. A. Walker, U. S. Bhatt et al., 2013. "Vegetation." In *Arctic Report Card 2012*. www.arctic.noaa.gov/reportcard/vegetation.html.

Epstein, H. E., M. K. Raynolds, D. A Walker et al., 2012. "Dynamics of Above-Ground Phytomass of the Circumpolar Arctic Tundra During the Past Three Decades." *Environmental Research Letters*, 7(1). doi:10.1088/1748-9326/7/1/015506.

Fabry, V. J., J. B. McClintock, J. T. Mathis and J. M. Grebmeier, 2009. "Ocean Acidification at High Latitudes: The Bellwether." *Oceanography*, 22(4):160–71.

Flanner, M. G., C. S. Zender, J. T. Randerson and P. J. Rasch, 2007. "Present-Day Climate Forcing and Response From Black Carbon in Snow." *Journal of Geophysical Research*, 112:D11202.

Gardner, A., G. Moholdt, J. Cogley et al., 2013. "A Reconciled Estimate of Glacier Contributions to Sea Level Rise: 2003–2009." *Science*, 340:852–57.

Ghatak, D. and J. Miller, 2013. "Implications for Arctic Amplification of Changes in the Strength of the Water Vapour Feedback." *Journal of Geophysical Research*, 118:7569–78.

Gillett, N. P., D. A. Stone, P. A. Stott et al., 2008. "Attribution of Polar Warming to Human Influence." *Nature Geoscience*, 1:750–54.

Greene, C. H. and B. C. Monger, 2012. "An Arctic Wild Card in the Weather." *Oceanography*, 25(2):7–9.

Gregory, J. M. and P. Huybrechts, 2006. "Ice Sheet Contributions to Future Sea-Level Change." *Philosophical Transactions of the Royal Society, Series A*, 364:1709–31.

Grinsted, A., J. C. Moore and S. Jevrejeva, 2009. "Reconstructing Sea Level from Paleo and Projected Temperatures 200 to 2100 AD." *Climate Dynamics*, 34 (4):461–72.

Hansen, J. et al., 2005. "Earth's Energy Imbalance: Confirmation and Implications." *Science*, 308:1431–35.

Hansen, J. and L. Nazarenko, 2004. "Soot Climate Forcing Via Snow and Ice Albedos." *Proceedings of the National Academy of Sciences*, 101:423–28.

Hansen, J., M. Sato, R. Ruedy et al., 2005. "Efficacy of Climate Forcings." *Journal of Geophysical Research*, 110:D18104.

Harper, J., N. Humphrey, W. T. Pfeffer, J. Brown and X. Fettweis, 2012. "Greenland Ice-Sheet Contribution to Sea-Level Rise Buffered by Meltwater Storage in Firn." *Nature*, 491:240–43.

Hewitson, B., A. C. Janetos, T. R. Carter, F. Giorgi, R. G. Jones, W.-T. Kwon, L. O. Mearns, E. L. F. Schipper, M. K. van Aalst et al., 2014. "Regional Context." In C. Field, V. Barros et al. (eds.), *Climate Change 2014: Impacts, Adaptation, and Vulnerability*. Contribution of Working Group II to the Fifth Assessment Report of the Intergovernmental Panel on Climate Change. In press.

Holland, M. M., C. M. Bitz, B. Tremblay and D. A. Bailey, 2008. "The Role of Natural Versus Forced Change in Future Rapid Summer Arctic Ice Loss." In E. T. DeWeaver et al. (eds.), *Arctic Sea Ice Decline: Observations, Projections, Mechanisms, and Implications*. Washington, DC: American Geophysical Union.

Howat, I. M. and A. Eddy, 2011. "Multidecadal Retreat of Greenland's Marine-Terminating Glaciers." *Journal of Glaciology*, 57(203):389–96.

Hudson, J. M. G. and G. H. R. Henry, 2009. "Increased Plant Biomass in a High Arctic Heath Community From 1981 to 2008." *Ecology*, 90:2657–63.

Hung, H. et al., 2005. "Temporal and Spatial Variabilities of Atmospheric Polychlorinated Biphenyls (PCBs), Organochlorine (OC) Pesticides and Polycyclic Aromatic Hydrocarbons (PAHs) in the Canadian Arctic: Results From a Decade of Monitoring." *Science of the Total Environment*, 342:119–44.

Hung, H. et al., 2010. "Atmospheric Monitoring of Organic Pollutants in the Arctic Under the Arctic Monitoring and Assessment Programme (AMAP): 1993–2006." *Science of the Total Environment*, 408:2854–73.

Jackson, J. M., W. J. Williams and E. C. Carmack, 2012. "Winter Sea-Ice Melt in the Canada Basin, Arctic Ocean." *Geophysical Research Letters*, 39:L03603.

Jacob, T., J. Wahr, W. T. Pfeffer and S. Swenson, 2012. "Recent Contributions of Glaciers and Ice Caps to Sea Level Rise." *Nature*, 482:514–18.

Jantunen, L. M. and T. F. Bidleman, 1995. "Reversal of the Air-Water Gas Exchange Direction of Hexachlorocyclohexanes in the Bering and Chukchi Seas: 1993 Versus 1988." *Environmental Science and Technology*, 29(4):1081–89.

Joughin, I., R. B. Alley and D. M. Holland, 2012. "Ice-Sheet Response to Oceanic Forcing." *Science*, 338:1172–76.

Kaufman, D. S., D. P. Schneider, N. P. McKay et al., 2009. "Recent Warming Reverses Long-Term Arctic Cooling." *Science*, 325(5945):1236–39.

Kelly, E. N., D. W. Schindler, P. V. Hodson, J. W. Short, R. Radmanovich and C. C. Nielsen, 2010. "Oil Sands Development Contributes Elements Toxic at Low Concentrations to the Athabasca River and Its Tributaries." *Proceedings of the National Academy of Sciences of the United States*, 107(37):16178–83.

Kelly, E. N., J. W. Short, D. W. Schindler, P. V. Hodson, M. Ma, A. K. Kwan and B. L. Fortin, 2009. "Oil Sands Development Contributes Polycyclic Aromatic Compounds to the Athabasca River and Its Tributaries." *Proceedings of the National Academy of Sciences of the United States*, 106(52):22346–51.

Kinnard, C., C. M. Zdanowicz, D. Fisher et al., 2011. "Reconstructed Changes in Arctic Sea Ice Over the Past 1,450 Years." *Nature*, 479:509–12.

Kirk, J. L., D. Muir, X. Wang, D. Antoniades, M. Douglas, M. Evans, T. Jackson, H. Kling, S. Lamoureux, D. S. S. Lim, R. Pienitz, J. Smol, K. Stewart and F. Yang, 2011. "Climate Change and Mercury Accumulation in Canadian High and Sub-Arctic Lakes." *Environmental Science and Technology*, 45(3):964–70.

Kjaer, K. H., S. A. Khan, N. J. Korsgaard et al., 2012. "Aerial Photographs Reveal Late–20th Century Dynamic Ice Loss in Northwest Greenland." *Science*, 337:569–73.

Kopp, R. E., F. Simons, J. Mitrovica et al., 2009. "Probabilistic Assessment of Sea Level During the Last Interglacial Stage." *Nature*, 462:863–67.

Krupnik, I. and D. Jolly (eds.), 2002. *The Earth Is Faster Now: Indigenous Observations of Arctic Environmental Change*. Arctic Research Consortium of the United States. Fairbanks, Alaska.

Kupiainen, K. and Z. Klimont, 2007. "Primary Emissions of Fine Carbonaceous Particles in Europe." *Atmospheric Environment*, 41(10):2156–70.

Kwok R., G. F. Cunningham, M. Wensnahan, I. Rigor, H. J. Zwally and D. Yi, 2009. "Thinning and Volume Loss of Arctic Sea Ice Cover: 2003–2008." *Journal of Geophysical Research Oceans*, 114:C07005.

Lamon, L. et al., 2009. "Modelling the Global Levels and Distribution of Polychlorinated Biphenyls in Air Under a Climate Change Scenario." *Environmental Science and Technology*, 43:5818–24.

Levermann, A., P. U. Clark et al., 2013. "The Multimillennial Sea-Level Commitment of Global Warming." *Proceedings of the National Academy of Sciences*. 10 (34):13745–50.

Lee, S. and S. Feldstein, 2013. "Detecting Ozone and Greenhouse Gas-Driven Wind Trends With Observational Data." *Science*, 339(6119):563–67.

Li, Y. F. et al., 2010. "Polychlorinated Biphenyls in Global Air and Surface Soil: Distributions, Air-Soil Exchange, and Fractionation Effect." *Environmental Science and Technology*, 44:2784–90.

Liu, J., J. A. Curry, H. Wang, M. Song and R. M. Horton, 2012. "Impact of Declining Arctic Sea Ice on Winter Snowfall." *Proceedings of the National Academy of Sciences*, 109:4074–79.

Lozier, S. M., 2010. "Deconstructing the Conveyor Belt." *Science*, 328(5985):1507–11.

Lund, D. C., J. Lynch-Stieglitz and W. B. Curry, 2006. "Gulf Stream Density Structure and Transport During the Past Millennium." *Nature*, 444:601–04.

Ma, J., H. Hung, C. Tian and R. Kallenborn, 2011. "Revolatilization of Persistent Organic Pollutants in the Arctic Induced by Climate Change." *Nature Climate Change*, 1:255–60.

Ma, J. and Z. Cao, 2010. "Quantifying the Perturbations of Persistent Organic Pollutants Induced by Climate Change." *Environmental Science and Technology*, 44:8567–73.

Macdonald, R. W., D. Mackay and B. Hickie, 2002. "Contaminant Amplification in the Environment: Revealing the Fundamental Mechanisms." *Environmental Science and Technology*, 36:456A–62A.

Macdonald, R. W., T. Harner and J. Fyfe, 2005. "Recent Climate Change in the Arctic and Its Impact on Contaminant Pathways and Interpretation of Temporal Trend Data." *Science of the Total Environment*, 342:5–86.

Macias-Fauria, M., B. C. Forbes, P. Zetterberg and T. Kumpula, 2012. "Eurasian Arctic Greening Reveals Teleconnections and the Potential for Structurally Novel Ecosystems." *Nature Climate Change*, 2:613–18.

Mahlstein, I. and R. Knutti, 2012. "September Arctic Sea Ice to Disappear Near 2° Global Warming Above Present." *Journal of Geophysical Research*, 117:D06104.

Mahlstein, I., G. Hegerl and S. Solomon, 2012. "Emerging Local Warming Signals in Observational Data." *Geophysical Research Letters*, 39:L21711.

Manney, G. L., M. L. Santee, M. Rex., N. J. Livesey et al., 2011. "Unprecedented Arctic Ozone Loss in 2011 Echoed the Antarctic Ozone Hole." *Nature*, 478:469–75.

Maslanik J. A., C. Fowler, J. Stroeve, S. Drobot, J. Zwally, D. Yi and W. Emery, 2007. "A Younger, Thinner Arctic Ice Cover: Increased Potential for Rapid, Extensive Sea-Ice Loss." *Geophysical Research Letters*, 34:L24501.

Maslanik, J., J. Stroeve, C. Fowler and W. Emery, 2011. "Distribution and Trends in Arctic Sea Ice Age Through Spring 2011." *Geophysical Research Letters*, 38: L13502.

Mastepanov, M., C. Sigsgaard, E. J. Dlugokencky et al., 2008. "Large Tundra Methane Burst During Onset of Freezing." *Nature*, 456:628–30.

Mathis, J. T., 2011. "The Extent and Controls on Ocean Acidification in the Western Arctic Ocean and Adjacent Continental Shelf Seas." In *Arctic Report Card 2011*. www.arctic.noaa.gov/report11.

Mathis, J. T., J. N. Cross and N. R. Bates, 2011. "Coupling Primary Production and Terrestrial Runoff to Ocean Acidification and Carbonate Mineral Suppression in the Eastern Bering Sea." *Journal of Geophysical Research*, 116:C02030.

Mathis, J. T., J. N. Cross, N. R. Bates, M. L. Lomas, S. B. Moran, C. W. Mordy and P. Stabeno, 2010. "Seasonal Distribution of Dissolved Inorganic Carbon and Net Community Production on the Bering Sea Shelf." *Biogeosciences*, 7:1769–87.

Mcguire, D., L. Anderson, T. R. Christensen et al., 2009. "Sensitivity of the Carbon Cycle in the Arctic to Climate Change." *Ecological Monographs*, 79(4):523–55.

Meyer, T. and F. Wania, 2008. "Organic Contaminant Amplification During Snowmelt." *Water Research*, 42:1847–65.

Meyer, T., Y. D. Lei, I. Muradi and F. Wania, 2009. "Organic Contaminant Release From Melting Snow. 1. Influence of Chemical Partitioning." *Environmental Science and Technology*, 43:657–62.

Milne, G. A., W. R. Gehrels, C. W. Hughes and M. Tamisiea, 2009. "Identifying the Causes of Sea-Level Change." *Nature Geoscience*, 2:471–78.

Mitrovica, J. X., N. Gomez and P. Clark, 2009. "The Sea-Level Fingerprint of West Antarctic Collapse." *Science*, 323(5915):753.

Monastersky, R., 2013. "Global Carbon Dioxide Levels Near Worrisome Milestone." *Nature*, 497:13–14.

Myers-Smith, I. H., B. C. Forbes, M. Wilmking et al., 2011. "Shrub Expansion in Tundra Ecosystems: Dynamics, Impacts and Research Priorities." *Environmental Research Letters*, 6:045509.

Myers-Smith, I. H., D. S. Hik, C. Kennedy, D. Cooley, J. F. Johnstone, A. J. Kennedy and C. J. Krebs, 2011. "Expansion of Canopy-Forming Willows Over the Twentieth Century on Herschel Island, Yukon Territory, Canada." *Ambio*, 40:610–23.

National Research Council of the National Academies, 2010. *America's Climate Choices: Panel on Advancing the Science of Climate Change, Board on Atmospheric Sciences and Climate, Division on Earth and Life Studies*. The National Academies Press. Washington, DC.

Nghiem, S. V., D. K. Hall, T. L. Mote, M. Tedesco, M. R. Albert, K. Keegan, C. A. Shuman, N. E. DiGirolamo and G. Neumann, 2012. "The Extreme Melt Across the Greenland Ice Sheet in 2012." *Geophysical Research Letters*, 39:L20502.

Nizzetto, L. et al., 2010. "Atlantic Ocean Surface Waters Buffer Declining Atmospheric Concentrations of Persistent Organic Pollutants." *Environmental Science and Technology*, 44:6978–84.

Orr, J. C. et al., 2005. "Anthropogenic Ocean Acidification Over the Twenty-First Century and Its Impact on Calcifying Organisms." *Nature*, 437:681–86.

Overland, J. E., 2011. "Potential Arctic Change Through Climate Amplification Processes." *Oceanography*, 24(3):176–85.

Overland, J., J. Key, B.-M. Kim, S.-J. Kim, Y. Liu, J. Walsh, M. Wang and U. Bhatt, 2012. "Air Temperature, Atmospheric Circulation and Clouds." In *Arctic Report Card 2012*. www.arctic.noaa.gov/report12.

Overland, J. E., K. R. Wood and M. Wang, 2011. "Warm Arctic-Cold Continents: Impacts of the Newly Open Arctic Sea." *Polar Research*, 30:15787.

Overland, J. E. and M. Wang, 2010. "Large-Scale Atmospheric Circulation Changes Are Associated With the Recent Loss of Arctic Sea Ice." *Tellus*, 62A:1–9.

Overland, J. E. and M. Wang, 2013. "When Will the Summer Arctic Be Nearly Ice Free?" *Geophysical Research Letters*, 40:2097–101.

Overland, J. E., M. Wang and S. Salo, 2008. "The Recent Arctic Warm Period." *Tellus*, 60A:589–97.

Perovich, D. K., J. A. Richter-Menge, K. F. Jones and B. Light, 2008. "Sunlight, Water, and Ice: Extreme Arctic Sea Ice Melt During the Summer of 2007." *Geophysical Research Letters*, 35:L11501.

Perovich. D., W. Meier, M. Tschudi, S. Gerland and J. Richter-Menge, 2012. "Sea Ice." In *Arctic Report Card 2012*. www.arctic.noaa.gov/report12.

Rahmstorf, S., 2006. "Thermohaline Ocean Circulation." In S. A. Elias (ed.), *Encyclopedia of Quaternary Sciences*. Amsterdam: Elsevier.

Rahmstorf, S., 2007. "A Semi-Empirical Approach to Projecting Future Sea-Level Rise. 2007." *Science*, 315:368–70.

Rahmstorf, S., G. Foster and A. Cazenave, 2012. "Comparing Climate Projections to Observations Up to 2011." *Environmental Research Letters*, 7:044035.

Raynolds, M. K., D. A. Walker, H. E. Epstein, J. E. Pinzon and C. J. Tucker, 2012. "A New Estimate of Tundra-Biome Phytomass From Trans-Arctic Field Data and AVHRR NDVI." *Remote Sensing Letters*, 3:403–11.

Regehr, E. V., C. M. Hunter, H. Caswell, S. C. Amstrup and I. Stirling, 2010. "Survival and Breeding of Polar Bears in the Southern Beaufort Sea in Relation to Sea Ice." *Journal of Animal Ecology*, 79:117–27.

Regehr, E. V., N. J. Lunn, S. C. Amstrup and I. Stirling, 2007. "Effects of Earlier Sea Ice Breakup on Survival and Population Size of Polar Bears in Western Hudson Bay." *Journal of Wildlife Management*, 71:2673–83.

Reid, D. G., R. A. Ims, N. M. Schmidt, G. Gauthier and D. Ehrich, 2012. "Lemmings." In *Arctic Report Card 2012*. www.arctic.noaa.gov/report12.

Rignot E., I. Velicogna, M. R. van den Broeke, A. Monaghan and J. Lenaerts, 2011. "Acceleration of the Contribution of the Greenland and Antarctic Ice Sheets to Sea Level Rise." *Geophysical Research Letters*, 38(5). doi:10.1029/2011GL046583.

Robinson, A., R. Calov and A. Ganopolski, 2012. "Multistability and Critical Thresholds of the Greenland Ice Sheet." *Nature Climate Change*, 2:429–32.

Romanovsky, V., N. Oberman, D. Drozdov, G. Malkova, A. Kholodov and S. Marchenko, 2011. "Permafrost." In "State of the Climate in 2010." *Bulletin of the American Meteorological Society*, 92(6):S152–53.

Rooney, R. C., S. E. Bayley and D. W. Schindler, 2012. "Oil Sands Mining and Reclamation Cause Massive Loss of Peatland and Stored Carbon." *Proceedings of the National Academy of Sciences of the United States*, 109(13):4933–37.

Ruppel, C. D., 2011. "Methane Hydrates and Contemporary Climate Change." *Nature Education Knowledge*, 3(10):29.

Russell, D. and A. Gunn, 2012. "Caribou and Reindeer." In *Arctic Report Card 2012*. www.arctic.noaa.gov/report12.

Sallenger, A. H., K. S. Doran and P. A. Howd, 2012. "Hotspot of Accelerated Sea-Level Rise on the Atlantic Coast of North America." *Nature Climate Change*, 2:884–8.

Sambrotto, R. N., C. Mordy, S. I Zeeman, P. J. Stabeno and S. A. Macklin, 2008. "Physical Forcing and Nutrient Conditions Associated With Patterns of Chl a and Phytoplankton Productivity in the Southeastern Bering Sea During Summer." *Deep Sea Research*, 11(55)(16–17):1745–60.

Scambos, T. A., C. Hulbe, M. Fahnestock and J. Bohlander, 2000. "The Link Between Climate Warming and Break-Up of Ice Shelves in the Antarctic Peninsula." *Journal of Glaciology*, 46(154):516–30.

Scambos, T. A., J. Bohlander, C. Shuman and P. Skvarca, 2004. "Glacier Acceleration and Thinning After Ice Shelf Collapse in the Larsen B Embayment, Antarctica." *Geophysical Research Letters*, 31(18). doi:10.1029/2004GL020670

Schaeffer, M., W. Hare, S. Rahmstorf and M. Vermeer, 2012. "Long-Term Sea-Level Rise Implied by 1.5 °C and 2 °C Warming Levels." *Nature Climate Change*, 2:867–70.

Schweiger A. J., R. W. Lindsay, S. Vavrus and J. A. Francis, 2008. "Relationships Between Arctic Sea Ice and Clouds During Autumn." *Journal of Climate*, 21:4799–810.

Screen, J. A. and I. Simmonds, 2010. "The Central Role of Diminishing Sea Ice in Recent Arctic Temperature Amplification." *Nature*, 464:1334–37.

Screen, J. A. and I. Simmonds, 2010. "Increasing Fall-Winter Energy Loss From the Arctic Ocean and Its Role in Arctic Temperature Amplification." *Geophysical Research Letters*, 37:L16707.

Sedláček, J., O. Martius and R. Knutti, 2011. "Influence of Subtropical and Polar Sea-Surface Temperature Anomalies on Temperatures in Eurasia." *Geophysical Research Letters*, 38(12):L12803.

Serreze, M. C., A. P. Barrett, J. C. Stroeve, D. N. Kindig and M. M. Holland, 2009. "The Emergence of Surface-Based Arctic Amplification." *The Cryosphere*, 3:11–9.

Serreze M. and R. Barry, 2011. "Processes and Impacts of Arctic Amplification." *Global and Planetary Change*, 77:85–96.

Shakhova, N. et al., 2010. "Extensive Methane Venting to the Atmosphere From Sediments of the East Siberian Arctic Shelf." *Science*, 327:1246–50.

Sharp, M., D. O. Burgess, J. G. Cogley, M. Ecclestone, C. Labine and G. J. Wolken, 2011. "Extreme Melt on Canada's Arctic Ice Caps in the 21st Century." *Geophysical Research Letters*, 38:L11501.

Sharp, M. and G. Wolken, 2012. "Glaciers and Ice Caps (Outside Greenland)." In "State of the Climate in 2011." *Bulletin of the American Meteorological Society*, 93 (7):133–4.

Sharp, M. and G. Wolken, 2011. "Glaciers and Ice Caps (Outside Greenland)." In *Arctic Report Card 2011*. www.arctic.noaa.gov/report11.

Sheng, J., X. Wang, P. Gong, D. Joswiak, L. Tian, T. Yao and K. Jones, 2013. "Monsoon-Driven Transport of Organochlorine Pesticides and Polychlorinated Biphenyls to the Tibetan Plateau: Three Year Atmospheric Monitoring Study." *Environmental Science and Technology*, 47(7):3199–208.

Shepherd, A., E. R. Ivins et al., 2012. "A Reconciled Estimate of Ice-Sheet Mass Balance." *Science*, 338(6111):1183–89.

Singarayer J. S., J. L. Bamber and P. J. Valdes, 2006. "Twenty-First Century Climate Impacts From a Declining Arctic Sea Ice Cover." *Journal of Climate*, 19:1109–25.

Skinner, L., 2012. "A Long View on Climate Sensitivity." *Science*, 337:917–19.

Smith, S. L., J. Throop and A. G. Lewkowicz, 2012. "Recent Changes in Climate and Permafrost Temperatures at Forested and Polar Desert Sites in Northern Canada." *Canadian Journal of Earth Sciences*, 49:914–24.

Spielhagen, R. F. et al., 2011. "Enhanced Modern Heat Transfer to the Arctic by Warm Atlantic Water." *Science*, 331:450–53.

Steinacher, M., F. Joos, T. L. Frolicher, G.-K. Plattner and S. C. Doney, 2009. "Imminent Ocean Acidification in the Arctic Projected With the NCAR Global Coupled Climate Carbon Cycle Climate Model." *Biogeosciences*, 6:515–33.

Stern, G. A., R. W. Macdonald, P. M. Outridge, S. Wilson et al., 2012. "How Does Climate Change Influence Arctic Mercury?" *Science of the Total Environment*, 414:22–42.

Strahan, S. E., A. R. Douglass and P. A. Newman, 2013. "The Contributions of Chemistry and Transport to Low Arctic Ozone in March 2011 Derived From Aura MLS Observations." *Journal of Geophysical Research: Atmospheres*, 118(3):1563–76.

Strauss, B., 2013. "Rapid Accumulation of Committed Sea-Level Rise From Global Warming." *Proceedings of the National Academy of Sciences*, 110(34):13699–700.

Stroeve, J. C., M. C. Serreze, M. M. Holland, J. E. Kay, J. Maslanik and A. P. Barrett, 2012. "The Arctic's Rapidly Shrinking Sea Ice Cover: A Research Synthesis." *Climatic Change*. doi:10.1007/s10584-011-0101-1.

Stroeve, J., M. M. Holland, W. Meier, T. Scambos and M. Serreze, 2007. "Arctic Sea Ice Decline: Faster Than Forecast." *Geophysical Research Letters*, 34:L09501.

Su, Y. et al., 2006. "Spatial and Seasonal Variations of Hexachlorocyclohexanes (HCHs) and Hexachlorobenzene (HCB) in the Arctic Atmosphere." *Environmental Science and Technology*, 40:6601–07.

Tamisiea, M. E. and J. X. Mitrovica, 2011. "The Moving Boundaries of Sea Level Change: Understanding the Origins of Geographic Variability." *Oceanography*, 24(2):24–39.

Timmermans, M.-L., A. Proshutinsky, I. Ashik, A. Beszczynska-Moeller, E. Carmack et al., 2012. "Ocean." In *Arctic Report Card 2012*. www.arctic.noaa.gov/report12.

Tingley, M. P. and P. Huybers, 2013. "Recent Temperature Extremes at High Northern Latitudes Unprecedented in the Past 600 Years." *Nature*, 496:201–05.

UNEP, 2011. *Integrated Assessment of Black Carbon and Tropospheric Ozone: Summary for Decision Makers*. UNEP/GC/26/INF/20. United Nations Environment Programme. Nairobi, Kenya.

UNEP, 2011. *Near-Term Climate Protection and Clean Air Benefits: Actions for Controlling Short-Lived Climate Forcers*. A UNEP Synthesis Report. United Nations Environment Programme. Nairobi, Kenya. www.unep.org/pdf/Near_Term_Climate_Protection_&_Air_Benefits.pdf.

UNEP/AMAP, 2011. *Climate Change and POPs: Predicting the Impacts*. Report of the UNEP/AMAP Expert Group. Secretariat of the Stockholm Convention. Geneva, Switzerland.

Vaks, A., O. S. Gutareva, S. F. M. Breitenbach, E. Avirmed, A. J. Mason, A. L. Thomas, A. V. Osinzev, A. M. Kononov and G. M. Henderson, 2013. "Speleothems Reveal 500,000-Year History of Siberian Permafrost." *Science*, 340 (6129):183–86.

Velders, G. J. M., A. R. Ravishankara, M. K. Miller, M. J. Molina et al., 2012. "Preserving Montreal Protocol Climate Benefits by Limiting HFCs." *Science*, 335:922–23.

Vongraven, D. and E. Richardson, 2011. "Biodiversity – Status and Trends of Polar Bears." In *Arctic Report Card 2011*. www.arctic.noaa.gov/report11.

Wang, M. and J. E. Overland, 2009. "A Sea Ice Free Summer Arctic Within 30 Years?" *Geophysical Research Letters*, 36:L07502.

Waugh, D., F. Primeau, T. Devries and M. Holzer, 2013. "Recent Changes in the Ventilation of the Southern Oceans." *Science*, 339(6119):568–70.

Weaver, A. J. and C. Hillaire-Marcel, 2004. "Global Warming and the Next Ice Age." *Science*, 304:400–02.

Whiteman, G., C. Hope and P. Wadhams, 2013. "Vast Costs of Arctic Change." *Nature*, 499:401–03.

Wong, F. et al., 2011. "Air-Water Exchange of Anthropogenic and Natural Organohalogens on International Polar Year (IPY) Expeditions in the Canadian Arctic." *Environmental Science and Technology*, 45:876–81.

Woodgate, R. A., T. J. Weingartner and R. Lindsay, 2012. "Observed Increases in Bering Strait Oceanic Fluxes From the Pacific to the Arctic From 2001 to 2011 and Their Impacts on the Arctic Ocean Water Column." *Geophysical Research Letters*, 39(24):24603.

Woodgate, R. A., T. J. Weingartner and R. W. Lindsay, 2010. "The 2007 Bering Strait Oceanic Heat Flux and Anomalous Arctic Sea-Ice Retreat." *Geophysical Research Letters*, 37:L01602.

World Glacier Monitoring Service, 2012. *Preliminary Glacier Mass Balance Data 2009 and 2010.* www.geo.uzh.ch/microsite/wgms/mbb/sum10.html.

Yamamoto-Kawai, M., F. A. McLaughlin, E. C. Carmack, S. Nishino and K. Shimada, 2009. "Aragonite Undersaturation in the Arctic Ocean: Effects of Ocean Acidification and Sea Ice Melt." *Science*, 326:1098–1100.

Yamamoto-Kawai, M., W. Williams, S. Nishino and F. McLaughlin, 2011. "Ocean Biogeophysical Conditions." In *Arctic Report Card 2011.* www.arctic.noaa.gov/report11.

Young, O. R., J. Deog Kim and Y. Hyung Kim (eds.), 2012. *The Arctic in World Affairs: A North Pacific Dialogue on Arctic Marine Issues.* Korea Maritime Institute. Seoul and East-West Center, Honolulu.

Zeng, H., G. Jia and H. Epstein, 2011. "Recent Changes in Phenology Over the Northern High Latitudes Detected From Multi-Satellite Data. *Environmental Research Letters*, 6:045508.

Chapter 11: Thoughts on Education, the Training of Arctic Scientists and Arctic Research

The primary sources for this chapter have been:

The Arctic Council Kiruna Declaration, 2013. Arctic Council Secretariat. Tromso, Norway. www.arctic-council.org.

The Arctic Council's Vision of the Arctic, 2013. Arctic Council Secretariat. Tromso, Norway. www.arctic-council.org.

Suggested Further Reading

Aksnes, D. W. and D. O. Hessen, 2009. "The Structure and Development of Polar Research (1981–2007): A Publication-Based Approach. *Arctic, Antarctic and Alpine Research*, 41(2):155–63.

Canada's Arctic Foreign Policy. Ottawa. (version as dated: 2013-06-03) www.international.gc.ca.

Communication From the Commission to the European Parliament and the Council:

European Commission, 2013. *European Union Strategy for the Arctic*. http://europa
.eu/rapid/press-release_SPEECH-13-329_en.htm.

The European Union and the Arctic Region, 2008. Commission of the European
Communities. Brussels, Belgium.

Krupnik, I. and D. Jolly (eds.), 2002. *The Earth Is Faster Now: Indigenous Observations
of Arctic Environmental Change*. Arctic Research Consortium of the United States.
Fairbanks, Alaska.

National Committee on Inuit Education, 2011. *First Canadians, Canadians First*.
Available from Inuit Tapiriit Kanatami, 75 Albert St., Suite 1101, Ottawa,
Ontario, Canada.

Chapter 12: The Long and the Short of It: Has the Arctic Messenger Been Noticed? What Can Be Done?

The primary sources for this chapter have been:

Diamond, J. A., 2005. *Collapse: How Societies Choose to Fail or Succeed*. New York:
Penguin.

GEA 2012: *Global Energy Assessment – Toward a Sustainable Future*. Cambridge, UK,
and New York and the International Institute of Applied Systems Analysis,
Laxenburg, Austria: Cambridge University Press.

Suggested Further Reading

Adly, J. E. and R. N. Stavins, 2012. "Climate Negotiators Create an Opportunity for
Scholars." *Science*, 337:1043–44.

Allen, M. R. and D. Frame, 2007. "Call Off the Quest." *Science*, 318:582–83.

Allen, M. R. and D. Frame et al., 2009. "The Exit Strategy." *Nature Reports Climate
Change*, 3:56–57.

Allen, M. R. and D. Frame et al., 2009. "Warming Caused by Cumulative Carbon
Emissions Towards the Trillionth Tonne." *Nature*, 458:1163–66.

Diringer, E., 2013. "A Patchwork of Emissions Cuts." *Nature*, 501:307–09.

Guardans, R., 2012. "Perspective: A Note on the Advantages of Large-Scale Monitoring as in the Global Monitoring Plan (GMP) on Persistent Organic Pollutants
(POPs)." *Atmospheric Pollution Research*, 3:369–70.

Grifo, F., M. Halpern and P. Hansel (contributors), 2012. *Heads They Win. Tails We
Lose*. Union of Concerned Scientists. Cambridge, Massachusetts.

Kelly, E. N., D. W. Schindler, P. V. Hodson, J. W. Short, R. Radmanovich and
C. C. Nielsen, 2010. "Oil Sands Development Contributes Elements Toxic at Low
Concentrations to the Athabasca River and Its Tributaries." *Proceedings of the
National Academy of Sciences of the United States*, 107(37):16178–83.

Kelly, E. N., J. W. Short, D. W. Schindler, P. V. Hodson, M. Ma, A. K. Kwan and
B. L. Fortin, 2009. "Oil Sands Development Contributes Polycyclic Aromatic
Compounds to the Athabasca River and Its Tributaries." *Proceedings of the
National Academy of Sciences of the United States*, 106(52):22346–351.

Krupnik, I. and D. Jolly (eds.), 2002. *The Earth Is Faster Now: Indigenous Observations
of Arctic Environmental Change*. Arctic Research Consortium of the United States.
Fairbanks, Alaska.

Laube, J. C., M. J. Newland and C. Hogan et al., 2014. "Newly Detected Ozone-
Depleting Substances in the Atmosphere." *Nature Geoscience*, 7:266–9.

Monastersky, R., 2013. "Global Carbon Dioxide Levels Near Worrisome
Milestone." *Nature*, 497:13–14.

Overland, J. E., M. Wang, J. E. Walsh and J. C. Stroeve, 2014. "Future Arctic Climate Changes: Adaptation and Mitigation Time Scales." *Earth's Future*, 2(2):68–74.

Sabin, P., 2013. *The Bet: Paul Ehrlich, Julian Simon, and Our Gamble Over the Earth's Future*. New Haven, CT: Yale University Press.

Shulman, S. (lead investigator), 2007. *Smoke, Mirrors and Hot Air*. Union of Concerned Scientists. Cambridge, Massachusetts.

Turco, R. P., O. B. Toon, T. P. Ackerman, J. B. Pollack and C. Sagan, 1983. "Nuclear Winter: Global Consequences of Multiple Nuclear Explosions." *Science*, 222 (4630):1283–92.

UNEP, 2006. *Strategic Approach to International Chemicals Management: Comprising the Dubai Declaration on International Chemicals Management, the Overarching Policy Strategy and the Global Plan of Action. Resolutions of the International Conference on Chemicals Management*. United Nations Environment Programme. Geneva, Switzerland.

Weaver, A., 2008. *Keeping Our Cool: Canada in a Warming World*. Toronto: Viking Canada.

Wright, R., 2001. *Nonzero: The Logic of Human Destiny*. New York: Vintage Books.

Credits

Tables

8.3: Adapted and reproduced with kind permission from AMAP's *Arctic Pollution 2002*, page 30

Figures

2.1: Adapted from AMAP archives

4.1: Adapted with kind permission from *AMAP Assessment 2009: Radioactivity in the Arctic*, figure 4.25

8.1: Adapted with kind permission from *AMAP: Climate Issues 2011*, page 61

8.2: Adapted with kind permission from *AMAP Assessment 2002: Persistent Organic Pollutants*, figure 2.2 (as modified from Wania and Mackay, 1996)

8.3: Adapted with kind permission from *AMAP Assessment 2009: Human Health in the Arctic*, figure 5.13

8.4: Adapted with kind permission from *AMAP Assessment 2011: Mercury in the Arctic*, figure 3.4

8.5: Adapted with kind permission from *AMAP Arctic Pollution 2011*, page 26

8.6: Adapted with kind permission from *AMAP Assessment 2011: Human Health in the Arctic*, figure 5.14

8.7: Adapted with kind permission from *AMAP Arctic Pollution 2009*, page 17

10.1: Adapted with kind permission from AMAP: *Snow, Water, Ice and Permafrost in the Arctic (SWIPA): Climate Change and the Cryosphere*, 2009, figure 9.17 (as based on Wang and Overland 2009)

10.2:   Adapted with kind permission from AMAP:
          The Greenland Ice Sheet in a Changing Climate (SWIPA), 2009,
          figure 1.2. Data source: giss UNEP.

10.3:   Adapted with kind permission from AMAP archives:
          Data from CDIAC: "Historical CO_2 Record From the
          Law Dome DE08, DE08–2, and DSS Ice Cores."
          D.M. Etheridge et al. Data link: http://cdiac.esd.ornl.gov/
          trends/co2/lawdome-data.html. Data page link:
          http://cdiac.esd.ornl.gov/trends/co2/modern_co2.html.

10.4:   Adapted with kind permission from AMAP: *Summary of
          the Greenland Ice Sheet in a Changing Climate*, 2009, page 12

10.5:   Adapted with kind permission from AMAP: *Snow, Water,
          Ice and Permafrost in the Arctic (SWIPA): Climate Change and
          the Cryosphere*, 2011, figure 2.4 (based on analysis by
          NASA's Goddard Institute for Space Studies
          [http://data.giss.nasa.gov/gistemp])

10.6:   Reproduced by kind permission of Dáithí A. Stone,
          Lawrence Livermore National Laboratory, Berkeley, California

10.7:   Adapted with kind permission from *AMAP Arctic
          Pollution 2011*, page 31

10.8:   Adapted with kind permission from AMAP: *The
          Greenland Ice Sheet in a Changing Climate* (SWIPA), 2009,
          figure 2.3. Source: UNEP Maps.

10.9:   Adapted with kind permission from AMAP: *Climate
          Issues 2011*, page 21

10.10:   Adapted with kind permission from AMAP: *Arctic
          Climate Issues 2011*, page 33

10.11:   Reproduced by kind permission of Dáithí A. Stone,
          Lawrence Livermore National Laboratory, Berkeley,
          California, following the methods used for tables
          18–5 through 18–9 of Cramer et al., 2014. Chapter 18
          of IPCC WGII AR5

Quotations

Chapters 2, 8 and 12:   Quotations of poetry by Dáithí Ó hÓgáin
                          at the beginning of chapters. Reproduced
                          by kind permission of Philomel Productions
                          and Caitríona Uí Ógáin.

Chapter 10:   Quotations from IPCC reports are footnoted
                          in text. They are reproduced with kind

Chapter 13:

Index